STUDY GUIDE

Larry M. Lewis
Salem State College

to accompany

MICROBIOLOGY
Principles and Explorations

Sixth Edition

Jacquelyn G. Black
Marymount University

JOHN WILEY & SONS, INC.

Cover Images: ©Russell Kightley, www.rkm.com.au

To order books or for customer service please, call 1-800-CALL WILEY (225-5945).

ISBN-13 978-0-471-48244-4
ISBN-10 0-471-48244-7

Printed in the United States of America

10 9 8 7 6 5 4 3 2 1

Printed and bound by Malloy Lithographing, Inc.

To the Student

Welcome to microbiology! I have taught this subject for many years and have used Dr. Black's fine textbook through many editions. It has been a privilege to update this Study Guide to accompany the sixth edition of ***Microbiology: Principles and Explorations***.

I understand that many students greatly benefit from a guide such as this. One of my nursing students shared with me that she felt that the microbiology course is much more difficult than her previous anatomy and physiology class. At first I found this difficult to understand. She then went on to say while her previous class only involved memorization, microbiology requires one to actually think. Indeed. It does require a lot of thought and is a lot more fun too.

I introduce each chapter with a story or insight from my own experience. An outline of each chapter follows. It may be useful to take the book to lecture if your instructor follows the book closely. A variety of self-tests and essay questions will both test and develop your understanding of the material.

Larry M. Lewis, Ph.D.
Professor of Biology
Salem State College
2005

Table of Contents

Chapter 1 Scope and History of Microbiology

INTRODUCTION

September 11, 2001 was the date stamped on envelopes addressed to NBC News anchor Tom Brokow and the editor of the New York Post. The letter stated:

This is next
Take Penicillin Now
Death to America
Death to Israel
ALLAH is Great

America was under biological attack from the bacterium Anthrax. Mail may never seem the same again. In the winter of 2004, the flu has come early and hard. We read about many children who have died of influenza this year and are reminded that over 36,000 Americans died of complications of this disease in 2003. Many of us have tried hard to get a flu vaccine in 2004 and failed. We remember how the new disease SARS brought fear to travelers to Asia and we all have concern about its possible return.

Clearly, microbiology is very important for students entering the world of health and medicine. If you are such a student, you might be most interested in the microbiological fields of immunology and chemotherapy—they will teach you how your body responds to microbial invaders and how certain drugs can be used to combat these microorganisms.

Or, if you are more interested in history, you might be excited to learn that concepts in modern-day microbiology stem from ideas presented in the Bible and by Greek and Roman scholars. Perhaps you are a student who likes to eat and drink more than you like to study. Microbiology could still catch your interest, since biochemical reactions carried out by microbes are used to make pickles, sauerkraut, yogurt and other dairy products, fructose used in soft drinks, the artificial sweetener aspartame, and, of course, beer and wine. Or, if you are determined to be the first scientist to create a human clone of yourself, you will want to study the fields of virology, genetics, and molecular biology, fields that are all included under the title of "microbiology."

It is important to appreciate that microbes are essential for the ecology of our world and our bodies. Consider that the song from the musical Carosel, 'You Never Walk Alone,' is literally correct for there are more bacteria living on and in your body than you have eukaryotic (human) cells. Microbes may even exist throughout the universe. In the early winter of 2004, a probe on Mars searched for evidence of past microbial life.

STUDY OUTLINE

I. **Why Study Microbiology?**
   A. Microbial relationships
      Microorganisms are essential to the web of life in every environment, including the human environment. Consider, bacteria invented photosynthesis and first began to produce atmospheric oxygen.
   B. Beneficial aspects of microbes
      Most microorganisms are beneficial. For example, they are important in food chains, in digestive functions, and in the food industry. Microbes degrade oil spills and synthesize antibiotics. Bacteria produce interferon and human growth hormones in genetic engineering.
   C. Microbes in research
      Microorganisms are easily studied because they have relatively simple structures, they can be grown in large numbers, and they can reproduce quickly. The study of microbes has already led to remarkable success in understanding life processes and disease control. Biologists will continue to use microbiology to help them solve future challenges.
   D. Bacteria live at great depths under ground, in hot springs, and have a total mass greater than all other living organisms combined.

II. **Scope of Microbiology**
   A. The microbes
      Major groups include bacteria, algae, fungi, viruses, and protozoa.
      Macroscopic helminthes and arthropod s are also considered.
   B. The microbiologists
      Microbiologists work universities, genetic engineering laboratories, industrial laboratories, and hospitals.

III. **Historical Roots**
   A. Biblical accounts
      Mosaic laws found in the Bible include basic sanitation like burying human waste.
   B. Ancient Greek and Roman men
      Hippocrates, Thucydides, Varro, and Lucretius are some who contributed to an understanding of diseases caused by microbes by careful observation of patients.
   C. Bubonic plague
      Killed tens of millions of people in Europe but did not affect the Jewish population as drastically, due to the cleaner sanitation practices of the Jewish people.
   D. Development of microscope
      Anton van Leeuwenhoek first observed living microorganisms including protozoa and bacteria. Robert Hooke first used the term "cell" to refer to the cell walls of plant cells. Carolus Linnaeus developed a general classification system for all living organisms. Matthias Schleiden and Theodor Schwann formulated the cell theory.

IV. **The Germ Theory of Disease**
   A. Spontaneous generation theory
      In the mid-nineteenth century, many believed that microorganisms arose from nonliving things.
   B. Early studies
      Many scientists used their experiments to try to disprove the spontaneous generation theory, including Francesco Redi and Lazzaro Spallanzani. But it was the experiments of Louis Pasteur and John Tyndall that finally disproved spontaneous generation to most scientists of the time.
   C. Pasteur's further contributions
      In addition to his work regarding spontaneous generation, Pasteur developed the technique of pasteurization, associated specific organisms with particular diseases, and contributed to the development of vaccines.
   D. Koch's contributions
      Robert Koch identified the bacterium that causes anthrax, developed techniques for studying cells in vitro, developed pure culture techniques, formulated Koch's postulates, identified the bacterium that causes tuberculosis, and worked with cholera, malaria, typhoid fever, sleeping sickness, and several other diseases.
   E. Work toward controlling infections
      Ignaz Philipp Semmelweis reduced puerperal fever by encouraging physicians to use more sanitary practices (for example, by washing their hands after performing an autopsy and before delivering a

baby). Joseph Lister, who is considered the father of antiseptic surgery, developed aseptic techniques, which reduced the incidence of infection in surgical wards that followed his techniques.

V. Emergence of Special Fields of Microbiology

A. Immunology

Immunology considers the host's response to microbial invasion. Microbiologists developed the first vaccines. Metchnikoff discovered phagocytosis in the 1880's.

B. Virology

Viruses pass through fine filters and are difficult to study. In 1935, Stanley crystallized tobacco mosaic virus and demonstrated they consisted of protein and RNA. In 1939, viruses were imaged with an electron microscope for the first time.

C. Chemotherapy

The science of chemotherapy develops drugs to treat disease. It began in the first century AD, as people used herbal medicines to treat diseases, and is still under investigation today, as scientists try to find newer and better antibiotics and antiviral agents.

D. Genetics and molecular biology

Research in this field has led to a more comprehensive understanding of the genetics of all organisms and of human antibodies.

E. Tomorrow's history

Today's discovery is tomorrow's history—with all of these special fields of microbiology still undergoing active research today, a history of microbiology could never be complete. While the period between 1874–1917 is called the Golden Age of Microbiology, focus on AIDS and genetic engineering promise even greater accomplishments in the future. For example, bacteriophages may one day replace antibiotics to treat bacterial diseases. On the other hand, the potential for bioterror looms a greater danger than ever.

F. Genomics

The Human Genome Project has identified the location and chemical sequence of all genes in the human genome. Over 113 human genes are derived from bacteria. Only 300 genes in people are not found in mice.

SELF-TESTS

Continue with this section only after you have read Chapter 1 of your textbook. Write your answers in the appropriate space provided. Correct answers to all questions can be found at the end of the Self-Test section. A score of 80 percent or better is good. If your score is less than 65, reread the chapter.

True/False

Mark T for True, F for False.

_____ **1.** Koch's postulates can be applied to the identification of all microbes, including viruses.

_____ **2.** Jon Needham actually was a proponent of the spontaneous generation theory.

_____ **3.** Leeuwenhoek is credited with observing the first microorganisms and making the first microscope.

_____ **4.** Fleming is credited with the development of porcelain filters used to remove bacteria from water.

_____ **5.** The CDC stands for The Center of Disease Control.

_____ **6.** John Tyndall's contribution to microbiology was in his demonstration of microbes on dust particles.

Multiple Choice

Select the best possible answer.

_____ **7.** Pasteur is credited with all of the following except:

a. construction of "swan-necked" vessels.
b. development of a vaccine for rabies.
c. development of the technique of pasteurization.
d. being the director of the Pasteur Institute in Paris, France.
e. all of the above are credited to him.

______ 8. One of the most important contributions of Robert Koch in his development of the germ theory of disease was the:
a. use of test animals in research.
b. use of the microscope.
c. development of the technique of pure culturing.
d. development of the Bunsen burner.

______ 9. Fungi can be characterized as:
a. photosynthetic organisms.
b. organisms lacking a cell wall.
c. organisms lacking a true nucleus.
d. organisms that absorb nutrients from their environment.

______ 10. Phycology is the study of:
a. molds.
b. bacteria.
c. viruses.
d. algae.

Matching

Select the answer from the right-hand side that corresponds to the term or phrase on the left-hand side of the page. An answer may be used more than once. In some cases, more than one answer is required.

Topic: Microbiology Subdivisions and Specialty Fields

______ 11. An organism's resistance to disease
______ 12. Development of new methods of disease prevention and detection
______ 13. The study of submicroscopic obligate intracellular parasites
______ 14. The study of microscopic details of microbial cells
______ 15. The naming and classification of microorganisms
______ 16. Blood typing
______ 17. The study of the principles of physics and how they apply to all living matter
______ 18. Improvements of food quality
______ 19. Preventing heavy economic livestock losses and ultrastructure
______ 20. The study of hookworms and tapeworms
______ 21. Metabolic activities of microbes
______ 22. The manufacture of fermented milk products
______ 23. Interactions between microbes and their environments

a. mycology
b. virology
c. microbial ecology
d. immunology
e. food microbiology
f. genetic engineering
g. microbial physiology
h. biophysics
i. microbial taxonomy
j. microbial morphology
k. food and dairy
l. microbiology
m. parasitology
n. veterinary microbiology
o. biotechnology
p. microbial ecology
q. microbial physiology
r. none of the above

Topic: Historical Highlights

______ 24. Was the first to describe anaerobes
______ 25. Showed that maggots did not arise spontaneously from decaying meat

a. Anton van Leeuwenhoek
b. Louis Pasteur
c. John Tyndall

______ 26. Demonstrated the role of yeasts in fermentation		d. Francesco Redi
______ 27. Is noted for the development and use of simple microscopes		e. Alexander Fleming
______ 28. Independently observed spores		f. Paul Ehrlich
______ 29. Was one of the first to recognize the true biological functions of microbes		
______ 30. Isolated penicillin		
______ 31. Is noted for the introduction of chemotherapy		

Fill-Ins

Provide the correct term or phrase for the following. Spelling counts.

32. The study of very small organisms that require a microscope to observe them is called ________________.
33. Single-celled and multicellular microscopic organisms with true nuclei and which absorb nutrients from their environment are ________________.
34. For their survival, ________________ are totally dependent on the cells of higher forms of life.
35. One way in which microorganisms can be used is in cleaning up the environment, or ________________.
36. The scientist who built a crude microscope in the 1600's and who coined the term "cell" is ________________.
37. The Dutch clothes merchant and amateur lens grinder who observed the first microorganisms is ________________.
38. The theory formulated by Matthias Schleiden and Theodor Schwann that states that cells are the fundamental units of life and carry out all the basic functions of living things is the ________________.
39. The theory that states that microbes can invade other organisms and cause disease is the ________________.
40. The theory of ________________ holds that life could and did appear spontaneously from nonliving or decomposing matter.
41. The Italian physician who demonstrated that maggots arise from fly eggs and not rotten meats is ________________.
42. "Swan-necked" flasks used to refute spontaneous generation were made by ________________.
43. ________________ can be used to destroy microbes that cause wine spoilage without altering the quality of the wine.
44. The scientist who formulated four postulates to associate a particular organism with a specific disease is ________________.
45. The German physician who recognized the connection between autopsies and puerperal fever is ________________.
46. ________________ developed a system with which to control surgical infections.
47. The method by which wound infections are prevented through the use of sterilized instruments and by which carbolic acid is applied to wounds is known as ________________.
48. Elie Metchnikoff discovered certain cells in the body that could ingest microbes and named them ________________, which literally means "cell-eating."
49. Martinus Beijerinck used the term "virus" to refer to specific ________________ (disease-causing) molecules incorporated into cells.
50. ________________ and ________________ determined the structure of DNA.
51. ________________, who coined the word "chemotherapy," developed the first successful treatment against syphilis.

52. ________________ are substances derived from one microorganism that kill or restrict the growth of other microorganisms.

53. Oswald Avery, Maclun McCarty, and Colin MacLeod discovered that the chemical material that produces heritable changes in bacteria is ________________.

54. The mold used by the American geneticists Edward Tatum and George Beadle to show how genetic information controls metabolism was ________________.

55. ________________ are molecules that the immune system produces to combat invading microbes and their toxic products.

56. The project designed to map the location of every gene in all human chromosomes is the ________________ project.

Critical Thinking

Write a paragraph in reponse to each of following below.

57. Why is the study of microbiology important?

58. What is the scope of microbiology?

59. What are some major events in the early history of microbiology?

60. How do we know that microbe cause disease?

61. What events mark the emergence of immunology, virology, chemotherapy, microbial genetics, and molecular biology as branches of microbiology?

Images

Match the type of organism to the correct image. *Place your answers in the spaces provided on page 9.*

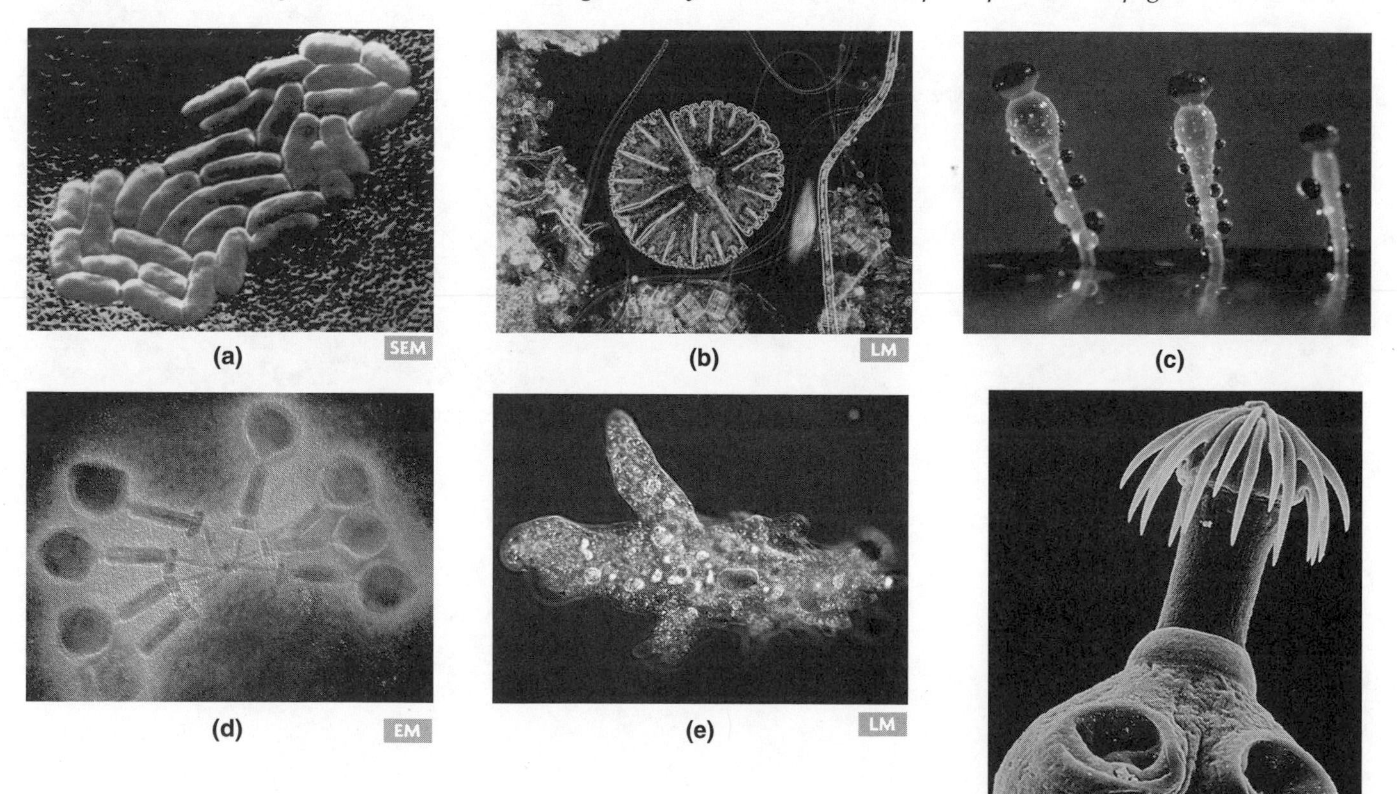

62. Head of a tapeworm ______

63. Green alga ______

64. Protozoan ______

65. Fungus ______

66. Virus ______

67. Bacterium ______

Match the position of the jar(s) to its correct description.

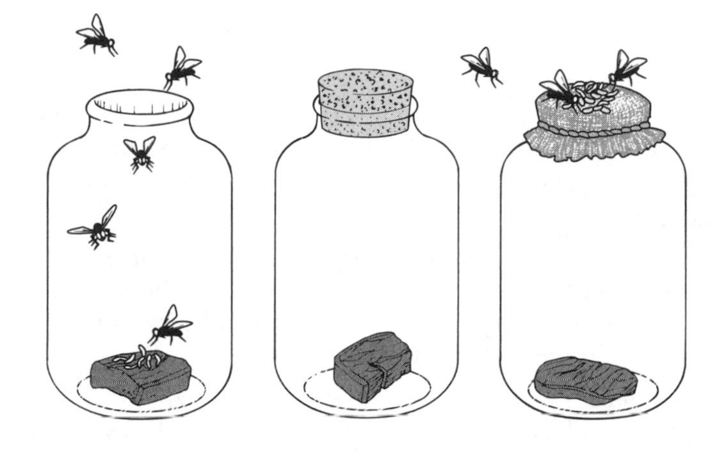

68. The control jar for the experiment. ______

69. Jar that is consistent with the view that eggs come from flies. ______

70. Jar that is consistent with the view that maggots can not arise without flies. ______

71. Does this experiment prove that bacteria do not arise by spontaneous generation? Explain.

ANSWERS

True/False

1. **False** Koch's postulates were written for those organisms that can be purified on artificial culture medium. Since that is not possible with viruses and other obligate intracellular parasites, the postulates cannot be applied directly.
2. **True** Needham, a British clergyman, still believed that microbes arose spontaneously. By boiling and sealing flasks of broth, oxygen (air) needed for their "spontaneous generation" was not present.
3. **False** Leeuwenhoek, although credited with discovering the first microbes, did not construct the first microscope. In fact, his microscopes were considered very crude compared with his contemporaries'.
4. **False** Fleming is credited with the identification of the first antibiotic, penicillin. Chamberland developed porcelain filters used to remove bacteria.
5. **True** The United States Center for Disease Control is located in Atlanta, Georgia, and is the largest such agency in the world.
6. **True** Tyndall used airtight light boxes into which he placed flasks of boiled infusions. Once all the dust settled (which could be visualized through the peepholes), the flasks would not become contaminated.

Multiple Choice

7. e; 8. c; 9. d; 10. d.

Matching

11. d; 12. d; 13. b; 14. j; 15. i; 16. d; 17. h; 18. e; 19. m; 20. l; 21. p; 22. k; 23. o; 24. b; 25. d; 26. b; 27. a; 28. c; 29. b; 30. e; 31. f.

Fill-Ins

32. microbiology; 33. fungi; 34. viruses; 35. bioremediation; 36. Robert Hooke; 37. Anton van Leeuwenhoek; 38. cell theory; 39. germ theory of disease; 40. spontaneous generation; 41. Francesco Redi; 42. Louis Pasteur; 43. pasteurization; 44. Robert Koch; 45. Ignaz Philipp Semmelweis; 46. Joseph Lister; 47. aseptic technique; 48. phagocytes; 49. pathogenic; 50. James Watson, Francis Crick; 51. Paul Erlich; 52. antibiotics; 53. DNA; 54. *Neurospora*; 55. antibodies; 56. human genome.

Critical Thinking

57. **Why is the study of microbiology important?**
Microorganisms have always been and will continue to be a major influence in our lives. Because of their relationship to our health and welfare, it is essential that we understand their characteristics. By studying their features we can also gain insight into the processes and relationships of all life forms.

58. **What is the scope of microbiology?**
Microbiology is often considered just a study of bacteria. However, it actually is a discipline that includes not only bacteria but viruses, algae, protozoa, fungi, and even the helminths. It even includes a great variety of diverse fields such as immunology and genetics that indirectly or directly relate to these microbes. There is tremendous diversity in this field of science.

59. **What are some major events in the early history of microbiology?**
The field of microbiology is as old as recorded time itself; however, it was never recognized as a science until late into the 19th century. The Greeks, Romans, and Jews all made major contributions to microbiology but only indirectly since microbes were not seen nor even believed to exist. These invisible organisms, however, frequently devastated major parts of the world as a result of epidemics such as the plague and smallpox and were often responsible for changing the course of history. It wasn't until the Dutch lens grinder Leeuwenhoek, using a very crude microscope, first observed the tiny "animalcules" that the relationship these organisms had to everyday activities could even be appreciated.

60. **How do we know that microbes cause disease?**
The germ theory of disease provides us with a mechanism to relate a particular bacterium to a particular disease. Before this relationship could be established, several important concepts had to be taken care of. First and foremost, the theory of spontaneous generation, which began in the days of Aristotle, had to be disproved. This theory represented a major stumbling block in the attempts to prove that a particular bacterium could cause a particular disease. If this theory were true, bacteria and hence diseases could arise at any given time and in any given place, spontaneously. Therefore, diseases could never be prevented nor controlled. Fortunately, several scientists including Pasteur and Tyndall using very simple equipment managed to demonstrate once and for all that bacteria could only arise from other living organisms. Once this was confirmed, the stage was set for Koch to develop a series of four postulates that could be used to prove that a particular microbe could cause a particular disease.

61. **What events mark the emergence of immunology, virology, chemotherapy, microbial genetics, and molecular biology as branches of microbiology?**
With the development of the germ theory, numerous specialized fields within microbiology began to emerge. The field of immunology began with the development of the smallpox vaccine by Jenner. Virology emerged as a separate field with the discovery of pathogenic, filterable agents by Beijerinck. Ehrlich and Fleming pioneered the field of chemotherapy with the discovery of sulfa drugs and antibiotics. Lastly, the field of genetics and molecular biology was greatly expanded through the efforts of Griffith who discovered genetic transformation and Avery who demonstrated that this genetic change was due to DNA.

Images

62. f; 63. b; 64. e; 65. c; 66. d; 67. a.

68. Left; 69. Right; 70. Middle and Right.

71. No, in part because this experiment was done with flies and not bacteria. In addition, the stopper in the middle jar might be blocking critical air or light needed for spontaneous generation. In addition, even the mesh on the right might not let in sufficient light for spontaneous generation of maggots to occur. It is not easy to design a perfect experiment!

Chapter 2 Fundamentals of Chemistry

INTRODUCTION

Why does a student of microbiology need to learn basic chemistry? The biology of microbes is largely the biology of cells, and cells are composed of chemicals just as a brick is composed of clay. Just as the clay of bricks imposes limitations on the nature of the brick, the chemicals of life will in part determine the biology of microbes.

The review of chemistry presented in this chapter will provide the basis for an understanding of the rest of the textbook. Some of you have already had a college level class in chemistry. For others, high school chemistry may seem like a distant memory. In any event, this chapter will review enough basic and organic chemistry to help you understand the metabolism of microbes and the basis of disease. If you can answer the Self-tests successfully in the guide, you will proceed through your class in microbiology with confidence.

STUDY OUTLINE

I. Why Study Chemistry?

An understanding of the basic principles of chemistry is necessary in order to understand the metabolic processes of all living things. With chemistry, we can understand both how microorganisms affect humans in disease processes and how they affect all life on earth.

II. Chemical Building Blocks and Chemical Bonds

A. Chemical building blocks

The smallest chemical unit of matter is the *atom*. Matter composed of one kind of atom is called an *element*. When two or more atoms combine chemically, they form a *molecule*. Molecules made up of atoms of two or more elements are called *compounds*.

B. Structure of atoms

Atoms are made up of protons, electrons, and neutrons. The number of protons in the atoms of a particular element is the *atomic number* of the element. *Ions*, or charged atoms, are produced when atoms gain or lose one or more electrons. The *atomic weight* is the sum of the number of protons and neutrons in an atom. Atoms of a particular element that contain different numbers of neutrons are called *isotopes*. A *gram molecular weight*, or *mole*, is the weight of a substance in grams equal to the sum of the atomic weights of the atoms in a molecule of the substance. An isotope with an unstable nucleus that tends to emit subatomic particles and radiation is called a *radioisotope*.

C. Chemical bonds

Chemical bonds form between atoms through interactions of electrons in their outer shells, forming molecules. *Ionic bonds* result from the attraction between ions that have opposite charges. *Covalent bonds* are bonds between atoms created when pairs of electrons are shared. *Hydrogen bonds* are relatively weak attractions that form between a hydrogen atom (carrying a partial positive charge) and an oxygen or nitrogen atom (carrying a partial negative charge).

D. Chemical reactions

Chemical reactions in living organisms either require energy (they are *endergonic*) to form chemical bonds (*anabolism*) or release energy (they are *exergonic*) as chemical bonds are broken (*catabolism*).

III. Water and Solutions

A. Water

Water has several properties that make it important to living things. Because it is a polar compound, it can act as a *solvent*, or dissolving medium. Water forms a thin, invisible, elastic membrane on its surface, a phenomenon known as *surface tension*. Water has a high *specific heat* (it can absorb or release large quantities of heat energy with little temperature changes). Finally, water provides the medium for most chemical reactions in cells and it participates in many of these reactions, like in *dehydration synthesis* (where the components of water are removed to form a larger product molecule) and in *hydrolysis* (where water is added to form simpler products).

B. Solutions and colloids

A *solution* is a mixture of two or more substances in which the molecules are evenly distributed and will not separate out on standing. In a solution, the medium in which substances are dissolved is the *solvent* and the dissolved substance is the *solute*. A *colloid* is a mixture formed by particles too large to form a true solution dispersed in a liquid.

C. Acids, bases, and pH

An *acid* is a hydrogen ion (H^+) donor. A *base* is a hydrogen ion acceptor or a hydroxyl ion (OH^-) donor. pH is a means of expressing the hydrogen-ion concentration, and thus the acidity, of a solution.

IV. Complex Organic Molecules

A. Basic characteristics

Organic chemistry is the study of complex organic molecules, or compounds that contain carbon. *Biochemistry*, a branch of organic chemistry, is the study of the chemical reactions that occur in living systems. The simplest organic compounds are *hydrocarbons*, chains of carbon atoms with associated hydrogen atoms. Other organic compounds may contain *functional groups*, or parts of molecules that generally participate in chemical reactions as a unit and give the molecule some of its chemical properties. The relative amount of oxygen in a functional group is significant; groups with little oxygen are said to be *reduced* and groups with relatively more oxygen are said to be *oxidized*.

B. Carbohydrates

Carbohydrates serve as the main source of energy for most living things. There are three groups of carbohydrates: *monosaccharides*, *disaccharides*, and *polysaccharides*. Monosaccharides consist of a carbon chain or ring with several alcohol groups and either an aldehyde or a ketone group. Several monosaccharides are *isomers* (they have the same molecular formula but have different structures and different properties). *Disaccharides* are formed when two monosaccharides are connected by the removal of water and the formation of a *glycosidic bond* (a sugar alcohol/sugar linkage). *Polysaccharides* are formed when many monosaccharides are linked by glycosidic bonds.

C. Lipids

Lipids constitute a chemically diverse group of substances that includes *fats*, *phospholipids*, and *steroids*. Fats contain the three-carbon alcohol glycerol and one or more fatty acids (which can be either *saturated* or *unsaturated*). Phospholipids, which are found in all cell membranes, are identical to fats except that a phosphoric acid is found in place of one fatty acid. *Steroids* have a four-ring structure and are quite different from other lipids.

D. Proteins

Proteins are composed of building blocks called *amino acids*—a protein actually is a polymer of amino acids joined by *peptide bonds*. Proteins have several levels of structure: the *primary structure* (the specific amino acid sequence), the *secondary structure* (the folding or coiling of amino acid chains into a particular pattern), the *tertiary structure* (the further bending and folding of the protein molecule into globular shapes or fibrous threadlike strands), and sometimes the *quaternary structure* (the association of several tertiary-structured chains). Most proteins can be classified by their major functions as *structural proteins* (contributing to the three-dimensional structure of cells, cell parts, and membranes) or as *enzymes* (protein *catalysts* that control the rate of chemical reactions in cells).

E. Nucleotides and nucleic acids
Nucleotides consist of three parts: a nitrogenous base, a five-carbon sugar, and one or more phosphate groups. *Nucleic acids* consist of long polymers of nucleotides, called *polynucleotides*, and contain the genetic information that determines all the heritable characteristics of a living organism. The two nucleic acids found in living organisms are *ribonucleic acid (RNA)* and *deoxyribonucleic acid (DNA)*.

SELF-TESTS

Continue with this section only after you have read Chapter 2 of your textbook. Write your answers in the appropriate space provided. Correct answers to all questions can be found at the end of the Self-Test section. A score of 80 percent or better is good. If your score is less than 65, reread the chapter.

True/False

Mark T for True, F for False.

_____ **1.** Water is actually a polar compound.

_____ **2.** An atomic particle that has mass but lacks an electrical charge is a neutron.

_____ **3.** Covalent bonds are strong bonds formed by the exchange of electrons resulting in the formation of cations and anions.

_____ **4.** Glucose and fructose are isomers because they have the same structure but differ in the number of atoms they possess.

_____ **5.** Amino acids always possess amino, carboxyl, and variable (R) groups.

Multiple Choice

Select the best possible answer.

_____ **6.** Which of the following is not true, concerning water?
a. Water acts as a polar compound.
b. Water has a low specific heat.
c. Water forms thin layers due to its surface tension.
d. Water forms hydrogen bonds.

_____ **7.** Disruption of a protein's structure due to changes in pH and/or high temperatures is called:
a. denaturation.
b. hybridization.
c. hydrolysis.
d. complementation.

_____ **8.** The rule of octets refers to:
a. formation of isotopes.
b. chemically stable orbits.
c. the ratio of protons to neutrons.
d. the number of electron shells an atom possesses.

_____ **9.** Lipids can be characterized by:
a. being soluble in water.
b. possessing large numbers of carbon and oxygen atoms and few hydrogen atoms.
c. forming amino acids on hydrolysis.
d. possessing ester bonds.

_____ 10. Select the most correct statement:
   a. Amino acids are all acidic and nonpolar molecules.
   b. Nucleic acids consist of long polymers of nucleotides.
   c. Fatty acids are always unsaturated, while phospholipids are always saturated molecules.
   d. ATP is an energy-carrying catalytic protein.
   e. All of the above are correct statements.

_____ 11. Which of the following does not relate to the nucleic acid DNA?
   a. Hydrogen bonding between nucleotide bases.
   b. Complementary base pairing.
   c. A five-carbon ribose sugar.
   d. Guanine, cytosine, thymine, and adenine.
   e. All of the above relate to DNA.

Matching

Select the answer from the right-hand side that corresponds to the term or phrase on the left-hand side of the page. An answer may be used more than once. In some cases, more than one answer may be required.

_____ 12. The loss of an electron in a reaction

_____ 13. The gain of an electron in a reaction

_____ 14. A polysaccharide

_____ 15. The subunit of protein

_____ 16. The type of association between different polypeptide chains in a protein

_____ 17. Involved in peptide bonds

_____ 18. Genetic information of a cell

_____ 19. Glucose

_____ 20. Contains a carboxyl group

a. reduction
b. starch
c. amino acid
d. quaternary
e. DNA
f. monomer
g. oxidation

Fill-Ins

Provide the correct term or phrase for the following. Spelling counts.

21. The smallest chemical unit of matter is the _________________.
22. If atoms of two or more different elements combine, they form a _________________.
23. The atomic number of an element is equal to the number of its _________________.
24. The atomic particles that whirl around the atomic nucleus along predictable paths known as orbitals are called _________________.
25. A cation is a _________________ charged ion.
26. Atomic weight is defined as the sum of the number of protons and _________________ in an atom.
27. Atoms of a particular element that contain different numbers of neutrons are called _________________.
28. Bonds that form as a result of the attraction between ions of opposite charges are _________________ bonds.
29. Carbon and some other atoms in _________________ bonds share pairs of electrons.
30. Since catabolic reactions release energy, they can be called _________________ reactions.
31. Water molecules possess a surface tension as a result of _________________ bonding.
32. When water is added to a reactant to form simpler products, the reaction is called a _________________ reaction.

33. In a solution, the medium in which solutes are dissolved is the ________________.

34. Particles too large to form true solutions can sometimes form ________________.

35. A substance which acts as a proton acceptor or a hydroxyl ion donor is a ________________.

36. Alcohols, aldehydes, ketones, and organic acids are all ________________ groups that contain oxygen.

37. Lactose is a combination of two sugars and is therefore a ________________.

38. The bond that links two monosaccharides together is a ________________bond.

39. Starch, glycogen, and cellulose are all polymers of the sugar ________________.

40. A type of lipid found in cell membranes that possesses a phosphate group is a ________________.

41. A lipid that exhibits four-ring structures is a ________________.

42. Proteins are composed of building blocks called ________________, which have at least one amino group and one acidic carboxyl group.

43. When proteins fold into a helix, they exhibit ________________ structure.

44. A protein catalyst is called an ________________.

45. ________________ are the monomers (building blocks) of nucleic acids.

46. DNA contains the sugar ________________, which contains one less oxygen atom than the RNA sugar ribose.

47. A nitrogenous base found in RNA but not in DNA is ________________.

Labeling

Questions 48–55: Select the best possible label for each of the molecules below from the following list: nucleotide, phospholipid, saturated fatty acid, monosaccharide, amino acid, disaccharide, steroid, unsaturated fatty acid.

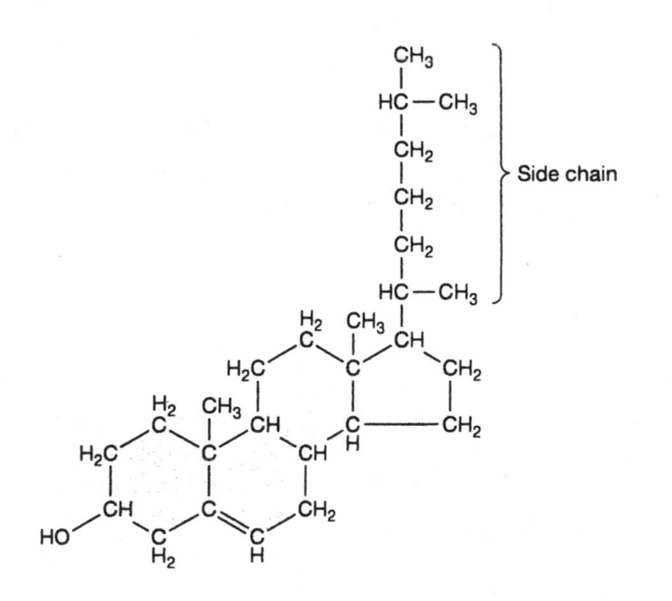

48. ________________

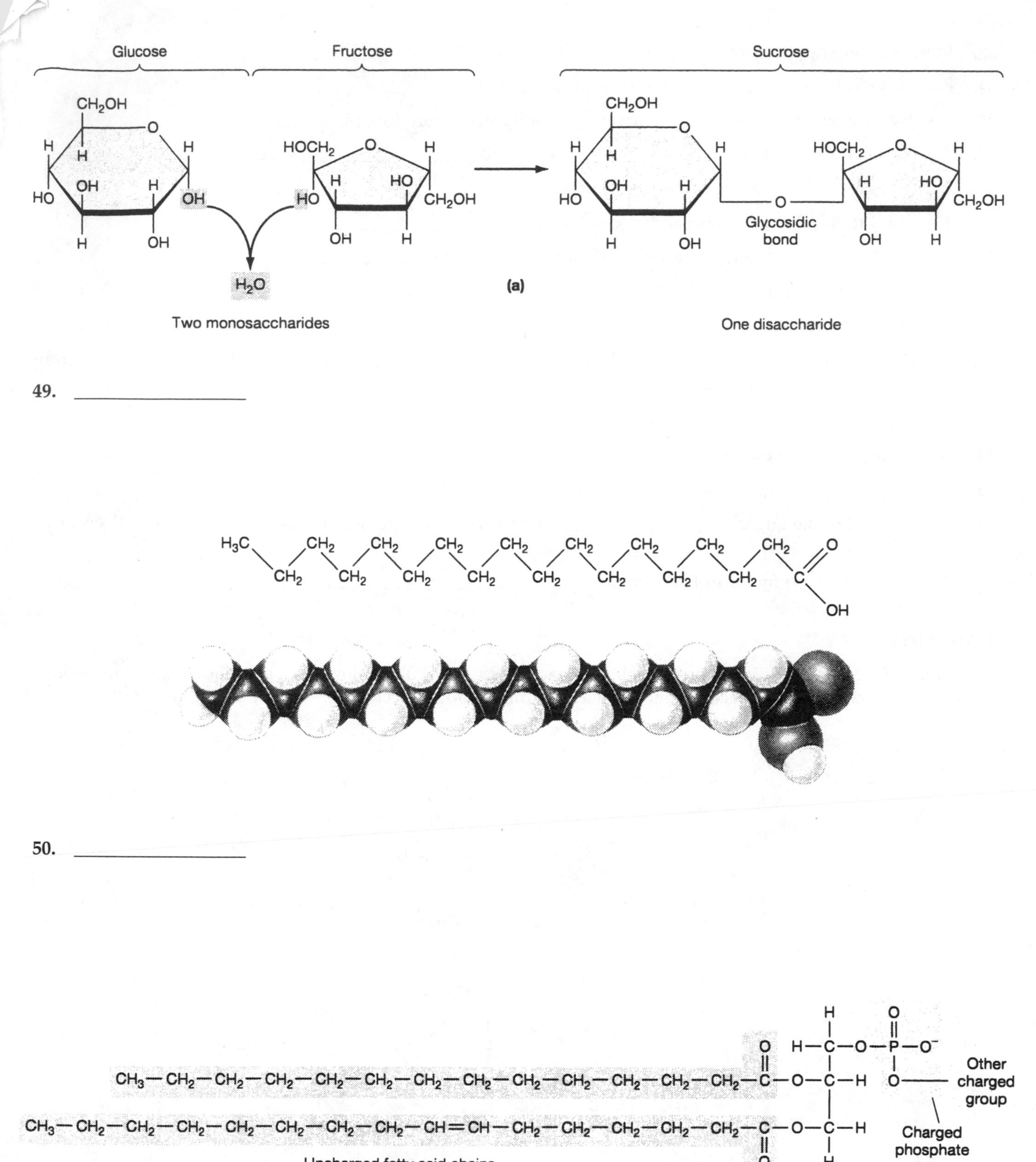
Glucose
Fructose
Sucrose
CH2OH
HOCH2
CH2OH
OH
HO
H
O
H2O
(a)
Glycosidic bond
Two monosaccharides
One disaccharide
H3C
CH2
C
O
OH
Uncharged fatty acid chains
Other charged group
Charged phosphate group
Glycerol portion

49. ____________________

50. ____________________

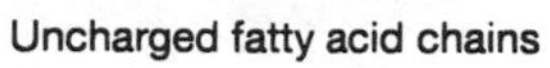

51. ____________________

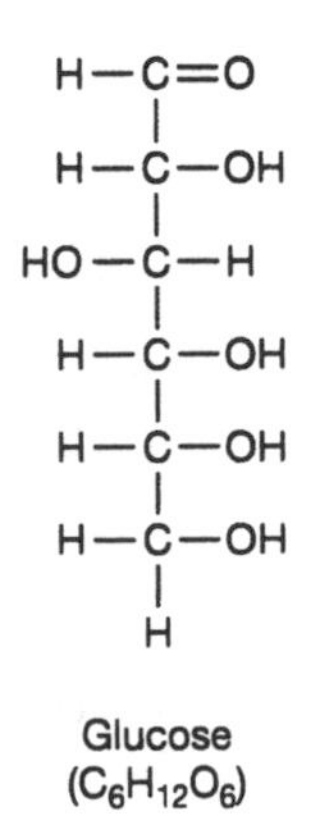

Glucose
($C_6H_{12}O_6$)

52. ____________________

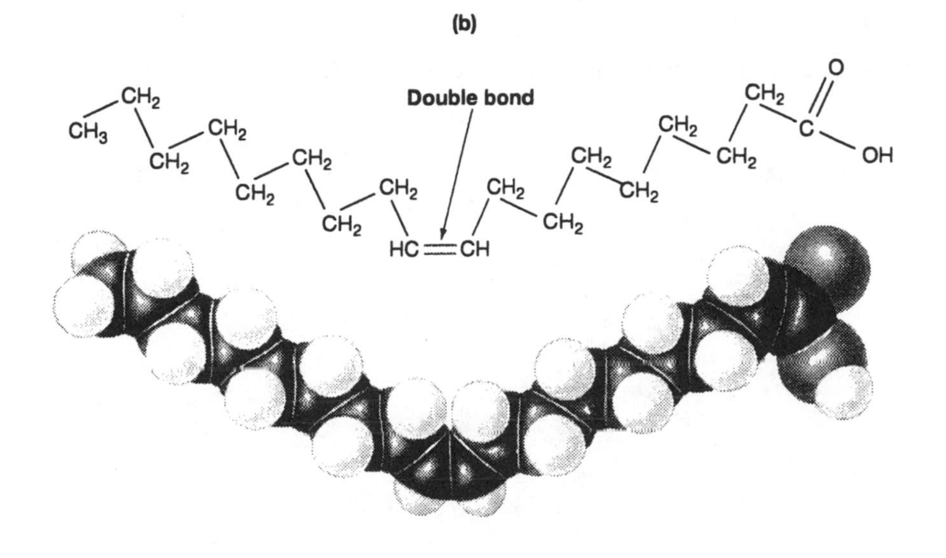

53. ____________________

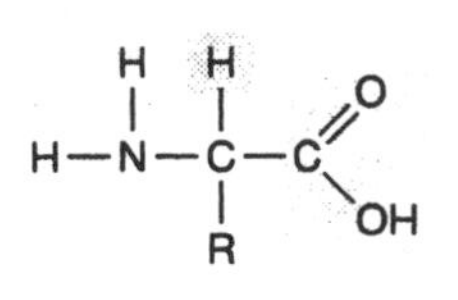

54. ____________________

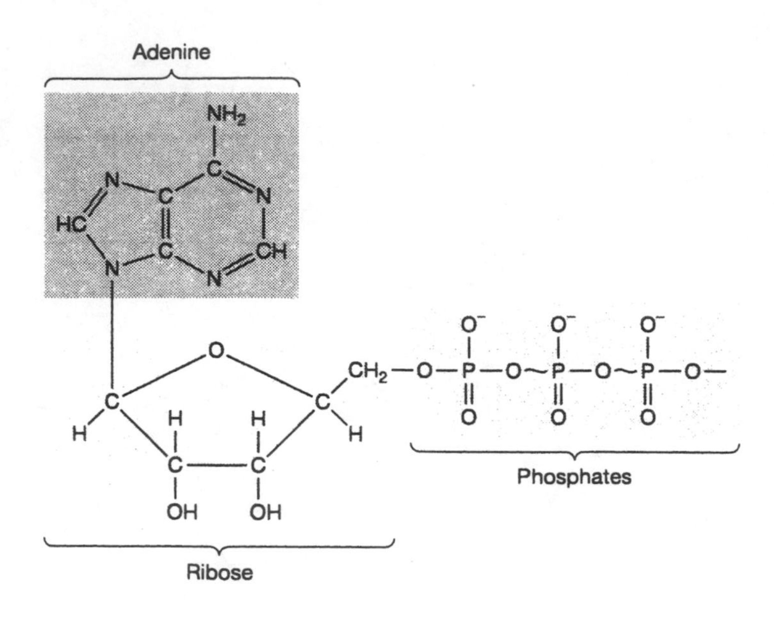

55. ________________

Critical Thinking

56. Why is knowledge of basic chemistry necessary to understanding microbiology?

57. What terms describe the organization of matter and what elements are found in living organisms?

58. What are the properties of chemical bonds and chemical reactions?

59. What properties of water, solutions, colloidal dispersions, acids, and bases make them important in living things?

60. What is organic chemistry and what are the major functional groups of organic molecules?

61. How do the structures and properties of carbohydrates contribute to their roles in living things?

62. How do the structures and properties of proteins, including enzymes, contribute to their roles in living things?

63. How do the structures and properties of simple lipids, compound lipids, and steroids contribute to their role in living things?

64. How do the structures and properties of nucleotides contribute to their role in living things?

ANSWERS

True/False

1. **True** Water molecules have a region of partial positive charge associated with the hydrogen atoms and a region of partial negative charge associated with the oxygen atoms. This allows water to act as an excellent dissolving medium.
2. **True** Neutrons possess a relative mass of one and occupy the nucleus, but do not possess a charge. Protons, however, possess a positive charge and have a mass of one.
3. **False** Covalent bonds are strong bonds that actually share electrons rather than exchange them. Ionic bonds exchange electrons and form ions.
4. **False** Glucose and fructose are isomers in that they have the same molecular formula but differ in their structure and properties.
5. **True** Amino acids are the building blocks of protein and consist of one amino group ($-NH_2$), one carboxyl group (–COOH), and one R or variable group which varies with each amino acid.

Multiple Choice

6. b; 7. a; 8. b; 9. d; 10. b; 11. c.

Matching

12. g; 13. a; 14. b; 15. c; 16. d; 17. c; 18. e; 19. f; 20. c.

Fill-Ins

21. atom; 22. compound; 23. protons; 24. electrons; 25. positively; 26. neutrons; 27. isotopes; 28. ionic; 29. covalent; 30. exergonic; 31. hydrogen; 32. hydrolysis; 33. solvent; 34. colloids; 35. base; 36. functional; 37. disaccharide; 38. glycosidic; 39. glucose; 40. phospholipid; 41. steroid; 42. amino acids; 43. secondary; 44. enzyme; 45. nucleotides; 46. deoxyribose; 47. uracil.

Labeling

48. steroid; 49. disaccharide; 50. saturated fatty acid; 51. phospholipid; 52. monosaccharide; 53. unsaturated fatty acid; 54. amino acid; 55. nucleotide.

Critical Thinking

56. **Why is knowledge of basic chemistry necessary to understanding microbiology?**
This is a common question asked by virtually all students. It would seem that chemistry should be studied only by chemists. That may have been true decades ago when basic structures of organisms were not understood, but today we need to know more about how those structures actually function. That involves chemistry. To understand how the cell membrane functions, it is necessary to know something about the molecules that make up the structure. To understand how our immune system works, it is necessary to know something about the proteins that form the antibodies. Without a basic knowledge of chemistry that is impossible. In most cases, only a moderate amount of information is required to understand even the most complex mechanisms and structures.

57. **What terms describe the organization of matter and what elements are found in living organisms?**
There are only a few basic terms that need to be understood. Elements are made up of invisible atoms which, in turn, consist of even smaller particles, namely, protons, neutrons, and electrons. The protons determine the atomic number of the atoms and therefore the nature of the element. They are found in the nucleus along with the neutrons, which together make up the atomic weight. The electrons are in energy levels or orbits found outside the nucleus and play a major role in how the atom combines with others. The number of protons which are positive in charge always equals the number of electrons which are negative in charge unless the atom has combined or reacted with other atoms. The first orbit is filled with two electrons, the second is filled with eight, as is the third orbit. Atoms whose outer electron shells are nearly full or nearly empty can either gain or lose a few electrons and thus react with other atoms.

58. **What are the properties of chemical bonds and chemical reactions?**
The loss or gain of electrons to arrive at completely full or empty orbits creates electrically charged atoms called ions. This allows atoms to form chemical bonds, in this case, called ionic bonds. Atoms that can accept or donate an equal number of electrons can also form bonds which are called covalent bonds. A third type of bond, the hydrogen or weak bond, is formed by an attraction of atomic particles rather than a donation or sharing of electrons. All three types of bonds hold the small and even the large organic molecules together. Chemical reactions allow for the breaking or forming of these chemical bonds and usually involve the release (exergonic) or the utilization (endergonic) of energy.

59. **What properties of water, solutions, colloidal dispersions, acids, and bases make them important in living things?**
Water is essential to all living organisms. Because it is a polar molecule, it acts as the perfect solvent or mixer. Water releases both a hydrogen ion and a hydroxyl ion so is neither an acid nor a base. Those molecules that release hydrogen ions can increase the acidity of a solution, and those that release hydroxyl ions can decrease the acidity. Chemists have devised the concept of pH to help measure the acidity of solutions. This is especially useful since living organisms can survive in only very narrow pH ranges.

60. **What is organic chemistry and what are the major functional groups of organic molecules?**
Molecules that contain carbon are organic chemicals whereas those that don't are inorganic. Organic molecules frequently have important reactive or functional groups which are characteristic of that molecule. For example, four significant groups of compounds—alcohols, aldehydes, ketones, and organic acids—have functional groups that contain oxygen. An alcohol has one or more hydroxyl groups (–OH). An aldehyde has a carbonyl group (–CO) at the end of the carbon chain; a ketone has a carbonyl group within the chain. An organic acid has one or more carboxyl groups (–COOH). One key functional group that does not contain oxygen is the amino group ($-NH_2$). Found in amino acids, amino groups account for the nitrogen in proteins.

61. **How do the structures and properties of carbohydrates contribute to their roles in living things?**
Carbohydrates act as the primary energy molecules for cells and consist of carbon chains with various attached functional groups. Monosaccharides such as glucose are simple sugars and are the easiest to metabolize. Disaccharides such as sucrose and lactose must first be split in half before they can be eaten. Polysaccharides are long chains of monosaccharides bonded in a chain and are used as storage sugars.

62. **How do the structures and properties of proteins, including enzymes, contribute to their roles in living things?**
Proteins act as structural and catalytic molecules for the cell. They are made of long chains of amino acids that can form into globular structures similar to a telephone cord that is coiled up by continuously twisting in opposite directions. Because of their functional amine and carboxyl groups, they are electrically charged as well. This allows them to form into important molecules such as membrane proteins, enzymes, and flagella. Because of their functional groups, they are, however, difficult to metabolize as food molecules and are not readily used for energy.

63. **How do the structures and properties of simple lipids, compound lipids, and steroids contribute to their role in living things?**
Lipids are long-chained and/or large cyclic organic molecules that are polarized, that is, they have electrically charged and uncharged ends. This allows them to function as effective insulating and storage molecules and as membranes, since their polarity allows them to orient automatically in water similar to a fishing bobber. The cyclic molecules can act as important membrane stabilizers and as chemical messengers.

64. **How do the structures and properties of nucleotides contribute to their role in living things?**
Nucleotides, because of their size and shape, act as energy carriers and as carriers of genetic information. The phosphate groups of adenosine triphosphate hold a large amount of energy and allow it to act as the energy currency molecule for the cell. The DNA nucleotides of adenine, thymine, guanine, and cytosine form double chains in the form of a helix and act to carry coded genetic information based on the sequences of the nucleotides within the chain. The RNA nucleotides of adenine, uracil, guanine, and cytosine form a single chain and act to decode the genetic information for the cell. The information coded by the nucleic acids allows for the formation of the wide variety of proteins needed by the cell.

Chapter 3 Microscopy and Staining

INTRODUCTION

In 1975, Roman Vishniac addressed a meeting of the International Society of Protozoology. Dr. Vishniac was 78 years old and began his talk by introducing his wife and telling the audience how much he loved her. It was an unusual and memorable introduction to a scientific meeting. Professor Vishniac's career was perhaps unique in the world of microscopy because Vishniac trained his camera on people as well as microbes. In addition to a vast library of images and videos of microbes, he is also famous for leaving us images of Europe's Jews in the 1930's. Of this work, Roman Vishniac said, "I was unable to save my people, only their memory". In a small effort to save my memory of Dr. Vishniac, I will present Dr. Vishniac's reflections on the first person to see living microbes and the role of the light microscope.

Dr. Vishniac studied the person who first discovered protozoa in 1677, Anton van Leewenhoek. Professor Vishniac described Holland in the 17th century and how Leewenhoek lost family members to the plague. In his grief he receives not flowers or food but a copy of *Micrographia* by Robert Hooke. While Leewenhoek was unable to read the English he could appreciate the images presented by Hooke and he devoted much of his life to the making of microscopes and the reporting his observations. Leewenhoek was also the first to report bacteria in 1683. The bacteria came from the plaque layer on his teeth. Controversy has surrounded his discovery of bacteria because Leeuwenhoek is known for making simple microscopes that may not be able to resolve bacteria. Professor Vishniac reported to have found a lost letter from Leewenhoek to a relative in which he mentioned a secret compound microscope that he kept behind a curtain. The letter included directions for his niece to destroy the secret microscope upon his death. This may explain why others could not image bacteria for over a hundred years. While the precise truth of the story remains unclear, the story points out the clear link between the history of microbiology and the history of the microscope.

Today, the light microscope is perhaps the most used and least understood instrument in clinical medicine. When we look at microbes through the light microscope we are faced with a problem not so much related to optics but to our nervous system. Consider, there are two basic types of objects in the world: amplitude and phase objects. Amplitude objects are the most common in our ordinary world. Amplitude objects interact with light by absorbing light intensity and thus appear darker than their surrounding. Consider looking at a black car on a bright day. It appears dark and easy to see because is reduces the amplitude or brightness of the light that strikes it. Occasionally, we encounter phase objects in our ordinary world. The author of this guide was in a building during the summer that had a lot of floor to ceiling glass. I made the unfortunate mistake of walking into a glass door. I learned first hand why the nose extends from the body. I had walked into a phase object and it hurt!

Light moves through air the way waves move through water. The distance from the crest on one wave to the next peak is called the wavelength. When light interacts with a phase object like clean glass, the height or amplitude of the wave is not affected. However, when light moves through a medium denser than air, like glass, it moves more slowly through the denser material compared with air. Thus, the wave that travels through the glass is slowed down or retarded relative to the next to it that moves through the air. Thus, the difference between the two waves is that the peaks of the waves will be offset with the retarded wave peaking slightly in front of the wave that went through the air. In order for the human eye and brain to understand a phase object, the phase object must retard the light more than a quarter of a wavelength. Most microbes and cellular organelles in eukaryotes are phase objects that retard light less than a quarter of a wave and thus while our microscopes can easily resolve them; our brains see little or nothing. This is part of the reason that we stain bacteria in the lab, when we apply colored stains to cells we are changing phase objects into amplitude objects. We also learn about the chemistry of the cells. The chapter will also describe optical ways to turn phase objects into amplitude objects, the phase contrast and Nomarski optical systems. Lastly, the chapter will explain how electron microscopes can resolve detail too small to be revealed with wavelengths of light.

STUDY OUTLINE

I. Microscopy Then and Now

Anton van Leeuwenhoek was probably the first person to see individual microorganisms. He saw them through simple single-lens microscopes that magnified objects 100 to 300 times.

II. Principles of Microscopy

A. Metric units

The units used in describing microorganisms are usually: the micrometer (μm), which is equal to 0.000001 m; the nanometer (nm), which is equal to 0.000000001 m; and the angstrom (Å), which is equal to 0.0000000001 m.

B. Properties of light: wavelength and resolution

The *wavelength* of light is the length of a light ray. *Resolution* refers to the ability to see two items as separate and discrete units rather than as a fuzzy, overlapped single image. The key to resolution is to get light of a short-enough wavelength to fit between the objects you want to see separately. The *resolving power (RP)* of a lens is a numerical measure of the resolution of the lens. You can calculate the RP of a lens if you know its *numerical aperture (NA)*, the widest cone of light that can enter a lens.

C. Properties of light: light and objects

If light strikes an object and bounces back, giving the object color, *reflection* has occurred. *Transmission* refers to the passage of light through an object. If light rays neither pass through nor bounce off an object but are taken up by the object, *absorption* has occurred. *Refraction* is the bending of light as it passes from one medium to another of different density. If light waves are broken up into bands of different wavelengths as they pass through a small opening, *diffraction* has occurred.

III. Light Microscopy

A. Basic features

Light microscopy refers to the use of any kind of microscope that uses visible light to make specimens observable. Over the years, several kinds of light microscopes have been developed, each adapted for making certain kinds of observations.

B. The compound light microscope

The compound light microscope, also known as the optical microscope or the light microscope, has more than one lens. The condenser of the compound light microscope causes light to be concentrated and transmitted directly through the specimen.

C. Darkfield microscopy

A darkfield microscope has a condenser that prevents light from being transmitted through the specimen. Instead, it causes light to reflect off the specimen at an angle. When these rays are gathered and focused into an image, a light object is seen on a dark background.

D. Phase-contrast microscopy

TA phase-contrast microscope has a condenser that accentuates small differences in the refractive index of various structures within a cell, causing objects in the cell to display different degrees of brightness.

E. Differential interference contrast (Nomarski) microscopy

Differential interference contrast microscopy uses differences in refractive index to visualize structures, producing a nearly three-dimensional image.

F. Fluorescence microscopy

In fluorescence microscopy, ultraviolet light is used to excite molecules so that they release light of different colors.

IV. Electron Microscopy

A. Basic features

The development of the electron microscope (EM) allowed subcellular structures to be visualized and studied. This type of microscope uses a beam of electrons, rather than a beam of light, and electromagnets, rather than glass lenses to produce an image.

B. Transmission electron microscope (TEM)

The transmission electron microscope (TEM) gives a better view of the internal structure of microbes than do other types of microscopes, magnifying objects up to 500,000X. Very thin slices of specimens are used.

C. Scanning electron microscope

The scanning electron microscope (SEM) is used to create images of the surfaces of specimens, magnifying objects up to 50,000X.

D. Scanning tunneling microscopy

The scanning tunneling microscope (STM) is used to create three-dimensional images and movies of individual molecules and atoms.

V. Techniques of Light Microscopy

A. Preparation of specimens for the light microscope

Wet mounts, in which a drop of medium containing the organisms is placed on a microscope slide, can be used to view living microorganisms. *Smears,* in which microorganisms from a loopful of medium are spread onto the surface of a glass slide, can be used to view killed organisms.

B. Principles of staining

A *simple stain* makes use of a single dye and reveals basic cell shapes and cell arrangements. A *differential stain* makes use of two or more dyes and distinguishes between two kinds of organisms or between two different parts of an organism.

C. The Gram stain

The Gram stain is a differential stain. Four groups of organisms can be distinguished with the Gram stain: (1) *gram-positive* organisms, which stain violet; (2) *gram-negative* organisms, which stain pink; (3) *gram-nonreactive* organisms, which do not stain or which stain poorly; and (4) *gram-variable* organisms, which stain unevenly.

D. The Ziehl-Neelsen acid-fast stain

The *Ziehl-Neelsen acid-fast stain* is used to detect tuberculosis- and leprosy-causing organisms of the genus *Mycobacterium*—these organisms stain bright red.

E. Special staining procedures

Negative staining stains the background around a specimen, leaving the specimen clear and unstained. *Flagellar staining* is a technique for observing flagella by coating the surfaces of flagella with a dye or a metal such as silver. *Endospore staining* uses the differential *Schaeffer-Fulton spore stain,* which makes endospores easier to visualize.

SELF-TESTS

Continue with this section only after you have read Chapter 3 of your textbook. Write your answers in the appropriate space provided. Correct answers to all questions can be found at the end of the Self-Test section. A score of 80 percent or better is good. If your score is less than 65, reread the chapter.

True/False

Mark T for True, F for False.

______ **1.** Leeuwenhoek constructed the first two-lens or compound microscope, which enabled him to observe tiny microbes.

______ **2.** Nomarski interference creates a microscopic image where the background is dark and the organism appears bright.

______ **3.** Immersion oil increases resolution but does not actually increase magnification.

______ **4.** The acid-fast stain allows for the observation and detection of mycobacteria.

______ **5.** Ocular micrometers are used to measure viruses when using the bright-field microscope, whereas a stage micrometer is used to measure bacteria.

Multiple Choice

Select the best possible answer.

_____ **6.** The total magnification of a microscope with the low power lens (10X) and ocular lens (15X) in position would be:
a. 25X.
b. 15X.
c. 150X.
d. 1500X.

_____ **7.** A major difference between the SEM and the TEM is that the SEM:
a. can resolve objects smaller than 20 nanometers.
b. requires less of a vacuum system than the TEM.
c. can create three-dimensional images.
d. does not require the use of any metal coating of the specimen.

_____ **8.** Heat fixation accomplishes all of the following except:
a. helps the dye to penetrate the cells
b. kills the bacteria on the slide.
c. decreases distortion of the cells prior to the addition of stains.
d. fixes the organisms to the slide.
e. All of the above are true.

_____ **9.** The condenser lens of a microscope:
a. increases the magnification.
b. generally can magnify an object ten times.
c. increases the light refraction
d. converges light beams onto the specimen.

_____ **10.** If a bacterium measures 0.3 micrometers, it would measure how many angstroms?
a. 300.
b. 30.
c. 3000.
d. 3.

_____ **11.** Select the most correct statement about the Gram stain:
a. Iodine acts as the decolorizer in the Gram stain.
b. Gram-positive cells retain the crystal violet dye and the gram-negative cells lose the dye following the decolorizer step.
c. To obtain the best Gram stain reaction, your cultures should be at least 48 hours old.
d. Bacteria become gram-variable because they can change from a gram-negative type of cell wall to a tough gram-positive type of cell wall.
e. All of the above are incorrect.

Matching

Select the answer from the right-hand side that corresponds to the term or phrase on the left-hand side of the page. An answer may be used more than once. In some cases, more than one answer may be required.

Topic: Compound Microscope Parts and Functions

_____ **12.** Coarse adjustment knob	**a.**	magnification
_____ **13.** Iris diaphragm lever	**b.**	concentration
_____ **14.** Ocular	**c.**	brightness control
_____ **15.** Objective (oil immersion)	**d.**	fine focusing adjustment
_____ **16.** Nosepiece	**e.**	none of the above

Topic: Microscopy

_____ **17.** Spore staining is applicable

_____ **18.** Ultraviolet light is used for illumination

_____ **19.** Demonstration of viruses

_____ **20.** Specimens are generally dead

_____ **21.** Shadow casting

a. bright-field microscopy

b. fluorescent microscopy

c. scanning electron microscopy

d. darkfield microscopy

e. transmission electron microscopy

Topic: Metric System

_____ **22.** 19 m

_____ **23.** 2900 mm

_____ **24.** 200 Å

_____ **25.** 1,000,000 nm

_____ **26.** 20,000 nm

_____ **27.** 10,000 pm

a. 2 mm

b. 100 mm

c. 190 pm

d. 10 nm

e. none of the above

Topic: Differential Staining Procedures

_____ **28.** Carbolfuchsin

_____ **29.** Malachite green

_____ **30.** Methylene blue

_____ **31.** Acid alcohol

_____ **32.** Crystal violet

_____ **33.** Safranin

_____ **34.** Gram's iodine

_____ **35.** Acetone-alcohol

a. counterstain in the Gram reaction

b. decolorizing agent in the acid-fast procedure

c. primary stain in the acid-fast procedure

d. counterstain in the acid-fast procedure

e. none of the above

Fill-Ins

Provide the correct term or phrase for the following. Spelling counts.

36. Bacteria can be measured in metric units called _________________.

37. Viruses should be measured in metric units called _________________.

38. The ability to see two items as separate items is known as _________________.

39. The numerical measure of the resolution of a lens is known as its _________________.

40. _________________ is the widest cone of light that can enter an objective lens.

41. When light passes through an object, this is known as _________________.

42. _________________ refers to the reemission of absorbed light as light of longer wavelengths.

43. To cut down on the refraction of light, microscopists use _________________, which has the same index of refraction of glass.

44. A compound microscope with a single eyepiece is said to be _________________.

45. The _________________ of a microscope controls light intensity.

46. A microscope adjustment knob that changes the distance between the lens and the specimen very slowly is the _________________.

47. _______________ describes a microscope in which objectives can be changed without major focusing adjustments.

48. In order to observe live, unstained organisms, you must use a _______________ microscope.

49. In _______________ microscopy, ultraviolet light is used to excite molecules so that they release light of a longer wavelength (a different color).

50. Fluorescent dyes used to stain organisms that cause tuberculosis and syphilis are called _______________.

51. In microscopy, the thickness of a specimen that is in focus at any one time is called the _______________ of _______________.

52. _______________ serve as sources of illumination for both transmission and scanning electron microscopes.

53. The spraying of a heavy metal such as gold at an angle in the preparation of electron microscope specimens is known as _______________ _______________.

54. _______________ _______________ accomplishes three things: (1) it kills the organisms; (2) it causes the organisms to adhere to the slide; and (3) it alters the organisms so that they more readily accept stains (dyes).

55. The application of one dye solution to a smear is representative of a _______________ staining procedure.

56. Iodine acts as a _______________ in the Gram stain to fix the basic dye into the cell wall.

57. Gram-positive organisms stain _______________.

58. Acid-fast organisms stain _______________

59. The decolorizer in the acid-fast procedure is _______________.

60. Stains that color the background and not the organism are called _______________ stains.

61. Certain bacteria form highly resistant structures called _______________; the Schaeffer-Fulton stain makes these structures easier to see.

Labeling

Questions 62–74: Select the best possible label for each part of the microscope from the following list: base, eyepiece (ocular lens), arm, condenser, coarse and fine focusing adjustment knobs, body tube, illuminator, camera attachment tube, stage, light source, specimen, light pathway, objective lens.

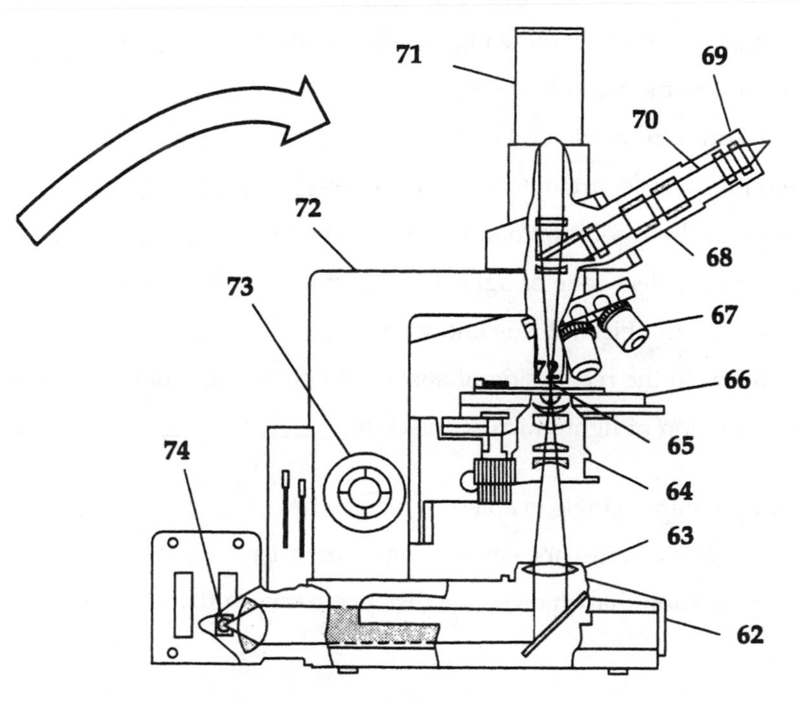

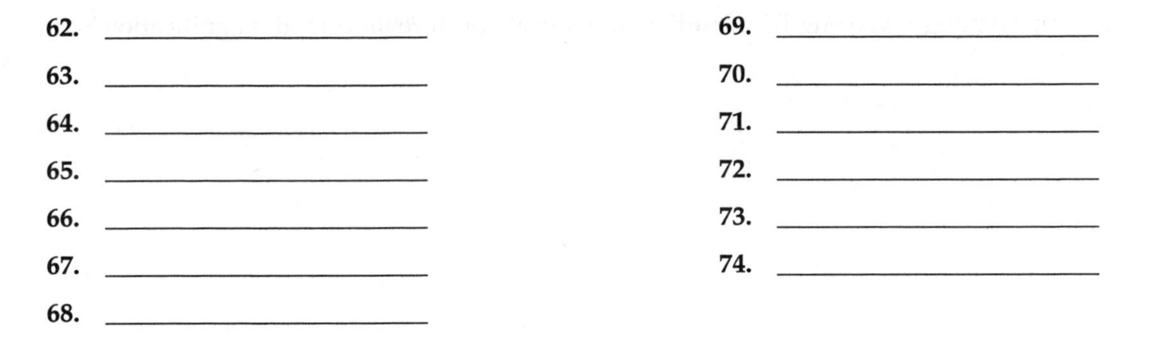

Critical Thinking

75. How is the evolution of microscopy instruments related to progress in microbiology?

76. Which metric units are most useful for the measurement of microbes?

77. What are the relationships among wavelength, resolution, numerical aperture, and total magnification?

78. How are the following properties of light related to microbiology: transmission, absorption, fluorescence, luminescence, phosphorescence, reflection, refraction, and diffraction?

79. What is the function of each part of a compound microscope?

80. What are the special uses and adaptations of brightfield, darkfield, phase-contrast, differential interference contrast, and fluorescence (UV) microscopes?

81. What are the principles of transmission and scanning electron microscopy? How do the advantages and limitations of electron microscopy compare with those of light microscopy?

82. What techniques are used to prepare and heighten contrast in specimens to be viewed with a light microscope?

83. What are the uses of the common types of microbial stains?

84. What are the functions and results of each of the steps in the Gram staining procedure?

ANSWERS

True/False

1. **False** Leeuwenhoek was the first to see microbes, but he constructed a very crude single lens microscope.
2. **False** Nomarski interference contrast produces images similar to phase contrast, where the image is virtually 3-D.
3. **True** Immersion oil has the same refractive index of glass and thus allows light to pass from the specimen to the lenses without refraction and loss of resolution.
4. **True** Mycobacteria have complex lipid components in their cell walls which allow them to resist the acid alcohol used in the acid-fast stain.
5. **False** Ocular and stage micrometers are both used to measure bacteria, but since viruses cannot be seen with a light microscope, they are not used for their measurement.

Multiple Choice

6. c; 7. c; 8. c; 9. d; 10. c; 11. b.

Matching

12. e; 13. c; 14. a; 15. a; 16. e; 17. a; 18. b; 19. c and e; 20. c and e; 21. e; 22. e; 23. e; 24. e; 25. e; 26. e; 27. d; 28. c; 29. e; 30. d; 31. b; 32. e; 33. a; 34. e; 35. e.

Fill-Ins

36. micrometers; 37. nanometers; 38. resolution; 39. resolving power; 40. numerical aperture; 41. transmission; 42. luminescence; 43. immersion oil; 44. monocular; 45. condenser; 46. fine adjustment; 47. parfocal; 48. phase-contrast; 49. fluorescence; 50. fluorochromes; 51. depth of field; 52. electrons; 53. shadow casting; 54. heat fixation; 55. simple; 56. mordant; 57. purple; 58. red; 59. acid alcohol; 60. negative; 61. endospores.

Labeling

62. base; 63. illuminator; 64. condenser; 65. specimen; 66. stage; 67. objective lens; 68. body tube; 69. eyepiece (ocular lens); 70. light pathway; 71. camera attachment tube; 72. arm; 73. coarse and fine focusing adjustment knobs; 74. light source.

Critical Thinking

75. **How is the evolution of microscopy instruments related to progress in microbiology?**
Although the activities of bacteria and other microbes were apparent for thousands of years, the organisms were completely invisible to the eye. Therefore, the existence of these organisms was understandably delayed until the advent of the microscope. However, the first microscopes built were probably used to observe objects already visible such as fly wings, ants, and plant leaves. Fortunately, the inquisitive mind of Anton van Leeuwenhoek developed a microscope to see and record the presence of these tiny animalcules. As new microscopic techniques were discovered, more and more information about the structure and nature of these organisms was obtained.

76. **Which metric units are most useful for the measurement of microbes?**
Since bacteria are too small to be seen or to be measured using the standard English system, a system of measurement called the metric system was developed. In this system, a meter is approximately equal to a yard and a millimeter is approximately equal to 1/16 of an inch. Both of these units can be visualized with the naked eye. The unit used to measure bacteria is the micrometer, which is 1/1000 of a millimeter, and the unit used to measure viruses is the nanometer, which is 1/1000 of a micrometer. Both of these units are too small for us to see but they do provide us with units to measure these very tiny organisms.

77. What are the relationships among wavelength, resolution, numerical aperture, and total magnification?
Most of our microscopes use light as the energy source. Since light must pass between two objects for them to be resolved, it becomes the limiting factor.

If the wavelength is too long to pass through them they will appear as one object. Therefore the smaller the objects, the harder it is to see or resolve them. Numerical aperture relates to the cone of light that enters the lens. The wider it is, the greater the resolving power of the lens.

78. How are the following properties of light related to microbiology: transmission, absorption, fluorescence, luminescence, phosphorescence, reflection, refraction, and diffraction?
Light exhibits a wide variety of properties as it passes through substances of different densities. Each of these properties affects how we can see objects viewed under a microscope. Transmission refers to the passage of light through an object. Absorption is defined as a process in which light rays are neither passed through nor reflected off an object but are retained. Fluorescence is the emission of light of one color when irradiated with another, shorter wavelength of light. Luminescence refers to a process in which absorbed light rays are reemitted at longer wavelengths. Phosphorescence is defined as continued emission of light by an object when light rays no longer strike it. Reflection is the bouncing of light as it passes from one medium to another medium of different density. Refraction refers to the bending of light as it passes from one medium to another medium of different density. Diffraction is defined as a phenomenon in which light waves, as they pass through a small opening, are broken up into bands of different wavelengths. Remember that the use of immersion oil in the oil immersion technique prevents the loss of light due to refraction.

79. What is the function of each part of a compound microscope?
All compound light microscopes have virtually the same components. Most of these are self-explanatory; however, be sure to understand the functions of the condenser lens, which converges light but does not magnify the object, and the iris diaphragm, which controls the amount of light.

80. What are the special uses and adaptations of brightfield, darkfield, phase contrast, differential interference contrast, and fluorescence (UV) microscopes?
The brightfield microscope, which is the most commonly used laboratory microscope, is characterized by a bright background and dark object and is best used to visualize stained specimens. The darkfield microscope exhibits a dark background and bright objects and is best used with live specimens. The phase contrast provides a contrast between the object and background and can be used with both stained and unstained specimens. Differential interference microscopes are very similar to phase contrast but provide a nearly three-dimensional image and are usually used with live specimens. Fluorescence microscopy involves the use of fluorescent dyes and is best used in clinical laboratories for diagnosis.

81. What are the principles of transmission and scanning electron microscopy? How do the advantages and limitations of electron microscopy compare with those of light microscopy?
Electron microscopes differ from light microscopes in five important ways. The electron scope is more expensive and difficult to use; it uses electrons as an energy source; it uses magnets to focus the electrons; and it always requires a viewing screen or photographic plate to view the object. The use of electrons with very short wavelengths provides greater magnification and resolution of the objects; however, specimen preparation is far more difficult.

The TEM blasts electrons at the specimen and provides a flat image of the object while the SEM scans the object with a thin beam, thus generating a 3-D image.

82. What techniques are used to prepare and heighten contrast in specimens to be viewed with a light microscope?
Wet mounts are the simplest techniques to use and only require a slide, cover slip, and a specimen. The organisms are live and unstained and generally hard to see. By air drying, heat fixing, and staining them with dyes, the organisms can be visualized.

83. What are the uses of the common types of microbial stains?
The acid-fast stain allows detection of mycobacteria which may cause leprosy or tuberculosis. The Schaeffer-Fulton spore stain allows visualization of resistant endospores produced by members of the genus *Bacillus* and *Clostridium*. Negative stains use acidic dyes that do not penetrate or damage the cells and thus can be used to visualize fragile organisms. Flagellar stains are used to exhibit the flagella of motile organisms.

84. What are the functions and results of each of the steps in the Gram staining procedure?

The Gram stain is the most important stain in microbiology. The first step is the use of crystal violet dye, which stains all organisms purple. The second step is the use of iodine, which acts as a mordant to fix the dye into the cell wall layers. All organisms are still purple. The third step is the alcohol decolorizer step which removes the dye from the gram negative cells. Alcohol dissolves the lipids found in the outer layer of gram negative cells. The final step is the addition of safranine, which provides a contrasting color to the colorless gram negative cells.

Chapter 4 Characteristics of Prokaryotic and Eukaryotic Cells

INTRODUCTION

Some years ago (1974), Christian du Duve received the Nobel Prize for his work on cell structure. When he received the award he said, "We are sick because our cells are sick". In our study of microbiology, it is often cells or cell products that make us sick. In his study of the enzymes of lysosomes, Dr. de Duve discovered that even a small defect in a single enzyme could prove fatal. In our study of microbes, seeming minor differences in the structure and chemistry of a cell may also prove very difficult for us. Thus, the details of the structure and chemistry of both prokaryotes and eukaryotes is a central concern for us.

Knowledge of cell structure and diversity will help us to understand microbiological principles that range from pathology to ecology and evolution. Here, our understanding of chemistry, microscopy, and staining that you just read about in previous chapters will be applied. After reading this chapter, you will be ready to understand how antibiotics affect microorganisms, how some microorganisms become resistant to antibiotics and disinfectants, and how your body responds to different microbial invaders. You will also be prepared to understand the relationships between organisms that is the basis for all life on Earth. In addition, I hope that you grasp a small measure of the beauty that underlies all biological form and organization.

STUDY OUTLINE

I. Basic Cell Types

A. Prokaryotic cells

Prokaryotic cells lack a nucleus and other membrane-bound structures.

B. Eukaryotic cells

Eukaryotic cells have a nucleus and membrane-bound organelles, or "little organs."

II. Prokaryotic Cells

A. Size, shape, and arrangement

Prokaryotes are among the smallest of all organisms—they are usually from 0.5 to 2.0 mm in diameter—and therefore have a large surface-to-volume ratio. Typically, prokaryotes display three basic shapes: spherical (*coccus*), rodlike (*bacillus*), and spiral (*vibrio, spirillum,* or *spirochete*). In addition, prokaryotic cells can be found in distinctive arrangements: division in one plane produces cells in pairs (indicated by the prefix *diplo-*) or in chains (*strepto-*); division in two planes produces cells in *tetrads*; division in three planes produces *sarcinae*; and random division planes produce grapelike clusters (*staphylo-*).

B. Overview of structure

Structurally, prokaryotic cells consist of the following: a cell membrane, an internal cytoplasm, and a variety of external structures.

C. Cell wall

The semirigid *cell wall* maintains the characteristic shape of the prokaryotic cell and prevents the cell from bursting when fluids flow into the cell. Its single most important component is *peptidoglycan*; other components include the *outer membrane* and the *periplasmic space*. Certain properties of cell walls result in different staining reactions. On the basis of these reactions, you can distinguish between *gram-positive, gram-negative*, and *acid-fast bacteria*. Some bacteria, like those of the genus *Mycoplasma*, lack a cell wall. While other bacteria, may exist in wall-deficient atrain under some conditions.

(1) Gram-positive bacteria have a thick peptidoglycan layer and teichoic acid polymers.

(2) Gram-negative bacteria have a thin peptidoglycan layer and an outer bilayer membrane with endotoxin.

(3) Acid-fast bacteria (mycobacteria) have a thick wall but with less peptidoglycan and more lipid than Gram-positive bacteria.

D. Cell membrane

The *cell membrane*, otherwise known as the *plasma membrane*, is a living membrane that forms the boundary between a cell and its environment. It is selectively permeable and is made up of lipids and proteins.

E. Internal structure

Internal structures found in bacterial cells include: (1) *ribosomes*, which serve as sites for protein synthesis; (2) a *nuclear region*, or *nucleoid*, which consists mainly of DNA but has some RNA and protein associated with it; (3) *internal membrane systems*, which are sometimes known as *chromatophores* or *mesosomes*; (4) *inclusions*, which are small bodies in the cytoplasm; and (5) *endospores*, which help the organism survive.

F. External structure

External structures of bacterial cells include: (1) *flagella*, which help bacteria to move in processes like *chemotaxis* or *phototaxis*; (2) *axial filaments*, which cause rigid spirochetes to rotate like a corkscrew; (3) *pili*, which are used to attach bacteria to surfaces; and (4) *glycocalyx* substances, which include capsules and slime layers.

III. Eukaryotic Cells

A. Overview of structure

Eukaryotic cells are larger and more complex than prokaryotic cells. They contain a variety of highly differentiated structures.

B. Plasma membrane

The *plasma membranes* of eukaryotic cells are similar to the cell membranes of prokaryotic cells, but they are less versatile and contain a greater variety of lipids.

C. Internal structure

Internal structures found in eukaryotic cells include: (1) *cytoplasm*, which is a semifluid substance consisting mainly of water; (2) a *cell nucleus*, which contains DNA, some RNA, and some proteins; (3) *mitochondria*, which are the powerhouses of eukaryotic cells; (4) *chloroplasts*, which carry out photosynthesis; (4) *ribosomes*, which provide sites for protein synthesis; (5) *endoplasmic reticulum*, which forms the vesicles that transport lipids and proteins to the Golgi apparatus; (6) *lysosomes*, which contain digestive enzymes that digest the substances in vacuoles; (7) *peroxisomes*, which oxidize amino acids in animal cells and fats in plant cells; (8) *vacuoles*, which store materials to be used for energy; and (9) a *cytoskeleton*, which supports and gives rigidity and shape to a cell.

D. External structure

External structures of bacterial cells include: (1) *flagella*, which help eukaryotic cells to move; (2) *cilia*, which allow organisms to move rapidly; (3) *pseudopodia*, which are temporary cytoplasmic extensions associated with *amoeboid movement*; and (4) *cell walls*, which are made up mainly of cellulose (algae) or chitin (fungi).

IV. Evolution by Endosymbiosis

A. The endosymbiont theory

The *endosymbiont theory* states that the organelles of eukaryotic cells arose from prokaryotic cells that developed *symbiotic* relationships with the eukaryote-to-be.

B. Evidence for the endosymbiont theory
(1) the mitochondria and chloroplasts are about the same size as prokaryotic cells;
(2) both the mitochondri and chloroplasts contain their own DNA that resembles the DNA of modern bacteria; and (3) protein synthesis in eukaryotic ribosomes is similar to the process found in bacteria.

V. Movement of Substances Across Membranes

A. Basic characteristics
The mechanism by which substances move across membranes can be a passive process (which would include *simple diffusion, facilitated diffusion*, or *osmosis*) or an active process (which would include *endocytosis* and *exocytosis*).

B. Simple diffusion
Simple diffusion is the net movement of particles from a region of higher to lower concentration. Diffusion through the phospholipid bilayer depends on: (1) the solubility of the diffusing substance in lipid; (2) the temperature; and (3) the difference between the highest and lowest concentration of the diffusing substance.

C. Facilitated diffusion
Facilitated diffusion is diffusion down a concentration gradient and across a membrane with the assistance of special pores or carrier molecules.

D. Osmosis
Osmosis is diffusion in which water molecules diffuse across a selectively-permeable membrane. Based on this definition of osmosis, if the fluid surrounding the cells is *isotonic* to the cells, no change in volume will occur; if the fluid is *hypertonic* to the cells, the cells will shrivel or shrink as water moves out of them into the fluid environment; and if the fluid is *hypotonic*, the cells will swell or burst as water moves from the environment into the cells.

E. Active transport
Active transport moves molecules and ions against concentration gradients from regions of lower concentration to those of higher concentration. This kind of transport requires the cell to expend energy.

F. Endocytosis and exocytosis
In *endocytosis*, invagination of the plasma membrane forms vesicles, moving substances into a cell. In *exocytosis*, vesicles inside a cell fuse with the plasma membrane and release their contents from the cell.

SELF-TESTS

Continue with this section only after you have read Chapter 4 of your textbook. Write your answers in the appropriate space provided. Correct answers to all questions can be found at the end of the Self-Test section. A score of 80 percent or better is good. If your score is less than 65, reread the chapter.

True/False

Mark T for True, F for False.

_____ **1.** All prokaryotic and eukaryotic cells possess nuclear material and a fluid mosaic plasma membrane.

_____ **2.** Spiral bacteria commonly exhibit multicellular forms such as chains, tetrads, and grapelike clusters.

_____ **3.** Gram-negative bacteria possess a complex lipopolysaccharide cell wall complete with a lipid bilayer, but lack the peptidoglycan component found in Gram-positive organisms.

_____ **4.** The nuclear membranes of all eukaryotic cells contain pores to allow for communication between the nucleus and the cytoplasm.

_____ **5.** Both mitochondria and golgi bodies found in eukaryotes contain DNA and can replicate independently.

_____ **6.** Hypertonic solutions may cause cells to lose water and shrink.

_____ **7.** When gram-negative bacteria are killed, their toxicity is also destroyed.

Multiple Choice

Select the best possible answer.

_____ 8. Which of the following would not be consistent with eukaryotic organisms?
a. membrane bound organelles.
b. presence of histones.
c. cell membranes lacking sterols.
d. paired chromosomes.

_____ 9. The peptidoglycan layer of the cell wall:
a. consists of lipopolysaccharide.
b. is formed from repeating molecules of gluNAc and murNAc.
c. represents the lipid bilayer.
d. lacks teichoic acids.

_____ 10. Dipicolinic acid is commonly associated with:
a. endospore coats.
b. lipopolysaccharide of Gram-negative bacteria.
c. peptidoglycan layer of Gram-positive bacteria.
d. mesosomes.
e. None of the above is true.

_____ 11. Bacterial flagella:
a. attach to the cell wall via the teichoic acids and calcium.
b. form a hook after leaving the cell.
c. are about the same size as eukaryotic flagella.
d. are composed of lipopolysaccharide units called flagellin.

Matching

Select the answer from the right-hand side that corresponds to the term or phrase on the left-hand side of the page. An answer may be used more than once. In some cases, more than one answer may be required.

Topic: Major Chemical Composition of Prokaryotic Structures

_____ 12. Cell wall
_____ 13. Cytoplasmic (cell) membrane
_____ 14. Capsule
_____ 15. Endospore
_____ 16. Pilus
_____ 17. Flagellum
_____ 18. Plasmid
_____ 19. Ribosome
_____ 20. Glycocalyx
_____ 21. Cytoplasmic granules
_____ 22. Nucleoid
_____ 23. Genome

a. *N*-acetylmuramic acid, lipopolysaccharides and protein
b. polysaccharide
c. calcium ions and dipicolinic acid
d. DNA
e. RNA and protein
f. glycogen or polyphosphate
g. lipid, protein
h. mainly protein

Topic: Prokaryotic Structure and Function

_____ 24. Nucleoid
_____ 25. Ribosome

a. locomotor organelle
b. regulates substances into and out of cells

______	**26.** Sex pili	**c.**	protects against extreme heat
______	**27.** Cell wall	**d.**	provides shape
______	**28.** Axial filament	**e.**	protects against phagocytosis
______	**29.** Capsule	**f.**	protects from drying; binds to form dental plaque
______	**30.** Cytoplasmic membrane	**g.**	extrachromosomal pieces of DNA
______	**31.** Slime	**h.**	site for protein synthesis
______	**32.** Protoplast	**i.**	contains one chromosome
______	**33.** Murein	**j.**	photosynthesis
______	**34.** Plasmid	**k.**	regulates depth for floating
______	**35.** Thylakoids	**l.**	none of the above
______	**36.** Gas vacuoles		
______	**37.** Endospores		

Topic: Eukaryotic Structure and Function

______	**38.** A major site of active transport	**a.**	plasma membrane
______	**39.** Consist mainly of protein	**b.**	cell wall
______	**40.** ATP production	**c.**	mitochondrion
______	**41.** Major site for photosynthesis	**d.**	Golgi apparatus
______	**42.** Structures bound to the surfaces of an endoplasmic reticulum	**e.**	ribosomes
		f.	nucleolus
______	**43.** Contain grana and stroma	**g.**	rough endoplasmic reticulum
______	**44.** Responsible for the form and shape of plant cells	**h.**	lysosomes
______	**45.** Contain a large number of hydrolytic enzymes	**i.**	peroxisomes
		j.	flagella
______	**46.** Involved in amino acid and carbohydrate utilization	**k.**	chloroplasts
		l.	none of the above
______	**47.** Regulates passage of materials into and out of cells		
______	**48.** Site of ribosome assembly		
______	**49.** Contain enzymes that convert hydrogen peroxide to water		

Fill-Ins

Provide the correct term or phrase for the following. Spelling counts.

50. Organisms that lack a defined nucleus are called ________________ cells.

51. ________________ bacteria vary widely in their cellular form, even within a single culture.

52. Cocci appearing in pairs are referred to as ________________.

53. Cocci appearing in grapelike clusters are referred to as exhibiting a ________________ arrangement.

54. A tough, interlinked component of the cell wall of bacteria that provides rigidity is the ________________.

55. The two repeating subunits of the peptidoglycan layer are _______________ and _______________.

56. The molecules in gram-positive cell walls that bacteriophages (viruses that infect bacteria) attach to are called _______________ molecules.

57. A toxic component of the outer layer of Gram-negative bacteria is _______________.

58. If a gram-positive organism loses its cell wall, the resulting structure is called a _______________.

59. If a gram-negative organism loses its cell wall, the resulting structure is called a _______________.

60. An antibiotic that affects the formation of a bacterial cell wall is _______________

61. The _______________ - _______________ model represents the current understanding of the cell or plasma membrane.

62. Nonpolar hydrocarbon ends of fatty acids are said to be _______________, or "water-fearing."

63. Sedimentation rates of structures such as ribosomes are expressed in terms of _______________ units.

64. The semifluid substance inside the cell membrane that is about four-fifths water and one-fifth substances dissolved or suspended in this water is called the _______________.

65. Chains of ribosomes are called _______________.

66. One of the major features differentiating prokaryotic cells from eukaryotic cells is the absence of a membrane-bound _______________.

67. Polyphosphate granules within the cytoplasm of bacteria are called _______________ or _______________ granules.

68. Members of the genus *Bacillus* and *Clostridium* produce resistant structures called _______________.

69. Flagella distributed all over the surface of bacteria are called _______________flagella.

70. Spirochetes may possess the ability to rotate like a corkscrew, due to the presence of _______________ _______________, or endoflagella.

71. Any substance containing polysaccharides found external to the cell wall is called the _______________.

72. The _______________ in eukaryotes provides support and shape to the cell.

73. The _______________ located inside a eukaryotic cell's nucleus serves as the site of ribosome assembly.

74. _______________ are proteins bound to the DNA of eukaryotic cells.

75. Mitochondria are characterized by extensive inner membrane folds called _______________.

76. The production of lipids in a eukaryotic cell is associated with the smooth _______________ _______________.

77. A eukaryotic organelle that contains digestive enzymes is a _______________.

78. A membrane that allows the passage of only a selected group of substances is said to be _______________ _______________.

79. Movement of molecules from a region of higher to lower concentration by means of a carrier protein is called _______________ _______________.

80. Cells immersed in an _______________ solution exhibit no change in their cell volume.

81. The energy-requiring process by which substances are chemically changed as they move across a membrane is known as _______________ _______________.

Labeling

Questions 82–87: Select the best possible label for each of the most common bacterial shapes from the following list: coccus, spirillum, bacillus, vibrio, coccobacillus, spirochete. *Place your answers in the spaces provided.*

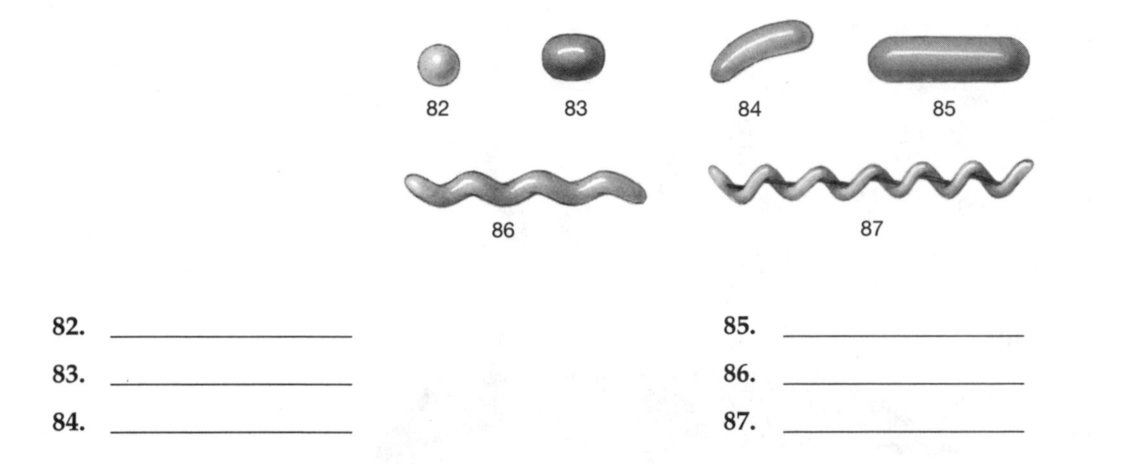

82. ____________________
83. ____________________
84. ____________________
85. ____________________
86. ____________________
87. ____________________

Questions 88–97: Select the best possible label for each structure of Prokaryotic cell from the following list: cell membrane, pilus (fimbrae), flagellum, capsule or slime layer, cell wall, inclusion, plasmid, cytoplasm, ribosomes, chromosomes.

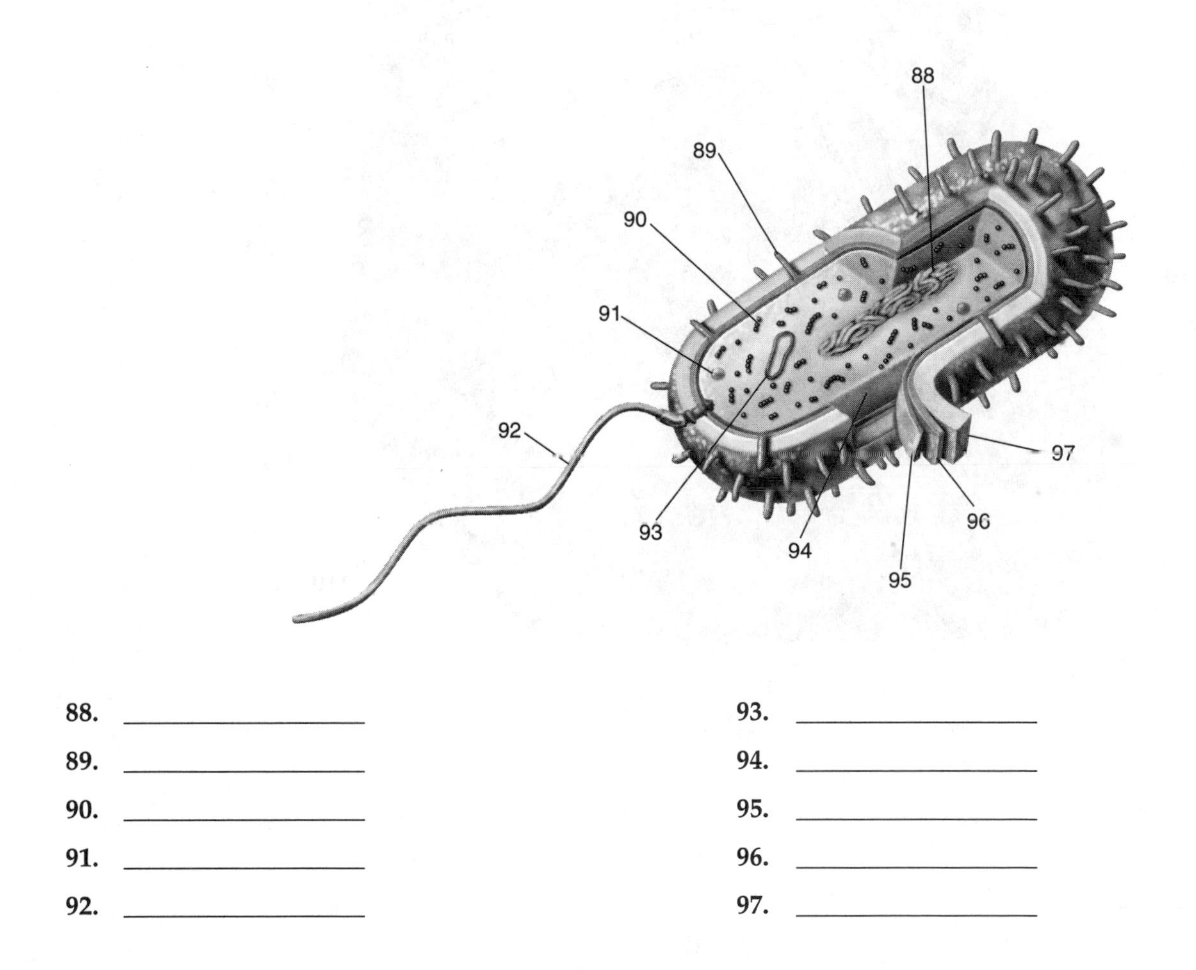

88. ____________________
89. ____________________
90. ____________________
91. ____________________
92. ____________________
93. ____________________
94. ____________________
95. ____________________
96. ____________________
97. ____________________

Labeling (cont'd)

Questions 98–117: Select the best possible label for each structure of a eukaryotic cell from the following list: ribosomes, nuclear envelope, chloroplast (in plants), nucleolus, secretory vesicle, cilia (in animals and protists), cytoplasm, nucleus, thylakoid, rough endoplasmic reticulum, mitochondrion, centrioles (in animals), microvilli (in animals), nucleoplasm, smooth endoplasmic reticulum, pinocytotic vesicle, lysosome (in animals), Golgi apparatus, nuclear pore, and nuclear chromatin. *Place your answers in the spaces provided on page 46.*

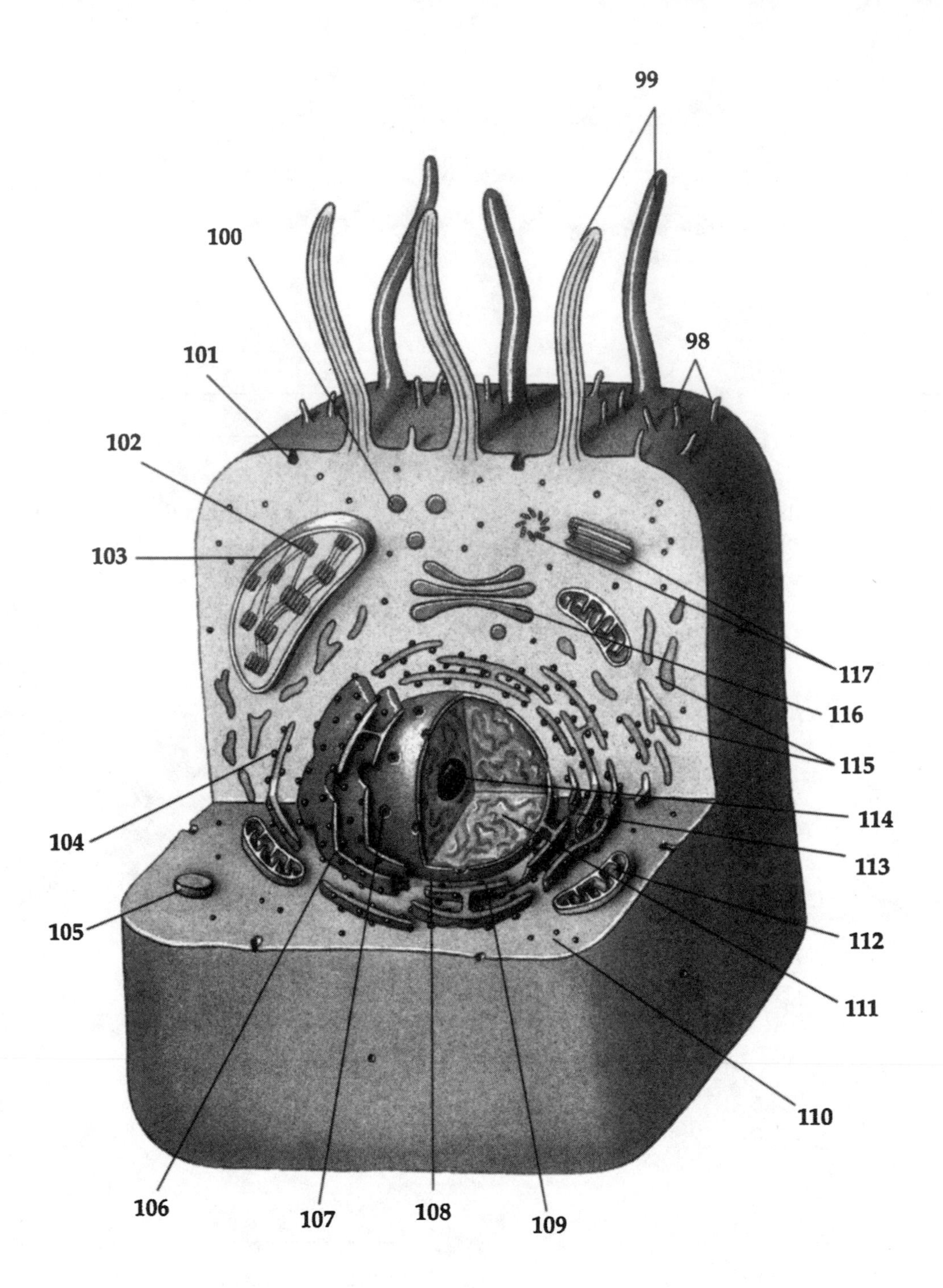

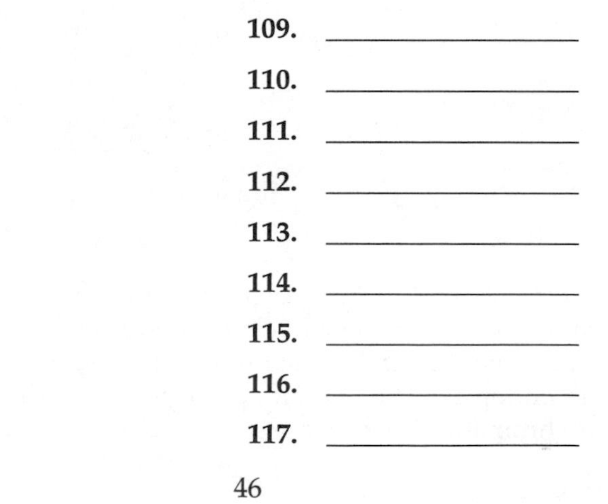

98.	____________________	109.	____________________
99.	____________________	110.	____________________
101.	____________________	111.	____________________
102.	____________________	112.	____________________
104.	____________________	113.	____________________
105.	____________________	114.	____________________
106.	____________________	115.	____________________
107.	____________________	116.	____________________
108.	____________________	117.	____________________

Critical Thinking

118. What are the characteristics of eukaryotic and prokaryotic cells?

119. How do prokaryotic cells differ in size, shape, and arrangement?

120. How are structure and function related in bacterial cell walls and cell membranes?

121. How are structure and function related in other bacterial components?

122. How are structure and function related in eukaryotic cell membranes?

123. How are structure and function related in other eukaryotic components?

124. How do passive transport processes function and why are they important? How does active transport function and why is it important?

125. How do exocytosis and endocytosis occur and why are they important?

ANSWERS

True/False

1. **True** These characteristics are shared by both types of organisms.
2. **False** Spiral organisms generally exist only as single cells. Bacilli often form chains and cocci can exhibit a variety of shapes.
3. **False** Gram-negative bacteria do possess a lipopolysaccharide layer but also possess a thin peptidoglycan layer.
4. **True** Eukaryotic cells have nuclear pores which allow RNA molecules to leave the nucleus and participate in protein synthesis.
5. **False** Both mitochondria and chloroplasts contain DNA and can replicate. Golgi bodies lack DNA and are incapable of division.
6. **True** Hypertonic solutions such as brine water may cause cells to shrink due to the osmotic "puling" effect of the excess salt outside the cells.
7. **False** Gram-negative bacteria have an endotoxin which is released when they die. Therefore killing them may only increase the toxicity.

Multiple Choice

7. b; 8. c; 9. b; 10. a; 11. b.

Matching

12. a; 13. g; 14. b; 15. c; 16. h; 17. h; 18. h; 19. d; 20. e; 21. b; 22. f; 23. d; 24. l; 25. h; 26. l; 27. d; 28. a; 29. e; 30. b; 31. f; 32. l; 33. l; 34. g; 35. j; 36. k; 37. c; 38. a; 39. j; 40. c; 41. k; 42. e; 43. k; 44. b; 45. h; 46. l; 47. a; 48. f; 49 i.

Fill-Ins

50. prokaryotic; 51. pleomorphic; 52. diplococci; 53. staphylococci; 54. peptidoglycan; 55. N-acetylglucosamine, N-acetylmuramic acid; 56. techoic acid; 57. lipid A; 58. protoplasm; 59. spheroplast; 60. penicillin; 61. fluid-mosaic; 62. hydrophobic; 63. Svedberg; 64. cytoplasm; 65. polyribosomes; 66. nucleus; 67. volutin, metachromatic; 68. endospores; 69. peritrichous; 70. axial filaments; 71. glycocalyx; 72. cytoskeleton; 73. nucleolus; 74. histones; 75. cristae; 76. endoplasmic reticulum; 77. lysosome; 78. selectively permeable; 79. facilitated diffusion; 80. isotonic; 81. group translocation.

Labeling

82. coccobacillus; 83. coccus; 84. bacillus; 85. vibrio; 86. spirillum; 87. spirochete. 88. chromosome; 89. pilus (fimbria); 90. ribosomes; 91. inclusion; 92. flagellum; 93. plasmid; 94. cytoplasm; 95. cell membrane; 96. cell wall; 97. capsule or slime layer. 98. microvilli (in animals); 99. cilia (in animals and protists); 100. secretory vesicle; 101. pinocytotic vesicle; 102. thylakoid; 103. chloroplast (in plants); 104. ribosome; 105. lysosome (in animals); 106. rough endoplasmic reticulum; 107. nuclear pore; 108. nuclear envelope; 109. nucleus; 110. cytoplasm; 111. nucleoplasm; 112. mitochondrion; 113. nuclear chromatin; 114. nucleolus; 115. smooth endoplasmic reticulum; 116. Golgi apparatus; 117. centrioles (in animals).

Critical Thinking

118. **What are the characteristics of eukaryotic and prokaryotic cells?**
Prokaryotic and eukaryotic cells have a number of similarities and differences between them. For example, both have cell membranes, ribosomes, and cytoplasm. Prokaryotes, however, have very complex cell walls not found within eukaryotes, but they lack the complex organelles found in eukaryotes (mitochondria and golgi bodies) which are as large or larger than the prokaryotic cell. Prokaryotes also lack a nuclear membrane that is found within eukaryotic cells.

119. **How do prokaryotic cells differ in size, shape, and arrangement?**
Prokaryotes, specifically the bacteria, have three basic shapes, namely, cocci, bacilli, and spirilli. Although we often see the bacteria in only two dimensions, try to visualize them as three-dimensional rods and cocci.

120. **How are structure and function related in bacterial cell walls and cell membranes?**
Bacterial cell walls surround the entire cell and are very complex. Notice that the cell wall construction directly relates to the staining properties of the cell. Those that are Gram-positive have a thick wall made al-

most entirely of peptidoglycan. The thick layer allows the cell to retain the purple dye. Gram-negative cells, on the other hand, have a thin inner layer of peptidoglycan and a relatively thick outer membrane of lipopolysaccharide material. This combination ensures that as the alcohol is added, any crystal violet dye that is adhering to these cells will be quickly removed since alcohol dissolves the lipid layer.

The lipopolysaccharide layer is also important because it can represent a toxic layer. Since it is a part of the cell, it is called the endotoxin.

Acid-fast bacteria also have a complex wall which includes lipids and waxy substances. These waxes greatly resist staining and especially decolorization steps. The presence of the waxes also explains why these cells do not readily mix in water when making a smear.

In contrast to the porous, protective, and lifeless bacterial cell wall, the cell membrane is very active and functional. The lipids of the membrane act primarily as a barrier while the proteins act as the active functional units, determining what moves in or out of the cell and providing sites for cellular respiration. Interestingly, the membrane is much like a chocolate-chip cookie, with the dough acting as the lipids (boring, tasteless barrier) and the chocolate chips acting as the proteins (active, tasty and functional units).

121. How are structure and function related in other bacterial components?

Bacterial cells have only a few other internal and external parts. The internal parts include the ribosomes that act as sites of protein synthesis and the inclusion granules that act as storage units for the cell. The nuclear material, which contains the genes, lacks a nuclear membrane and consists of only one circular chromosome. This is similar to a large ball of string that is connected end to end. The external structures include the incredibly thin but long flagella used for motion and the glycocalyx which acts as a slime layer or capsule external to the cell wall. A thick capsule can protect the cell by resisting engulfment by white blood cells. Finally, some bacteria have short, stiff bristles called pili that can be used for surface contact or for contacting other cells in the process of gene transmission.

122. How are structure and function related in eukaryotic cell membranes?

Eukaryotic cells are generally larger in structure and far more complex than prokaryotes. The cell membrane of these cells is very similar to that found in prokaryotes except that they often possess sterols, probably to aid in stabilizing or strengthening the membrane. Since eukaryotes have specialized organelles for respiration and secretion, the cell membrane functions primarily to move material in and out of the cell.

123. How are structure and function related in other eukaryotic components?

Eukaryotes possess specialized organelles such as mitochondria which are used to produce energy, golgi bodies which are used for packaging and secreting substances, and endoplasmic reticula which are used for internal transport of materials. The nucleus is very complex and possesses pairs of chromosomes rather than the single chromosomes found with prokaryotes. This allows eukaryotes a greater degree of variation and genetic change. Numerous other structures may be found within the cytoplasm, each with a particular function fitted to its structure.

External structures include flagella and cilia, both of which are considerably larger and more complex than the flagella of bacteria.

124. How do passive transport processes function and why are they important? How does active transport function and why is it important?

All living cells must move food material into their cytoplasm and move wastes out. This essential process is taken up by the cell membrane and specifically by the proteins found scattered within the lipid layer. Passive, non-energy requiring mechanisms such as osmosis, diffusion, and facilitated diffusion move the majority of materials. All operate on the basis of diffusion, in other words, if food is more abundant outside the cell (high concentration), it moves inside (low concentration) and vice versa for waste materials. Osmosis operates as a pulling force because the membrane is very selective. It, in effect, acts like a sponge. Facilitated diffusion involves the action of carrier membrane proteins that merely select which substances to bring in or take out of the cell. Osmosis can create environmental conditions that can greatly affect cells. Keep in mind that a condition that adversely affects one microbe may in fact be favorable to another. For example, freshwater organisms placed in marine waters would be adversely affected as well as marine organisms placed in fresh water.

125. How do exocytosis and endocytosis occur and why are they important?

Endocytosis and exocytosis are two energy-requiring mechanisms that move material through cell membranes; however, these only occur in eukaryotic cells. Phagocytosis is the most important example of endocytosis and involves the engulfing or "eating" of large particles such as cells or bacteria. Exocytosis is the reverse mechanism where a cell releases digested particles following phagocytosis

Chapter 5 Essential Concepts of Metabolism

INTRODUCTION

People can be thought of as a funny kind of machine. We may be viewed as a type of machine that is aware of its existence but not of the parts that make it go. Consider an imaginary automobile that is aware of itself and that it can move, but is not aware of the parts of its engine or fuel system. Both people and microbes are really like that. Consider, we share with microbes a considerable set of biochemical reactions that as a sum is called metabolism. Yet we are born aware of none of the chemicals or reactions that occur. We are even born with no knowledge of the chemicals called enzymes that make the chemistry of our body possible and regulate the rates of the different reactions.

In this chapter we will learn the principles and specific chemical reactions of man and microbe. You may even leave with insight into nutrition and the problem of obesity that plagues so many of us. More importantly, an understanding of the metabolism of microbes is critical to an understand of the unity and diversity of microbial life.

STUDY OUTLINE

I. Metabolism: An Overview

A. Metabolism

Metabolism is the sum of all chemical processes carried out by living organisms, including anabolism and catabolism.

B. Electron transfer

Electron transfers allow energy to be captured in high-energy bonds. *Oxidation* is the loss or removal of electrons; *reduction* is the gain of electrons.

C. Autotrophy

Autotrophs, or "self-feeders," use inorganic carbon dioxide to synthesize organic molecules. These include *photoautotrophs* and *chemoautotrophs*.

D. Heterotrophy

Heterotrophs, or "other-feeders," use ready-made organic molecules obtained from other organisms. These include *photoheterotrophs* and *chemoheterotrophs*.

E. Relationship of respiration with photosynthesis

In *respiration*, glucose is broken down to carbon dioxide, water, and energy in the presence of oxygen. In *photosynthesis*, carbon dioxide is reduced to an organic substance containing more energy in the presence of light energy and hydrogen from water (or other compounds).

II. Enzymes

A. General characteristics

Enzymes are protein *catalysts*: substances that remain unchanged as they speed up chemical reactions. Enzymes allow these chemical reactions to proceed at temperatures that cells can stand.

B. Properties of enzymes

Every enzyme has an *active site*, or a binding site, that allows the enzyme to form a loose association with its *substrate*, or the substance on which the enzyme acts. When the enzyme and the substrate join at the enzyme's active site, an *enzyme-substrate complex* forms. This allows the enzyme to catalyze the formation of the product. Enzymes speed up chemical reactions by lowing the activation energy.

C. Properties of coenzymes and cofactors

Many enzymes can act as catalysts only if *coenzymes* or *cofactors* are present. A coenzyme is a nonprotein organic molecule bound to or loosely associated with an enzyme. A cofactor is usually an organic ion that often improves the fit of an enzyme with its substrate.

D. Enzyme inhibition

In *competitive inhibition*, a nonsubstrate molecule competes with the substrate for an enzyme's active site. In *noncompetitive inhibition*, a molecule attaches to an enzyme at an *allosteric site*, changing the shape of the active site so that the substrate can no longer bind. In *feedback inhibition*, an end product of the pathway binds to and inactivates the enzyme that catalyzed the first reaction in the pathway.

E. Factors that affect enzyme reactions

Temperature, pH, and the concentrations of substrate, product, and enzyme affect the rates of enzyme reactions.

III. Anaerobic metabolism: Glycoloysis and Fermentation

A. Glycolysis

Glycolysis is an anaerobic metabolic pathway used to break glucose down to pyruvic acid. This process produces some ATP.

B. Fermentation

Fermentation is the anaerobic metabolism of the pyruvic acid produced in glycolysis. Types of fermentation include homolactic-acid fermentation and alcoholic fermentation.

IV. Aerobic Metabolism: Respiration

A. Role of glycolysis

Aerobes obtain some energy from glycolysis but mainly use glycolysis to prepare molecules for a much more productive process: aerobic respiration, which allows aerobes to obtain far more of the energy available in glucose.

B. The Krebs cycle

The *Krebs cycle* is a sequence of enzyme-catalyzed chemical reactions that metabolizes 2-carbon units called acetyl groups to carbon dioxide and water. It is also called the *tricarboxylic acid (TCA) cycle* and the *citric acid cycle*.

C. Electron transport and oxidative phosphorylation

Electron transport is a process in which pairs of electrons are transferred between cytochromes and other compounds to oxygen. *Oxidative phosphorylation* is a process in which the energy of electrons is captured in high-energy bonds as phosphate groups combine with ADP to form ATP. *Chemiosmosis* is a process of energy capture in which a proton gradient is created by means of electron transport and then used to drive the synthesis of ATP.

D. Significance of energy capture

If a glucose molecule undergoes glycolysis and fermentation, two net ATPs are created. If a glucose molecule undergoes glycolysis and aerobic respiration, 38 net ATPs are created. Aerobes therefore grow more rapidly than anaerobes in environments with ample oxygen.

V. Metabolism of Fats and Proteins

A. General features

Glucose is the major source of energy for most organisms, but other organic substances, like fats and proteins, can also be used as sources of energy.

B. Metabolism of fats

Fats are catabolized through a process known as *beta oxidation*—a metabolic pathway that breaks fatty acids down into 2-carbon acetyl-CoA pieces. These acetyl-CoA molecules are then oxidized via the Krebs cycle.

C. Metabolism of proteins

Proteins are first hydrolyzed into amino acids by *proteolytic enzymes*. The amino acids are then *deaminated*. The resulting molecules enter glycolysis, fermentation, or the Krebs cycle.

VI. Other Metabolic Processes

A. Photoautotrophy

Photoautotrophs are autotrophs that obtain energy from light. Photoautotrophs carry out *photosynthesis*, a process that captures light energy and uses the energy to manufacture carbohydrates from carbon dioxide.

B. Photoheterotrophy

Photoheterotrophs are heterotrophs that obtain energy from light but require organic substrates as carbon sources.

C. Chemoautotrophy

Chemoautotrophs, or *chemolithotrophs*, are autotrophs that obtain energy by oxidizing simple inorganic substances such as sulfides and nitrites.

VII. Uses of Energy

A. Biosynthetic activities

The intermediates of the energy-yielding pathways are sometimes involved in *biosynthetic pathways*, or energy-requiring processes in which proteins, carbohydrates, and lipids are synthesized.

B. Membrane transport and movement

Microorganisms also use energy to transport substances across their own membranes and to move.

C. Bioluminescence

Some bacteria exhibit bioluminescense and may live on the surface of fish and squid in the sea. The Microtox Acute Toxicity Test exploits these bacteria and uses their light output as a measure of how toxic a chemical is.

SELF-TESTS

Continue with this section only after you have read Chapter 5 of your textbook. Write your answers in the appropriate space provided. Correct answers to all questions can be found at the end of the Self-Test section. A score of 80 percent or better is good. If your score is less than 65, reread the chapter.

True/False

Mark T for True, F for False.

_____ **1.** Autotrophic organisms rarely cause human disease.

_____ **2.** Enzymes increase the amount of activation energy needed to initiate a reaction.

_____ **3.** The glycolysis cycle is required by anaerobes in the process of fermentation but it is not used by aerobic microbes in the process of respiration.

_____ **4.** NAD and FAD operate in the Krebs cycle to remove high energy electrons and hydrogen protons for transfer to the electron transport system.

_____ **5.** Because anaerobes produce only two ATP per food molecule (versus 38 ATP in aerobic systems), they grow very slowly and probably require many days to reach any appreciable population size.

_____ **6.** Photosynthesis in green plants differs for that in bacteria in that the bacteria generally do not produce oxygen as a byproduct.

Multiple Choice

Select the best possible answer.

_____ **7.** Which of the following is true concerning enzymes?

a. Enzymes are generally only used once in a reaction.
b. Enzymes always end with the suffix -ole.
c. Enzymes consist of large lipopolysaccharide molecules.
d. Enzymes form an enzyme-product complex before completing their reaction.
e. None of the above is true.

______ 8. The Krebs cycle is characterized by all the following except:
a. substrate-level energy capture.
b. oxidation of carbon.
c. removal of electrons by coenzymes.
d. formation of pyruvate as an end product.

______ 9. In the "dark" reaction of photosynthesis:
a. electrons in chlorophyll are activated by sunlight.
b. photolysis occurs.
c. carbon dioxide is reduced to form glucose.
d. NADP is reduced.
e. None of the above occurs.

______ 10. An amphibolic pathway:
a. generates energy in photolithotrophs.
b. can capture energy or synthesize substances needed by the cell.
c. is required to metabolize fats to fatty acids.
d. provides the surfactants needed for movement.

Matching

Select the answer from the right-hand side that corresponds to the term or phrase on the left-hand side of the page. An answer may be used more than once. In some cases, more than one answer may be required.

Topic: Nutritional Aspects of Bacterial Growth

______ 11. Use light energy and CO_2 as carbon sources
______ 12. Use light energy and organic compounds as carbon sources
______ 13. Use chemical energy and CO_2 as carbon sources
______ 14. Use chemical energy and organic compounds as carbon sources
______ 15. Use inorganic compounds as an energy source and organic compounds as carbon sources

a. lithoheterotrophs
b. photoautotrophs
c. chemoheterotrophs
d. photoheterotrophs
e. chemoautotrophs

Topic: Enzymes and Reactions

______ 16. Reaction in which there is a gain of electrons
______ 17. Conversion of glucose to pyruvic acid
______ 18. The apoenzyme and coenzyme
______ 19. Conversion of pyruvic acid to ethanol
______ 21. Nonprotein organic molecule bound to an enzyme
______ 22. Reaction in which there is a loss of electrons
______ 23. NAD
______ 24. Conversion of pyruvic acid to CO_2 and H_2O
______ 25. The addition of P_i to a molecule
______ 25. The protein part of an enzyme
______ 26. FAD

a. holoenzyme
b. coenzyme
c. aerobic respiration
d. fermentation
e. glycolysis
f. oxidation
g. reduction
h. apoenzyme
i. phosphorylation

Topic: Fermentation Pathways

______ 27. Butanol
______ 28. Succinic acid
______ 29. Propionic acid

a. butyric–butylic fermentation
b. homolactic-acid fermentation
c. alcoholic fermentation

_____ 30. Ethyl alcohol

_____ 31. Lactic acid

_____ 32. Acetone

_____ 33. Acetic acid

_____ 34. CO_2

d. mixed-acid fermentation

e. propionic acid fermentation

f. butanediol fermentation

Fill-Ins

Provide the correct term or phrase for the following. Spelling counts.

35. Reactions that require energy to synthesize complex molecules from simpler ones are known as _______________ reactions.

36. _______________ can be defined as the loss or removal of electrons from a molecule.

37. _______________ can be defined as the gain of electrons.

38. A _______________ pathway is a series of chemical reactions in which the product of one reaction serves as the reactant for the next.

39. Chemically, enzymes are composed of _______________.

40. The energy required to start a chemical reaction is called the reaction's _______________ energy.

41. The point at which a substrate attaches to its specific enzyme is known as the enzyme's _______________ site.

42. An _______________-_______________ complex results when a substrate collides with the active site of its enzyme.

43. Enzymes are usually named by adding the suffix _______________ to the name of the substrate on which they act.

44. Enzymes that act within the cells that produce them are called intracellular enzymes or _______________.

45. A nonprotein substance that acts with the enzyme to catalyze a reaction is a _______________.

46. The acronym _______________ stand for nicotinamide adenine dinucleotide.

47. Sulfa drugs exert their action on bacterial cells by _______________ inhibition.

48. If an inhibitor binds at a site other than the active site, altering the shape of the active site, it is called _______________ inhibition.

49. Most human enzymes have an optimum _______________ and an optimum _______________ at which they catalyze a reaction most rapidly.

50. The addition of a phosphate group, often from ATP to another molecule, is called _______________.

51. The common end product of the glycolysis cycle is _______________ _______________.

52. Pyruvic acid is metabolized in the absence of oxygen due to the metabolic process of _______________.

53. The cycle that metabolizes two carbon units to carbon dioxide and water is the _______________ cycle.

54. The molecule that enters the Krebs cycle from glycolysis is _______________.

55. Three types of carriers involved in _______________ _______________ are cytochromes, quinones, and iron-sulfur proteins.

56. The process by which fatty acids are catabolized is known as _______________ _______________.

57. The removal of amino groups from amino acids is called _______________.

58. The splitting of water molecules into protons, electrons, and oxygen molecules by light energy and membrane proteins is called _______________.

59. Bacteria that can oxidize inorganic substances for energy are known as ________________.

60. Photosynthesis requires ________________ for energy, but biosynthetic reactions do not.

61. Enzymes that extend through the membrane to aid in active transport of molecules are called ________________.

Labeling

Questions 62–65: Select the best possible label for each step in the chemical reaction graphed below: products, activation energy without enzyme, reactants, activation energy with enzyme

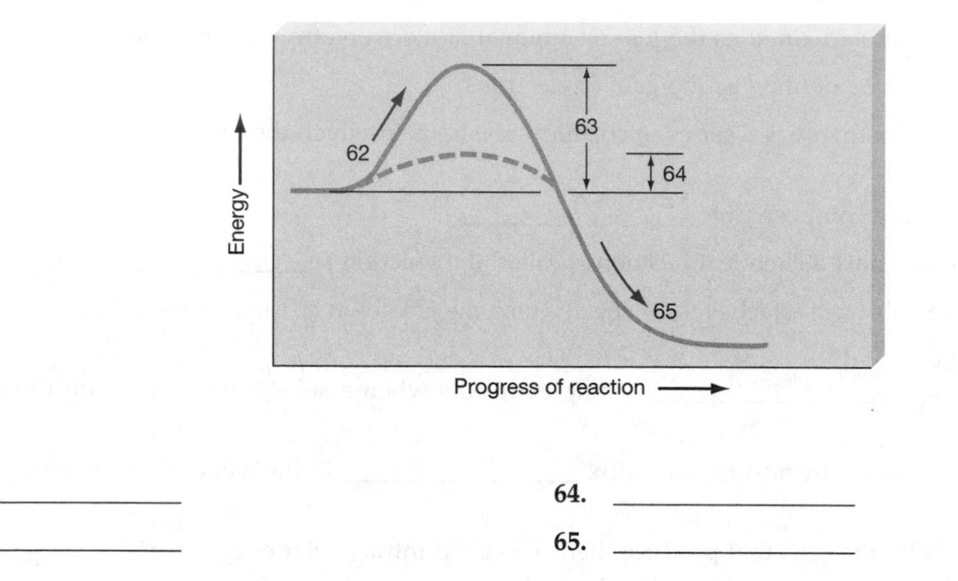

62. ________________

63. ________________

64. ________________

65. ________________

Questions 66–70: Select the best possible label for each of the parts of the following reaction: enzyme-substrate complex, active sites, product, enzyme, and substrate.

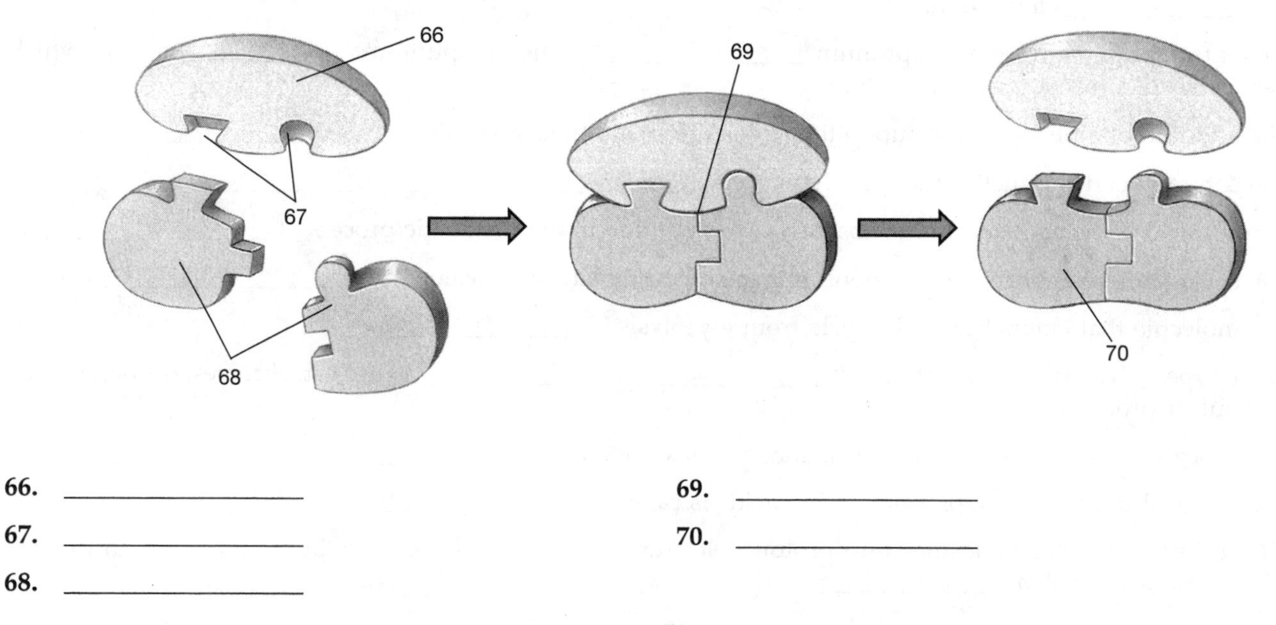

66. ________________

67. ________________

68. ________________

69. ________________

70. ________________

Questions 71–81: Select the best possible label for each step in the metabolism of the major classes of biomolecules: electron transport chain, fatty acids + glycerol, acetyl-CoA, Krebs cycle, ATP, amino-acid catabolism, beta-oxidation, pyruvate, glucose and other sugars, amino acids, and glycolysis.

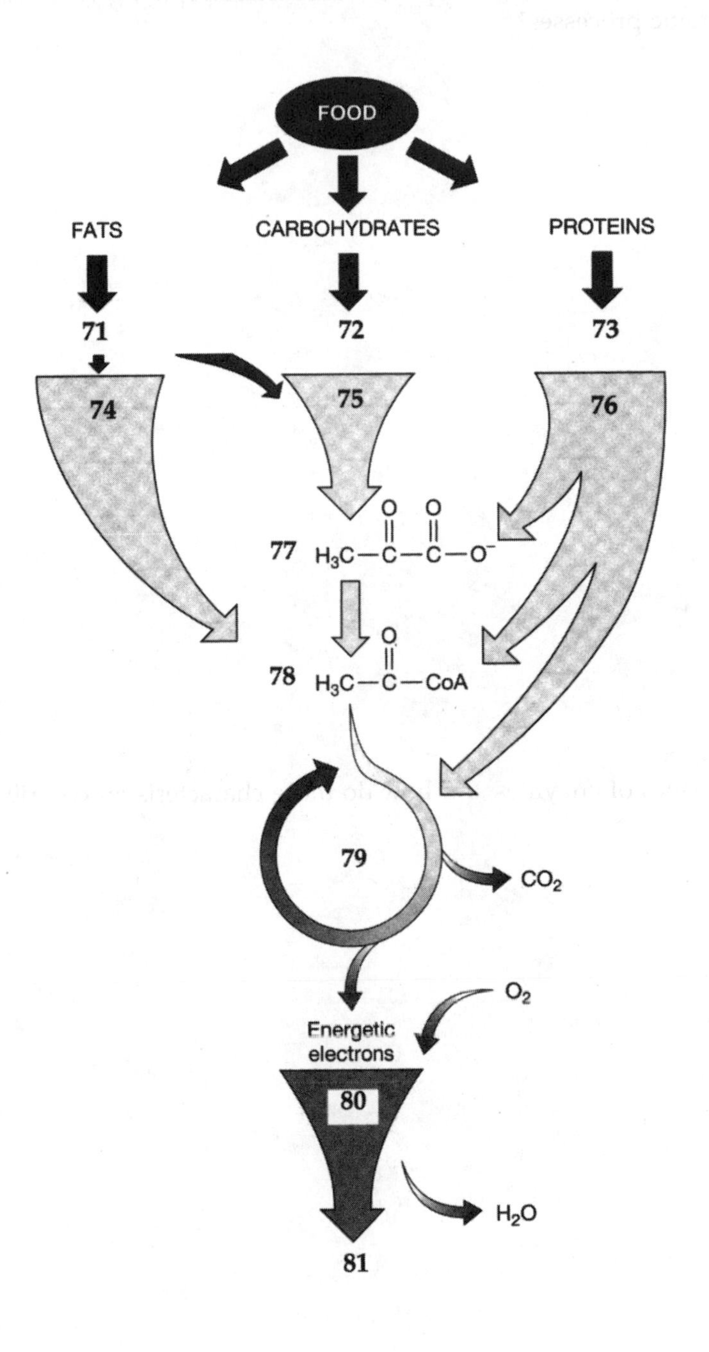

71. ____________________

72. ____________________

73. ____________________

74. ____________________

75. ____________________

76. ____________________

77. ____________________

78. ____________________

79. ____________________

80. ____________________

81. ____________________

Critical Thinking

82. How do the following terms relate to metabolism: autotrophy, heterotrophy, oxidation, reduction, photoautotrophy, photoheterotrophy, chemoautotrophy, chemoheterotrophy, glycolysis, fermentation, aerobic metabolism, and biosynthetic processes?

83. What are the characteristics of enzymes and how do those characteristics contribute to their function?

84. What are the main steps and significance of glycolysis and fermentation?

85. What is the significance of the Krebs cycle?

86. What are the roles of electron transport and oxidative phosphorylation in energy capture?

87. How do microorganisms metabolize fats and proteins for energy?

88. What are the main steps and significance of photosynthesis in microbes?

89. How do photoheterotrophy and chemoautotrophy differ?

90. How do bacteria carry out biosynthetic activities?

91. How do bacteria use energy for membrane transport and for movement?

ANSWERS

True/False

1. **True** Autotrophic or self-feeding metabolism especially photosynthesis, is an important means of energy capture. Since they are free-living, they usually do not cause disease.
2. **False** Enzymes decrease the activation energy needed for reactions thus facilitating them.
3. **False** Glycolysis is an essential metabolic cycle for both systems.
4. **True** Both NAD and FAD are coenzymes which function to accept electrons from molecules in the Krebs cycle and transfer them to coenzymes of ETS in the process of oxidative phosphorylation.
5. **False** Although anaerobes may grow more slowly than aerobes under optimum conditions for both, they still can reach maximum population densities in about the same time as aerobes.
6. **True** Photosynthetic bacteria differ from green plants in that bacterial chlorophyll absorbs longer wavelengths of light than plant chlorophylls, the bacteria use hydrogen compounds other than water for reducing carbon dioxide, and they do not release oxygen as a byproduct.

Multiple Choice

7. e; 8. d; 9. c; 10. b.

Matching

11. b; 12. d; 13. e; 14. c; 15. a; 16. g; 17. e; 18. a; 19. d; 20. b; 21. f; 22. b; 23. c; 24 i; 25. h; 26. b; 27. a; 28. d; 29. e; 30. c, d; 31. b; 32. a; 33. d, e; 34. a, c, d, e, f.

Fill-Ins

35. anabolic; 36. oxidation; 37. reduction; 38. metabolic; 39. protein; 40. activation; 41. active; 42. enzyme-substrate; 43. -ase; 44. endoenzyme; 45. coenzyme (or cofactor); 46. NAD; 47. competitive; 48. noncompetitive; 49. temperature, pH; 50. phosphorylation; 51. pyruvic acid; 52. fermentation; 53. Krebs; 54. acetyl-CoA; 55. electron transport; 56. beta oxidation; 57. deamination; 58. photolysis; 59. chemolithotrophs; 60. light; 61. permeases.

Labeling

62. reactants; 63. activation energy without enzyme; 64. activation energy with enzyme; 65. products. 66. enzyme; 67. active sites; 68. substrate; 69. enzyme-substrate complex; 70. product. 71. fatty acids + glycerol; 72. glucose and other sugars; 73. amino acids; 74. beta oxidation; 75. glycolysis; 76. amino-acid catabolism; 77. pyruvate; 78. acetyl-CoA; 79. Krebs cycle; 80. electron transport chain; 81. ATP.

Critical Thinking

82. **How do the following terms relate to metabolism: autotrophy, heterotrophy, oxidation, reduction, photoautotrophy, photo-heterotrophy, chemoautotrophy, chemoheterotrophy, glycolysis, fermentation, aerobic metabolism, and biosynthetic processes?**
Metabolism is the sum total of all chemical reactions in a cell which includes catabolism or energy-producing reactions and anabolism or energy-requiring reactions. This energy is found in the electrons of all foods used by organisms. By removing them (oxidation) and adding them to other compounds (reduction) they can obtain the energy they need.

Various mechanisms are used to obtain this energy. Certain organisms, called autotrophs, can use simple chemicals (chemoautotrophy) or the sun (photoautotrophy) as energy. Other organisms, called heterotrophs, can use ready-made organic molecules and are then called heterotrophs. Examples of heterotrophs include photoheterotrophs and chemoheterotrophs.

As this chapter will describe, various pathways and cycles are used such as fermentation and respiration in order to extract the energy found in foods.

83. What are the characteristics of enzymes and how do those characteristics contribute to their function?
Enzymes are amazing organic catalysts. They are highly specific and exhibit a "lock and key" type of mechanism as they react with a substrate. Keep in mind that they are made of protein and are therefore genetically coded in the DNA.

Some enzymes require helpers or coenzymes to react with substrates. These coenzymes are simple chemicals in comparison to the enzyme itself.

At certain times, it is necessary to control the enzyme activity. One way is to compete with or block the active site, sort of like a chastity belt. Another is to distort the active site by reacting with an inhibitor (noncompetitive inhibition). A final way is to turn the enzyme production system off at the DNA or gene level.

84. What are the main steps and significance of glycolysis and fermentation?
Fermentation is a very simple catabolic system used by a great number of organisms to produce their energy. Fermentors produce very little energy per food (net 2 ATP), which means they are very inefficient. Therefore they must eat a lot of food to fulfill their energy needs. As a result, they also produce a lot of wastes which are organic acids, gases, and/or alcohols. The only cycle they usually use is glycolysis. Notice that the beginning substrate is usually glucose and that the end product is pyruvic acid. All fermentors use this common pathway. Once they reach pyruvic acid they each go to different final organic end products. Many of these final end products are quite useful in industry.

85. What is the significance of the Krebs cycle?
While fermentation is a simple process utilizing only glycolysis, aerobic respiration is more complex involving three processes, namely, glycolysis, the Krebs cycle, and the electron transport system. Because more systems are used, aerobes produce much more energy per food (36-38 ATP) and are therefore more efficient. They don't need to eat as much food and they also produce fewer waste products. Aerobes, for the most part, are not as useful in industry as anaerobes.

As mentioned, glycolysis is used initially to produce pyruvic acid which is then converted to acetyl COA. This then enters the Krebs cycle. The cycle produces only a few, but important, products. A small amount of energy is produced, carbon dioxide is released, and a lot of high energy electrons are extracted for use in the third cycle.

86. What are the roles of electron transport and oxidative phosphorylation in energy capture?
Once electrons are extracted from molecules within the Krebs cycle by the help of the coenzymes, NAD and FAD, they can enter the electron transport system. This system allows for the production of tremendous amounts of ATP or energy currency. The system operates similar to water running down hill.

Once the electrons have been "relieved" of their energy they must be accepted by a final electron acceptor. The acceptor in aerobic respiration is oxygen, hence the term oxidative phosphorylation.

87. How do microorganisms metabolize fats and proteins for energy?
All foods contain energy. Even doughnuts, which are made of starch and sugar, contain enormous quantities of energy. This energy is, in fact, found in the electrons of each sugar molecule. All that living organisms have to do is use a system which will allow them to extract the energy from those electrons and convert it to ATP.

Although sugars are the best and easiest to use for energy, even fats and proteins can be used. Fats can be broken down by beta-oxidation systems to form acetyl-COA which can enter the Krebs cycle. Proteins are more difficult to metabolize and must first be deaminated and then altered before they can enter the cycles.

88. What are the main steps and significance of photosynthesis in microbes?
All photosynthetic microbes are harmless since they are autotrophic; however, they are very significant in the environment. All of these organisms possess a variety of pigments that are used in photosynthesis. Some, especially those found around thermal pools, can really enhance the beauty of the environment.

Photosynthetic bacteria differ from plants in four ways. First, their chlorophyll absorbs in a different range; second, they often use organic compounds to reduce CO_2; third, many release H_2S or sulfuric acid; and fourth, they do not release oxygen as a byproduct. Most photosynthetic bacteria are also strict anaerobes. The cycles used by bacteria in photosynthesis are, however, essentially the same as those found in green plants.

89. How do photoheterotrophy and chemoautotrophy differ?
Photoheterotrophs and chemoautotrophs are all harmless but important environmental organisms. Photoheterotrophs use the sun's energy and organic carbon molecules while chemoautotrophs use very simple inorganic chemicals for their carbon and energy source. The latter have recently been found around deep sea volcanic vents.

90. How do bacteria carry out biosynthetic activities?

Once organisms have produced energy, they can utilize it for the biosynthesis of building material for the cell wall and cell membrane, for cellular movement, and for growth. Some pathways called amphibolic pathways can produce energy and also construct building blocks at the same time.

91. How do bacteria use energy for membrane transport and for movement?

Membrane transport involves permease enzymes which allows substances to be concentrated inside the cell or perhaps for storage. Movement occurs by expending energy to move flagella or axial filaments. The rings of a flagellum are especially designed to rotate within the wall and membrane when energy is applied.

Chapter 6 Growth and Culturing of Bacteria

INTRODUCTION

December 7, 1941, a tragic day in the history of the United States. The Japanese attacked Pearl Harbor and started World War II and perhaps the first essential commodity to become in short supply as a result was something called agar. Agar is used to grow microbes and is essential for the diagnosis of disease and the selection of appropriate medicines for many microbial infections. Agar is derived from red seaweeds and before World War II it was imported from Japan. The late Dr. Harold Humm fought back by establishing an American agar industry in New England. He hired college students to brave the cold North Atlantic waters and harvest the red algae needed to grow and characterize bacteria and other microbes.

Today, we are even more dependent on the materials and methods to grow and study microbes. We require the methods culture methods to identify pathogenic microbes and select appropriate drugs. Many students in your class may not be alive if not for these techniques. This chapter will explore the care and feeding of bacteria and related forms.

STUDY OUTLINE

I. Growth and Cell Division

A. Microbial growth defined

Microbial growth is defined as the increase in the number of cells, which occurs by cell division.

B. Cell division

In bacteria, cell division usually occurs by *binary fission*, a process in which a cell duplicates its components and divides into two cells. Bacterial cell division can also sometimes occur by *budding*, a process in which a small new cell develops from the surface of an existing cell and then separates from the parent cell.

C. Phases of growth

The phases that form the *standard bacterial growth curve* include: (1) the *lag phase*, in which organisms do not increase significantly in number but are metabolically active; (2) the *log phase*, in which cells divide at an exponential or logarithmic rate; (3) the *stationary phase*, in which new cells are produced at the same rate that old cells die, leaving the number of live cells constant; and (4) the *decline (death) phase*, in which cells lose their ability to divide and thus die.

D. Measuring bacterial growth
Bacterial growth is measured using different methods, including: (1) *serial dilution* (in which successive 1:10 dilutions are made from the original sample) and *standard plate counts* (in which plated dilutions are left to grow and resulting bacterial colonies are counted); (2) *direct microscopic counts* (in which cells are counted in a known volume of medium that fills a specially calibrated counting chamber on a microscopic slide); and (3) *filtration* (in which the bacterial population is estimated when a known volume of air or water is drawn through a filter with pores too small to allow passage of bacteria).

II. Factors Affecting Bacterial Growth

A. Physical factors
Physical factors affecting bacterial growth include pH, temperature, oxygen concentration, moisture, hydrostatic pressure, osmotic pressure, and radiation.

B. Nutritional factors
Nutrition factors affecting bacterial growth include the availability of carbon, nitrogen, sulfur, phosphorus, trace elements, and, in some cases, vitamins.

III. Sporulation

A. Sporulation
Sporulation, or the formation of endospores, generally occurs during the stationary phase of bacterial growth in response to environmental, metabolic, and cell cycle signals.

B. Germination
Germination, in which a spore returns to its vegetative state, occurs in three phases: *activation, germination proper,* and *outgrowth.*

C. Cycles
Bacterial cells capable of sporulation display two cycles—the *vegetative cycle* and the *sporulation cycle.*

D. Other spore-like structures in bacteria
Certain bacteria form resistant *cysts,* or spherical, thick-walled cells that resemble endospores. Other filamentous bacteria form asexually reproduced *condida,* or chains of aerial spores with thick outer walls.

IV. Culturing Bacteria

A. Methods of obtaining pure cultures
Pure cultures, or cultures containing only a single species of organism, are obtained by isolating the progeny of a single cell. These cells are isolated through the *streak plate method* or the *pure plate method.*

B. Culture media
There are different types of culture media, including *synthetic media, defined synthetic media,* and *complex media.* Commonly found in culture media are yeast extracts, caesein hydrolysates, and/or serum. These media are sometimes selective, differential, and/or enrichment mediums. Care must be taken to control the oxygen content in any of these media. Once an organism has been isolated on any medium, it can be maintained indefinitely in a pure culture called a *stock culture;* to avoid the risk of contamination and to reduce the mutation rate, stock culture organisms also should be kept in a *preserved culture.*

V. Living, But Nonculturable Organisms

A. Most bacteria cannot be cultured.
B. They are identified by samples of their DNA.

SELF-TESTS

Continue with this section only after you have read Chapter 6 of your textbook. Write your answers in the appropriate space provided. Correct answers to all questions can be found at the end of the Self-Test section. A score of 80 percent or better is good. If your score is less than 65, reread the chapter.

True/False

Mark T for True, F for False.

______ **1.** Although organisms may differ in the length of the lag phase, all organisms have the same generation time.

______ **2.** Generally, microbes divide at an arithmetic rate.

______ **3.** Most bacteria that cause disease are in the neutrophilic category.

______ 4. An obligate anaerobe means it prefers an anaerobic condition but can tolerate a small amount of air.

______ 5. All organisms require a carbon and nitrogen source along with trace elements such as sulfur and phosphorous.

______ 6. Nutritional complexity reflects a deficiency in biosynthetic enzymes.

Multiple Choice

Select best possible answer.

______ 7. Select the most correct statement in relation to serial dilution.
a. Diluted samples are transferred to nutrient broth tubes.
b. The number of colonies on the plate is multiplied by the denominator of the dilution factor.
c. The test can accurately measure live and dead cells.
d. Countable plates should contain between 10 and 30 colonies.

______ 8. Which of the following methods is not used to determine bacterial numbers?
a. turbidity.
b. most probable number.
c. serial dilution.
d. direct microscopic counts.
e. All of the above are used.

______ 9. Temperatures can control bacteria because:
a. freezing will kill all bacteria.
b. hot temperatures (above 80 °C) will denature bacterial protein.
c. refrigerator temperatures stop the growth of all bacteria.
d. no bacteria can live above 180 °F.

______ 10. Endospores:
a. are generally formed for protection and reproduction.
b. are formed only when conditions become unfavorable.
c. contain dipicolinic acid and calcium.
d. contain laminated layers of peptidoglycan called the exosporium.

______ 11. Select the most incorrect statement concerning culturing of bacteria.
a. A synthetic medium consists of unidentifiable ingredients such as those found in beef extract.
b. Agar is a solidifying agent that melts at about 96 °C.
c. The streak plate method uses agar plates and a wire inoculating loop.
d. An enrichment medium contains ingredients such as blood, which can enhance the growth of certain organisms.
e. All of the above are true.

Matching

Select the answer from the right-hand side that corresponds to the term or phrase on the left-hand side of the page. An answer may be used more than once. In some cases, more than one answer may be required.

Topic: Physical Factors Affecting Growth

______ 12. Requires small amounts of O_2

______ 13. An organism with a low optimum temperature

______ 14. An organism that grows best at a pH of 9

______ 15. An organism that grows best at a pH of 5.4

______ 16. An organism that grows best at temperatures between 25° and 40 °C

a. psychrophile
b. acidophile
c. microaerophile
d. thermoduric
e. facultative anaerobe
f. mesophile

_____ 17. An organism that ordinarily lives as a mesophile but can metabolize for short periods at higher temperatures

_____ 18. A yeast cell that grows in the presence or absence of O_2

_____ 19. A carbon dioxide-loving bacterium

_____ 20. An organism requiring high atmospheric pressure

g. capnophile

h. alkaliphile

i. barophile

Fill-Ins

Provide the correct term or phrase for the following. Spelling counts.

34. A bacterial cell duplicates its components and divides into two cells by _______________ _______________.

35. Organisms that have adapted to medium, thus allowing for rapid exponential growth, are in the _______________ phase.

36. The time required for one cell to divide into two is the doubling time, or the _______________ time.

37. If bacteria could divide together and have the exact same generation time, they would exhibit _______________.

38. During the decline phase, many cells exhibit unusual shapes, which is caused by _______________.

39. The visible growth of bacteria on an agar medium is called a _______________.

40. A method used to measure bacterial growth that requires the use of a series of dilution tube is called _______________ _______________.

41. The _______________ _______________ count is a rapid procedure but it does not distinguish between living and dead bacteria.

42. _______________, or a cloudy appearance, is an indication of bacterial growth.

43. Organisms that can tolerate very low pH conditions are called _______________.

44. _______________ are organisms that prefer temperatures above 50 °C.

45. The term _______________ means that the organism can tolerate one environmental condition but still live in another.

46. An enzyme that can break H_2O_2 into H_2O and O_2 is _______________.

47. _______________ _______________are killed by free oxygen.

48. Organisms that grow best in the presence of small amounts of oxygen are called _______________.

49. Obligate anaerobes are killed by a highly reactive form of oxygen called _______________.

50. When cells lose water and their membranes shrink away from the cell wall, they are undergoing _______________.

51. Organisms that have many special nutritional needs are _______________.

52. Enzymes that are produced by the cell to operate outside the cell are _______________ enzymes.

53. The process of endospore formation is known as _______________.

54. The two bacterial genera that produce endospores are _______________ and _______________

55. A method of obtaining a pure culture of bacteria by means of an agar plate and an inoculating loop is called the _______________ _______________ method.

56. A culture medium that contains known specific kinds and amounts of chemicals is a ________________ ________________ medium.

57. Organisms that cause hemolysis can be detected on a ________________ ________________ medium.

58. A medium that encourages the growth of some organisms but suppresses others is a ________________ medium.

59. An ________________ medium contains special nutrients that allow the growth of a particular organism that might not otherwise be present in sufficient numbers to allow it to be isolated and identified.

60. Isolated organisms can be maintained in a pure culture called a ________________ culture.

61. A preserved culture maintained to keep its characteristics as originally defined is a ________________ culture.

62. A specially calibrated counting chamber used with direct microscopic counts is called a ________________ counter.

63. Trace elements such as copper and zinc often serve as ________________ in enzymatic reactions.

64. Microbes such as *E. coli* are able to manufacture vitamin ________________ in the human intestinal tract.

Labeling

Questions 65–68: Select the best possible label for each step in the bacterial growth curve seen below: decline or death phase, log phase, lag phase, stationary phase.

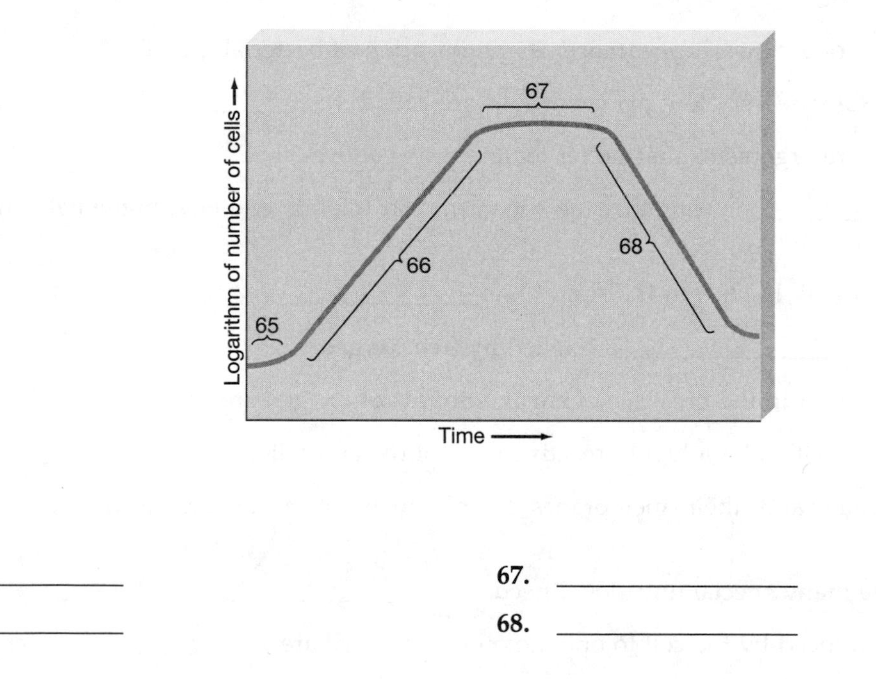

65. ________________

66. ________________

67. ________________

68. ________________

Questions 69–80: Select the best possible label for each step in the vegetative and sporulation cycles: axial nucleoid, binary fission, spore, core, double membrane, peptidoglycan fragments, spore membrane, spore coat, cortex, endospore septum, spore membrane, cell membrane.

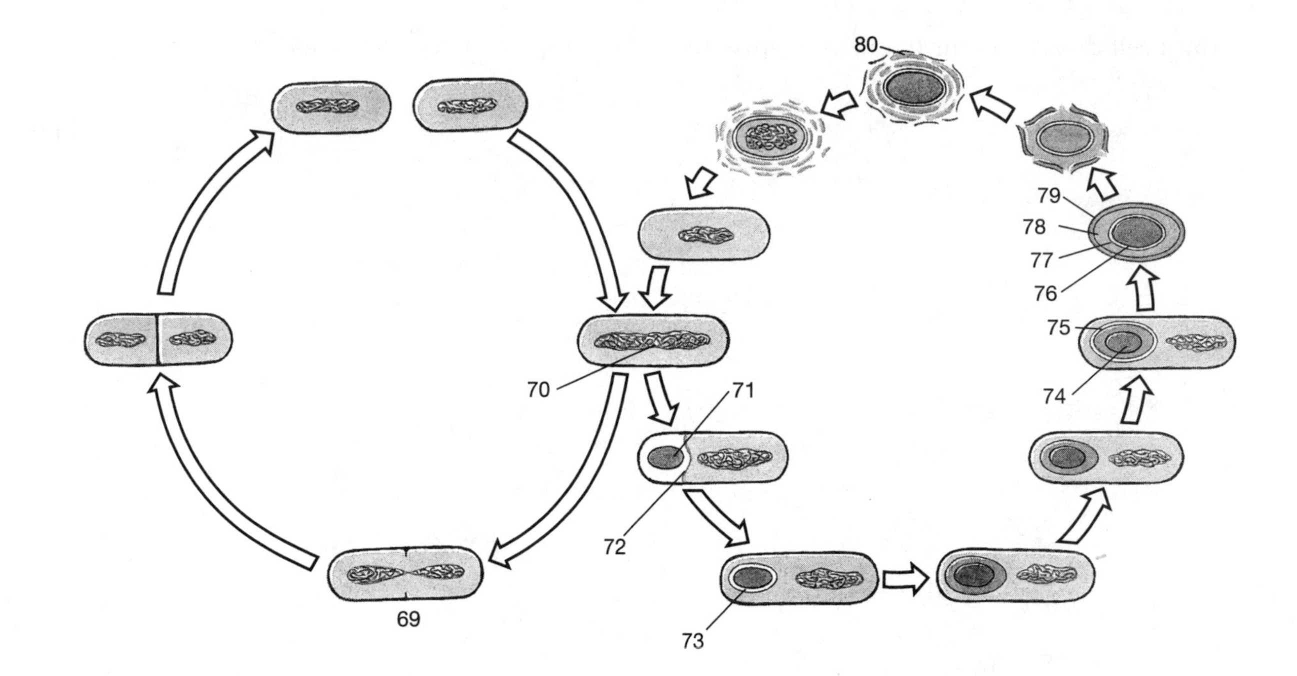

69. ____________________
70. ____________________
71. ____________________
72. ____________________
73. ____________________
74. ____________________
75. ____________________
76. ____________________
77. ____________________
78. ____________________
79. ____________________
80. ____________________

Critical Thinking

81. How is growth defined in bacteria?

82. How does cell division occur in microorganisms?

83. What are the phases of growth in a bacterial culture?

84. How is bacterial growth measured?

85. How do physical factors affect bacterial growth?

86. How do nutritional factors affect bacterial growth?

87. What occurs in sporulation and what is its significance?

88. What methods are used to obtain a pure culture of an organism for study in the laboratory?

89. How are different nutritional requirements supplied by various media?

ANSWERS

True/False

1. **False** Organisms vary not only in the length of the lag phase but also in the log phase. Generation times for the most bacteria is 20 minutes to 20 hours.
2. **False** Bacteria all divide at an exponential or logarithmic rate due to binary fission.
3. **True** Neutrophiles have an optimum pH near neutrality which is about pH 5.4 to 8.5. Since the pH of the human body is within that range it is understandable that most pathogens would be in this group.
4. **False** The term obligate means that the organism must have the environmental condition specified. Therefore an obligate anaerobe can only tolerate complete anaerobic conditions.
5. **True** All organisms do require all of these nutrients; however, they can be supplied in either an inorganic or organic form.
6. **True** Organisms with fewer enzymes have complex nutritional requirements because they lack the ability to synthesize many of the substances needed for growth.

Multiple Choice

7. b; 8. e; 9. b; 10. c; 11. a.

Matching

12. c; 13. a; 14. h; 15. b; 16. f; 17. d; 18. e; 19. g; 20. i; 21. a, d; 22. e; 23. c; 24. b; 25. b; 26. h; 27. d; 28. j; 29. g; 30. a; 31. c; 32. i; 33 f.

Fill-Ins

34. binary fission; 35. log; 36. generation; 37. synchronous growth; 38. involution; 39. colony; 40. serial dilution; 41. direct microscopic; 42. turbidity; 43. acidophiles; 44. thermophiles; 45. facultative; 46. catalase; 47. obligate anaerobes; 48. microaerophiles; 49. superoxide; 50. plasmolysis; 51. fastidious; 52. extracellular; 53. sporulation; 54. *Bacillus, Clostridium*; 55. streak plate; 56. defined synthetic; 57. blood agar; 58. selective; 59. enrichment; 60. stock; 61. reference; 62. Petroff-Hausser; 63. cofactors; 64. K.

Labeling

65. lag phase; 66. log phase; 67. stationary phase; 68. decline or death phase; 69. binary fission; 70. axial nucleoid; 71. core; 72. endospore septum; 73. double membrane; 74. spore; 75. spore coat; 76. spore membrane; 77. cell membrane; 78. cortex; 79. spore coat; 80. peptidoglycan fragments.

Critical Thinking

81. **How is growth defined in bacteria?**
Growth is defined as the orderly increase in the quantity of all the components of an organism. Since bacteria have a limited increase in size, their growth is measured by an increase in the number of cells.

82. **How does cell division occur in microorganisms?**
Cell division in bacteria occurs primarily by binary fission. That is accomplished by duplicating its components, forming a transverse septum, and splitting into two cells. Yeast cells and some bacteria may also divide by budding, in which a new cell forms from the surface of an existing cell.

83. **What are the phases of growth in a bacterial culture?**
There are four major phases of growth, namely, the lag, log, stationary, and decline or death phase. During the lag phase, the organisms are metabolically active but are not increasing in number. The log phase exhibits the rapid logarithmic growth pattern based on the organisms' generation time or the time needed to divide one cell into two at regular intervals. The formula for calculating the generation times is given as $n = \log B_t - \log B/\log 2$, where n = number of generations; B_t = number of bacteria at the end of a selected period of time, t; and B_0 = number of bacteria at zero time.

A simpler but less accurate formula for determining the number of bacteria that can reproduce in a period of time is 2^N where N = the total number of generations an organism has made in that given time. For example, if an organism can divide once every 30 minutes and is given 10 hours to divide, the total number of bacteria in the population would be 2^{20} or about 1,000,000 bacteria (assuming you have started with only one cell). Now consider a glass of milk left on the counter during the day and think of how many bacteria could be generated! Awe inspiring.

The third phase is the stationary phase which occurs because many of the organisms begin to die off due to the production of wastes, lack of food, and overcrowding. The decline phase occurs as the number of cells dying is far greater than those being reproduced. Once this phase is reached, the entire population may soon die off.

84. How is bacterial growth measured?

Methods to measure bacterial growth include the colony count, which is accomplished by transferring a known dilution of bacteria onto an agar plate and counting the colonies that arise. Other methods include direct microscopic counts, most probable numbers, filtration, turbidity, and others.

85. How do physical factors affect bacterial growth?

Various physical factors affect the growth of bacteria either through its effect on the cell's enzymes, on its metabolic systems, or on its structure such as the cell wall or membrane. However, because of the wide diversity of organisms, many have evolved to survive under almost any condition.

Acidity and alkalinity of the medium effect microbial growth and most organisms have an optimum pH range, with some preferring an acidic, some an alkaline, and some a neutral range. Temperature also affects organisms, with some preferring a high temperature, some a low temperature, and some a moderate one.

It is important to remember that pathogens seem to prefer a neutral pH and a moderate temperature range.

Oxygen also affects microbes, with some preferring aerobic conditions, some anaerobic conditions, some with a reduced amount of air, and some without any preference. Other factors that affect microbes include water and its osmotic effect and hydrostatic pressure.

86. How do nutritional factors affect bacterial growth?

Nutritional factors also greatly affect the growth of microbes. All require a carbon source supplied either organically (sugars) or inorganically (CO_2); a nitrogen source supplied either organically (amino acids) or in organically (ammonia); and several other elements such as sulfur, phosphorous, and potassium. Vitamins are also required but some can be synthesized by the organism.

Organisms' nutritional requirements are determined by the type of enzymes they possess. Various bioassay techniques can be used to determine the needs they require. Organisms can adjust to their environment by producing exoenzymes accordingly, by making enzymes to metabolize another available nutrient, or by adjusting their metabolic activities.

87. What occurs in sporulation and what is its significance?

Certain bacteria in the genus *Bacillus* and *Clostridium* produce highly resistant structures called endospores. The sporulation process occurs continuously to ensure survival under almost any condition. When favorable conditions arise, the endospores undergo germination and the cells return to the normal or vegetative state.

88. What methods are used to obtain a pure culture of an organism for study in the laboratory?

Various methods are used to grow and pure culture microbes. The most common method is the streak plate, which involves spreading bacteria on a solid medium by means of an inoculating loop. The pour plate is a second method where organisms are pipetted into a dish and cooled liquid agar is added. The individual organisms are then free to develop into isolated colonies within the gel.

89. How are different nutritional requirements supplied by various media?

Various culture media have been devised for microbial cultivation. Media may be constructed with known or defined contents such as glucose and water, or unknown, undefined (complex) contents such as beef extract or "hot dogs." Complex media are the most common to use since they provide all the elements needed for growth.

Specialized media include selective or inhibitory media designed to grow only certain organisms, differential media designed to distinguish one organism from another, and enrichment media designed to enhance the growth of certain "fussy" organisms.

Organisms can be maintained on stock cultures for routine work or as preserved cultures to store and save for long periods of time.

Chapter 7 Microbial Genetics

INTRODUCTION

In the first part of the 20th century, a young man sought to understand the 'secret of life'. James Watson went to England and in early 1950's began to work the laboratory of Francis Crick. The answer to the question had different aspects. What was the structure of the gene? How did genes work? What is the relationship between the structure of genes and how they are expressed in the proteins that provide the structure for organisms as diverse as bacteria, elephants, molds, and men? In 1953, Watson and Crick published a paper that was to change our world. The structure and function of DNA was revealed. The science of molecular biology began.

An understanding of the structure of and function of DNA is central to an understanding of the biology of microbes and the applications that exploit bacteria to reveal if chemicals can cause cancer in humans. It is also critical to an understanding of biotechnology presented in the next chapter.

STUDY OUTLINE

I. Overview of Genetic Processes

A. Basis of heredity

Heredity is the transmission of information from an organism to its offspring in the form of chromosomes or genes.

B. Nucleic acids in information storage and transfer

All the information for the structure and functioning of a cell is stored in the cell's DNA. This information is used both to guide the replication of DNA in preparation for cell division and to direct protein synthesis.

II. DNA Replication

A. DNA arrangement

Pairs of helical DNA strands are held together by the base pairing of adenine with thymine and of cytosine with guanine. When these two strands combine by base pairing, they do so in an *antiparallel* fashion.

B. Replication forks

The *replication fork* is the site at which the two strands of the DNA double helix separate during replication and new complementary DNA strands form.

C. Mechanism of DNA polymerase

The *DNA polymerase* enzyme moves along behind each replication fork, synthesizing new DNA strands complementary to the original ones.

D. Semiconservative replication
In *semiconservative replication,* a new DNA double helix is synthesized from one strand of parent DNA and one strand of new DNA.

III. Protein Synthesis

A. Transcription
Transcription is the synthesis of RNA from a DNA template. It involves the use of the enzyme RNA polymerase.

B. Kinds of RNA
Three kinds of RNA are involved in protein synthesis: *ribosomal RNA, messenger RNA, and transfer RNA.*

C. Translation
Translation is the synthesis of protein from the information in mRNA.

IV. Regulation of Metabolism

A. Significance of regulatory mechanisms
Bacterial cells, as well as the cells of all organisms, have developed mechanisms to turn reactions on and off in accordance with their needs. Cells can therefore use energy to synthesize substances in the amounts needed and then shut off these processes before *wasteful* excesses are produced.

B. Categories of regulatory mechanisms
Regulatory mechanisms can either: (1) regulate enzymes directly, or (2) regulate the synthesis of enzymes by turning on or off genes that code for particular enzymes.

C. Feedback inhibition
In feedback inhibition, the end product of a biosynthetic pathway directly inhibits the first enzyme in the pathway.

D. Enzyme induction
Enzyme induction is a mechanism whereby the genes coding for enzymes needed to metabolize a particular nutrient are activated by the presence of that nutrient.

E. Enzyme repression
Enzyme repression is a mechanism by which the presence of a particular metabolite represses the genes coding for enzymes used in its synthesis.

V. Mutations

A. Types of mutations and their effects
Point mutations are mutations in which one base is substituted for another at a specific location in a gene. These mutations may or may not change the amino acid sequence in a protein. *Frameshift mutations* are mutations resulting from the deletion or insertion of one or more bases. These mutations alter all the three-base sequences beyond the deletion or insertion.

B. Phenotypic variations
Phenotypic variations frequently seen in mutated bacteria include alternations in colony morphology, colony color, or nutritional requirements.

C. Spontaneous versus induced mutations
Spontaneous mutations occur in the absence of any agent known to cause changes in DNA. *Induced mutations* are produced by agents called *mutagens,* or agents that increase the mutation rate above the spontaneous mutation rate.

D. Chemical mutagens
Chemical mutagens act at the molecular level to alter the sequence of bases in DNA. They include *base analogs, alkylating agents, deaminating agents,* and *acridine derivatives.*

E. Radiation as a mutagen
Radiation such as X-rays and ultraviolet rays can act as mutagens, causing, for example, pyridine dimers to form.

F. Repair of DNA damage
Many bacteria and other organisms have enzymes that can repair certain kinds of damage to DNA. These enzymes work in mechanisms known as *light repair* and *dark repair,* which repair damage caused by dimers.

G. The study of mutations
Microorganisms are especially useful in studying mutations because of their short generation time and the relatively small expense of maintaining large populations of mutant organisms for study. The *fluctuation test* and *replica plating,* for example, have already used microorganisms to demonstrate that mutations are spontaneous.

H. The Ames test
The *Ames test* is used to determine whether a particular substance is mutagenic, based on its ability to induce mutations in auxotrophic *bacteria.*

SELF-TESTS

Continue with this section only after you have read Chapter 7 of your textbook. Write your answers in the appropriate space provided. Correct answers to all questions can be found at the end of the Self-Test section. A score of 80 percent or better is good. If your score is less than 65, reread the chapter.

True/False

Mark T for True, F for False.

_____ **1.** When two DNA strands combine by base pairing, they do so in a head-to-head or parallel fashion.

_____ **2.** Although the first codon of the mRNA always codes for methionine, it acts as the "start" signal.

_____ **3.** Actually several ribosomes "read" an mRNA at the same time.

_____ **4.** The operon theory was first proposed by Watson and Crick in 1951.

_____ **5.** All DNA mutations will ultimately result in the death of the cell.

Multiple Choice

Select the best possible answer.

_____ **6.** Feedback inhibition:
a. allows the cell to conserve energy.
b. acts to inhibit the active site by competitive inhibition.
c. involves control by the end product.
d. All of the above are correct.
e. Only **a** and **c** are correct.

_____ **7.** In DNA replication:
a. the old strands of DNA are first destroyed.
b. a DNA-dependent RNA polymerase forms the parent strands.
c. replication is semiconservative.
d. DNA synthesis is continuous to ensure accurate synthesis.

_____ **8.** Bacteria are ideal for studies on regulatory mechanisms because:
a. they reproduce at a very rapid rate.
b. they can be grown inexpensively.
c. they mutate at a fairly rapid rate.
d. All of the above are good reasons.

_____ **9.** In respect to the *lac* operon:
a. the absence of lactose results in the inhibition of the repressor.
b. the repressor binds to the operator and turns on the *lac* genes.
c. lactose acts as the inducer when it is present.
d. the inducer induces the repressor gene to turn off the operator.

_____ **10.** All of the following are true concerning mutagenic agents except:
a. they act at the genotypic level.
b. alkylating agents can add methyl groups to nitrogenous bases.
c. once chemicals mutate the DNA, it cannot be repaired.
d. X-rays can easily break chemical bonds.

Concept Question

11. How can a mutation in the DNA affect cells polysaccharide capsule?

Matching

Select the answer from the right-hand side that corresponds to the term or phrase on the left-hand side of the page. An answer may be used more than once. In some cases, more than one answer may be required.

Topic: Mutations

______	**12.** 5-bromouracil	**a.**	deaminating agent
______	**13.** Nutritionally deficient mutant	**b.**	ionizing form of radiation
______	**14.** Quinacrine	**c.**	base analog
______	**15.** X-ray	**d.**	replica plating
______	**16.** Nitrous acid	**e.**	ultraviolet light
______	**17.** Nonmutant form of bacteria	**f.**	auxotroph
______	**18.** Salvador Luria and Max Delbrück	**g.**	fluctuation test
______	**19.** Joshua Lederberg and Esther Lederberg	**h.**	acridine derivative
______	**20.** Dimer formation	**i.**	prototroph
______	**21.** Ames test	**j.**	histidine mutant detections

Fill-Ins

Provide the correct term or phrase for the following. Spelling counts.

22. Genetic information is contained in molecules of ________________.
23. The DNA of an organism comprises its ________________.
24. Eukaryotic genes with different information at the same locus are called ________________.
25. A permanent alteration in the DNA is a ________________.
26. When DNA serves as a pattern in the formation of mRNA, it acts as a ________________.
27. The pairing of certain nucleotide bases in DNA to other nucleotides is called ________________ ________________ ________________.
28. The process of mRNA formation from the DNA template is called ________________.
29. When RNA messages are decoded into proteins, the process is known as ________________.
30. DNA replication creates two moving strands called ________________ ________________.
31. Replication processes that form one strand of old or parent DNA and one of new DNA are called ________________ ________________.
32. An enzyme necessary to form one strand of old or parent DNA and one of new DNA is called ________________ ________________.
33. There are three kinds of RNA: ________________ RNA, ________________ RNA, and ________________ RNA.
34. Codons such as UCC, UAG, GGU would be found in ________________.
35. If an mRNA codon sequence such as UAA or UAG appears during the reading cycle, it acts as the ________________ ________________.

36. An RNA molecule that brings the appropriate amino acid to the ribosome for packaging is the ________________.

37. Each tRNA molecule has a three-nucleotide base region that is complementary to a particular mRNA codon called an ________________.

38. Several ribosomes can be attached at different points along an mRNA molecule to form a ________________.

39. Enzymes that are synthesized continuously are said to be ________________.

40. A substance that binds to and inactivates the repressor associated with an operon is called an ________________.

41. A series of closely-associated genes that relate to the synthesis of specific structures and to their regulation are known as ________________.

42. When a needed metabolite (such as glucose) is in adequate supply within the cell and the activity of the operon is then inhibited, the effect is called ________________.

43. A genetic change that affects the DNA would be a ________________ change.

44. A mutation that results in a deletion or an insertion of single bases is a ________________ mutation.

45. A nutritionally-deficient mutant organism is an ________________ organism.

46. Normal nonmutant forms of bacteria are called ________________, or wild types.

47. A molecule similar to a nitrogenous base found in DNA is a ________________ ________________.

48. Mutagenic agents (such as nitrous acid) that can remove amino groups from nitrogenous bases are known as ________________ agents.

49. Ultraviolet light creates thymine-thymine ________________.

50. DNA repair that occurs in the presence of visible light can be called light repair or ________________.

51. A test commonly used to detect antibiotic-resistant organisms is the ________________ ________________ test.

52. A test that is based on the ability of bacteria to mutate and used to screen chemicals for mutagenic properties is the ________________ test.

Labeling

Questions 53–58: Select the best possible label for each step in the information transmission from DNA to protein: translation, DNA, reverse transcription, replication, transcription, RNA.

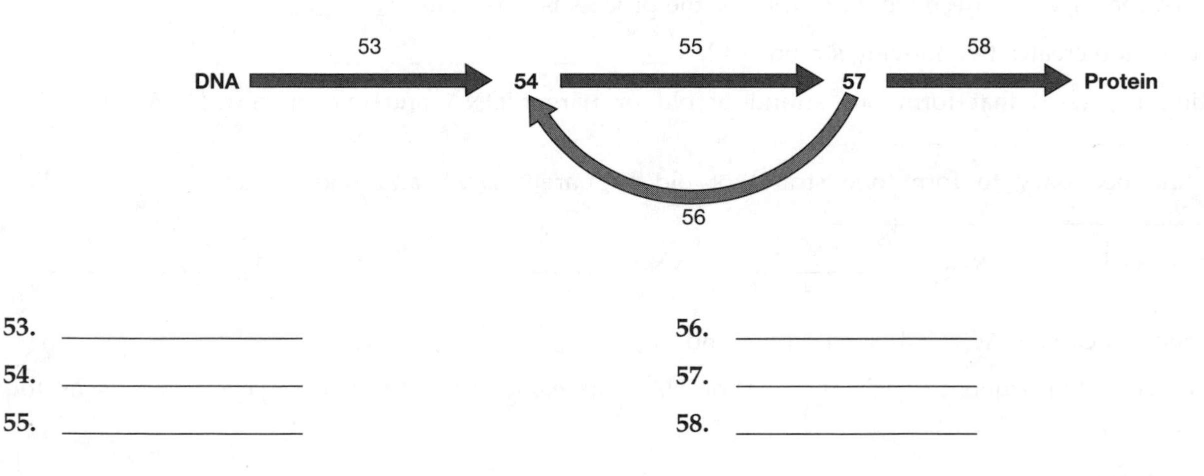

53. ________________

54. ________________

55. ________________

56. ________________

57. ________________

58. ________________

Questions 59–64: Select the best possible label for each part of the molecules involved in amino acid transfer, amino acid binding site, codon, amino acid, mRNA, acceptor arm, anticodon.

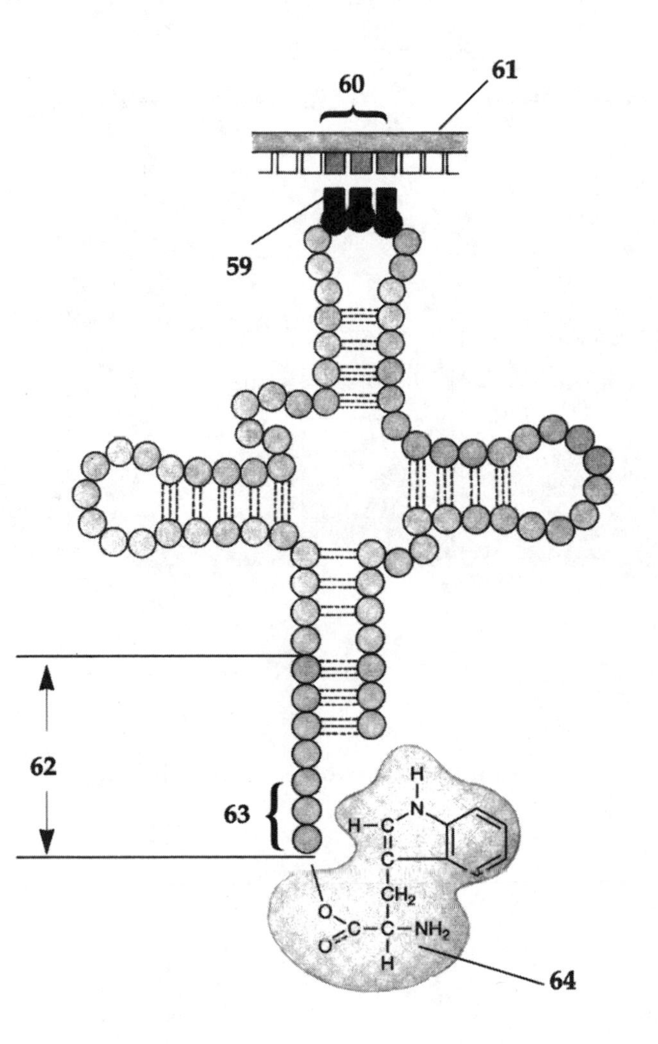

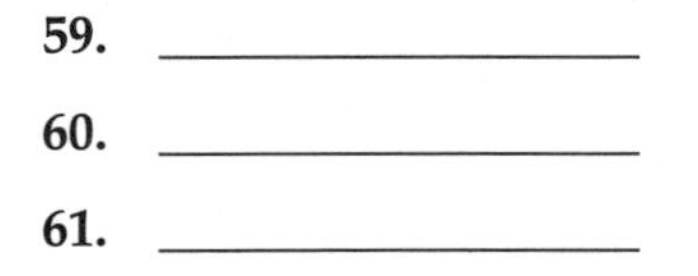

59. ____________________

60. ____________________

61. ____________________

62. ____________________

63. ____________________

64. ____________________

Questions 65–68: Select the best possible label for each step in dark repair of DNA: (a) a DNA polymerase synthesizes new DNA, (b) ligase reattaches new and old DNA, (c) endonuclease breaks DNA, (d) exonuclease removes defective segment.

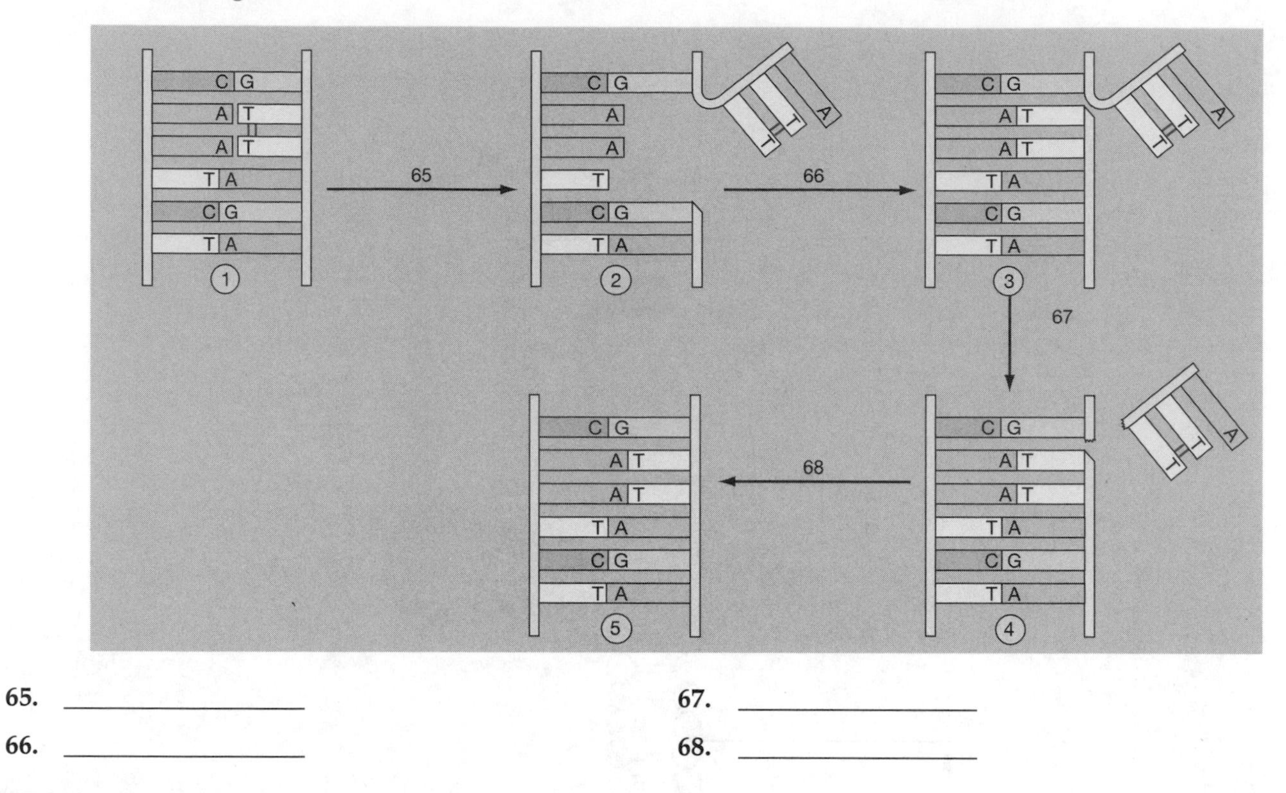

65. ____________________

66. ____________________

67. ____________________

68. ____________________

Critical Thinking

69. How are genes, chromosomes, and mutations involved in heredity in prokaryotic organisms?

70. How do nucleic acids store and transfer information?

71. How is DNA replicated in prokaryotic cells?

72. What are the major steps in protein synthesis?

73. How do mechanisms that regulate enzyme activity differ from those that regulate gene expression?

74. What happens in feedback inhibition, enzyme induction, and enzyme repression?

75. What changes in DNA occur in mutations, and how do mutations affect organisms?

76. How do the fluctuation test, replica plating, and the Ames test make use of bacteria in studying mutations?

ANSWERS

True/False

1. **False** DNA strands combine by base pairing in a head-to-tail or antiparallel fashion.
2. **True** Start and stop signals are part of the coded message in mRNA. The first RNA code is AUG which codes for methionine and acts as the start signal.
3. **True** Several ribosomes called polyribosomes read the mRNA strand at the same time, attaching to the end of mRNA that corresponds to the beginning of a protein.
4. **False** The operon theory was first proposed by the French scientists, Jacob and Monod in 1961. Watson and Crick developed our knowledge of DNA and the genetic code.
5. **False** Mutations always change the genotype but it may not result in the death of the cell. If the mutation is a minor change in the DNA with no change in the amino acid specified by the mRNA there may not be any effect on the organism.

Multiple Choice

7. c; 8. d; 9. c; 10. c.

Concept Question

11. Mutations affect the DNA which would then affect the mRNA and its ability to make specific proteins. If the affected protein is an enzyme that is required to construct the polysaccharide capsule, the result could be a defective capsule.

Matching

12. f; 13. h; 14. b; 15. a; 16. i; 17. g; 18. d; 19. e; 20. j.

Fill-Ins

22. DNA; 23. genotype; 24. alleles; 25. mutation; 26. template; 27. complementary base pairing; 28. transcription; 29. translation; 30. replication forks; 31. semiconservative replication; 32. RNA polymerase; 33. ribosomal, messenger, transfer; 34. RNA; 35. terminator codon; 36. tRNA; 37. anticodon; 38. polyribosome; 39. constitutive; 40. inducer; 41. operons; 42. catabolite repression; 43. genotypic; 44. framshift; 45. auxotrophic; 46. phototrophs; 47. base analog; 48. deaminating; 49. dimers; 50. photoreactivation; 51. replica plating; 52. Ames.

Labeling

53. replication; 54. DNA; 55. transcription; 56. reverse transcription; 57. RNA; 58. translation; 59. anticodon; 60. codon; 61. mRNA; 62. acceptor arm; 63. amino acid binding site; 64. amino acid.65. (c) ligase attaches new and old DNA; 66. (a) a DNA polymerase synthesizes new DNA; 67. (d) endonuclease breaks DNA; 68. (b) exonuclease removes defective segment.

Critical Thinking

69. **How are genes, chromosomes, and mutations involved in heredity in prokaryotic organisms?**
All information necessary for life is stored in an organism's genetic material, its DNA. Heredity involves the transmission of this information from an organism to its progeny. This is accomplished through the duplication and transfer of genes that consist of linear-coded sequences of DNA. The genes make up large genetic units called chromosomes. Prokaryotes carry one circular chromosome while eukaryotes possess numerous pairs of chromosomes. The genetic information is transmitted during binary fission in prokaryotes. Mutations are permanent alterations in the DNA that can be transmitted to the progeny and appear to account for much of the variations seen in organisms.

70. **How do nucleic acids store and transfer information?**
Deoxyribonucleic acid (DNA) makes up the chromosomes of all cells. The information is stored within the DNA in a coded message determined by the linear arrangement of the nucleotides adenine, guanine, cytosine, and thymine. This information is used to replicate the DNA in cell division and to provide information for protein synthesis. The mechanism of complementary base pairing ensures the accurate coding and decoding of the information.

71. **How is DNA replicated in prokaryotic cells?**
In the replication of DNA in a prokaryote, DNA strands separate, and replication begins at a replication fork on each strand. As synthesis proceeds, each strand of DNA serves as a template for the replication of its partner; complementary strands of the DNA double helix are antiparallel. Because synthesis of new DNA can take place in only one direction, the process must be discontinuous along one strand. Short segments are formed and then spliced together. Each new cell can undergo subsequent replications. But no matter how many daughter cells form in a population, only two of the cells in each generation will have one of the original strands from the mother cell (in other words, the process is semiconservative, where each chromosome receives a strand of parent DNA and one newly-synthesized strand of DNA).

72. **What are the major steps in protein synthesis?**
The process of protein synthesis involves two major steps. In transcription, a messenger RNA molecule is transcribed or formed from a section of the DNA. In translation, this message is decoded by ribosomes into specific proteins such as enzymes or structural units such as flagella.
The entire process of protein synthesis is directed by three types of RNA.
The necessary amino acids for this construction are brought to the ribosome by transfer RNA molecules. Using an anticodon system the tRNA's are matched to codons in the mRNA to insure an exact sequencing of amino acids as directed by the mRNA.
This process is somewhat analogous to the master computer at a college. This computer could represent the DNA of a cell. Information requested about a single class from the campus could be transcribed onto *floppy disks* which would act as the mRNA. These disks could then be translated by placing them into a *disk drive* of a small personal computer which would act as the ribosome. The disk would be decoded onto paper or onto a *printed tape* which would act as the finished product or protein produced by the cell.

73. **How do mechanisms that regulate enzyme activity differ from those that regulate gene expression?**
Regulatory mechanisms are essential to turn metabolic systems on and off when necessary for the cell. This

ensures that the cell does not waste energy producing more than what it needs and ensures the organism of essential materials.

Two basic regulatory mechanisms occur in prokaryote cells. The first involves regulation of enzyme activity already in the cell. This frequently involves feedback inhibition by the end product. The second involves regulation of enzyme production by affecting the action of the genes.

74. What happens in feedback inhibition, enzyme induction, and enzyme repression?

Feedback inhibition involves regulation of existing metabolic enzymes associated with a biochemical pathway. This is accomplished by the end product binding to an enzyme inhibitor or allosteric site which temporarily turns off the enzyme. Enzyme induction involves the activation of genetic units responsible for the production of a particular enzyme. Enzyme repression involves the inhibition of genetic units responsible for enzyme production. These two systems appear to affect genetic units called operons and effectively control enzyme synthesis. The presence of the nutrient or absence of the end product appears to act as the conditions that lead to the gene expression.

75. What changes in DNA occur in mutations and how do they affect organisms?

Permanent changes in the DNA affect the genotype and are called mutations. These may or may not be expressed in the phenotype or what is actually seen. These mutations may be either point mutations which affect only a single nucleotide, or frameshift mutations which involve the insertion or deletion of nucleotides.

76. How do the fluctuation test, replica plating, and the Ames test make use of bacteria in studying mutations?

Microbes are excellent organisms for genetic studies because they reproduce in very large numbers, are easy to handle, are inexpensive to produce, and can express their genetic information and/or changes very quickly. As a result several tests have been developed to take advantage of these features. The fluctuation and replica plating tests use bacteria to demonstrate the spontaneous nature of mutations. The Ames test is based on the ability of bacteria to mutate and is used to screen chemicals for potential mutagenic properties.

Chapter 8 Gene Transfer and Genetic Engineering

INTRODUCTION

The history of science can be a strange subject. Revolutions can arise from unusual places. A name that is missing from most histories of genetic engineering is Bernard O. Dodge. He worked at the New York Botanical Garden from 1928 to 1947 as a plant pathologist. Fungi were the major focus of his work, including a genus of bread mold, *Neurospora*. He published over 160 scientific papers in his life but has been hardly noticed in textbooks. On the other hand, two biologists, George Beadle and Edward Tatum (see page 18 in your textbook), noticed his work. They continued Dodge's work on this lowly fungus and received the Nobel Prize for discovering the link between genes and proteins. This discovery was critical for the development of genetic engineering described in this chapter. Many people today whose lives are enabled by products like genetically engineering human insulin produced with the help of bacteria may not appreciate the link between genes, proteins, and microbes. By the end of this chapter, you will gain an understanding of the methods that will produce many of the drugs developed in the future.

B. O. Dodge died in 1960. At his funeral, Beadle and Tatum came to deliver the eulogy. They said: "We are amateur professionals, but he was a professional amateur." Dodge represented a transition in biology. A transition from a discipline engaged in because of love and passion to a new profession, filled with all the rewards and trappings of the modern world. It is not possible, nor is it even desirable, to go back to the simpler era, but it is a good idea to have a little knowledge of the people and places that made the future of genetic engineering possible.

STUDY OUTLINE

I. **Nature and Significance of Gene Transfer**
   Gene transfer is the movement of genetic information between organisms by transformation, transduction, or conjugation. This is significant because gene transfer greatly increases the genetic diversity of organisms.

II. **Transformation**
   A. Discovery of transformation
      Bacterial transformation is a change in a bacterium's characteristics through the transfer of naked DNA. Frederick Griffith discovered this process in 1928 as he was performing experiments with pneumococcal infections in mice.

B. Mechanism of transformation

At a certain stage in a cell's growth cycle, a protein called *competence factor* is released into the cell's surroundings. This factor may facilitate them entry of naked DNA from the environment. Once the DNA has entered the cell, endonucleases allow the DNA to replace parts of the cell's normal DNA.

C. Significance of transformation

Transformation may contribute to the genetic diversity of organisms. It can also be used to study the locations of genes on a chromosome and to insert DNA from one species into the DNA of another species.

III. Transduction

A. Discovery of transduction

Transduction is the transfer of genetic material from one bacterium to another by a bacteriophage. Joshua Lederberg and Norton Zinder discovered this phenomenon in 1952 as they were studying *Salmonella*.

B. Mechanism of transduction

Transduction is usually a process associated with *temperate phages* whose DNA can be incorporated into the host's DNA in the form of a *prophage*. This occurs as *specialized transduction*, in which only specific genes are transferred by the phage, and as *generalized transduction*, in which any bacterial gene can be transferred.

C. Significance of transduction

First, transduction alters the genetic characteristics of recipient cells. Second, transduction can be used to demonstrate close evolutionary relationships between prophages and host bacterial cells. Third, transduction of prophages suggests a similar possible mechanism for the viral origin of cancer. Fourth, transduction probably brought genes from other hosts into human hosts. And fifth, transduction provides a way to study gene linkages.

IV. Conjugation

A. Discovery of conjugation

Conjugation is the transfer of genetic information from one bacterial cell to another by means of conjugation pili. Joshua Lederberg discovered this process in 1964 as he was performing experiments with *E. coli.*

B. Mechanism of conjugation

Conjugation can occur in a number of ways, including through the transfer of F plasmids, through high-frequency recombinations, and through the transfer of F′ plasmids.

C. Significance of conjugation

Conjugation is significant because it contributes to genetic variation. It also may represent an evolutionary stage between the asexual processes of transduction and transformation and the sexual process of eukaryotes.

V. Gene Transfer Mechanisms Compared

In transformation, less than one percent of the DNA in one bacterial cell is transferred to another and the transfer involves only one chromosomal DNA. In transduction, a few genes are transferred to large fragments of the chromosome and a bacteriophage is always involved in the transfer. In conjugation, highly variable quantities of DNA are transferred and a plasmid is always involved in the transfer.

VI. Plasmids

A. Characteristics of plasmids

Most plasmids are circular, double-stranded, extrachromosomal pieces of DNA that are self-replicating. Plasmids usually carry genes that code for functions not essential for cell growth.

B. Resistance plasmids

Resistance plasmids are plasmids that carry genes that provide resistance to specific antibiotics or to toxic metals. They generally contain two components: a *resistance transfer factor (RTF)* and one or more *resistance (R) genes*.

C. Transposons

The ability of a genetic sequence to move from one location to another is called *transposition*. The sequence itself is known as a *transposable element*. A *transposon* is a transposable element that contains the genes for transposition and one or more other genes as well.

D. Bacteriocinogens

Bacteriocins are growth-inhibiting proteins. Bacteriocin production is directed by a plasmid known as a *bacteriocinogen*.

VI. Genetic Engineering

A. Definition

Genetic engineering is defined as the use of various techniques to purposefully manipulate genetic material in order to alter the characteristics of an organism in a desired way.

B. Genetic fusion

Genetic fusion is a genetic engineering technique that allows the transposition of genes from one location on a chromosome to another location. It can also involve the deletion of a DNA segment, resulting in the coupling of portions of two operons. The major applications of genetic fusion are in research studies on the properties of microbes.

C. Protoplast fusion

Protoplast fusion is a genetic engineering technique in which genetic material is combined by removing the cell walls of two different types of cells and allowing the resulting protoplasts to fuse. By mixing two strains in this fashion, each of which has a desirable characteristic, new strains that have both characteristics can be produced.

D. Gene amplification

Gene amplification is a genetic engineering technique in which plasmids or bacteriophages carrying a specific gene are induced to reproduce at a rapid rate within host cells. In this way, plasmids that carry genes for antibiotic synthesis can be induced to create large quantities of antibiotics. Enzymes, amino acids, vitamins, and nucleotides are some of the other substances that can be created in large amounts, thanks to gene amplification.

E. Recombinant DNA Technology

Recombinant DNA technology allows DNA to be formed that contains information from two different organisms. It requires the use of a *vector* (a self-replicating carrier such as a phage or a plasmid) to carry the DNA segments from the donor to the recipient. Recombinant DNA is useful medically, since with it scientists can modify bacterial cells to produce substances useful to humans, making certain treatments potentially safer, cheaper, and available to more patients. Recombinant DNA is also useful industrially, since fermentation processes used in making wine, antibiotics, and other substances might be greatly improved by the use of recombinant DNA. And, lastly, recombinant DNA is useful agriculturally, since bacteria can be engineered to control insects that destroy crops and seeds can be engineered to produce plants that have high yields and are resistant to herbicides.

F. Hybridomas

Genetic engineering provides for the creation of *hybridomas*, or hybrid cells that are formed from the fusion of a cancer cell with another cell, usually an antibody-producing white blood cell. This kind of hybridoma can be used to produce pure, specific antibodies against any antigen to which the original white blood cell was sensitized.

G. Weighing the risks and benefits of recombinant DNA

In the past, some scientists were concerned that recombinant DNA techniques could result in the formation of new and especially virulent human pathogens. Presently, however, most scientists agree that DNA techniques offer significant benefits and exceedingly small risks to humans.

SELF-TESTS

Continue with this section only after you have read Chapter 8 of your textbook. Write your answers in the appropriate space provided. Correct answers to all questions can be found at the end of the Self-Test section. A score of 80 percent or better is good. If your score is less than 65, reread the chapter.

True/False

Mark T for True, F for False

_____ **1.** Transformation can only occur between a living recipient and a dead donor.

_____ **2.** An important significance of transduction is that it suggests a mechanism for the viral origin of cancer.

_____ **3.** Hfr's allow the transference of the largest numbers of genes from donors to recipient cells.

_____ **4.** Hybridomas are formed by opening the cell walls of similar bacteria and allowing them to fuse.

_____ **5.** Lysogenized bacteria contain viral prophages, which can act as "timebombs."

_____ **6.** Recombinant DNA technology as applied to industry ensures easy isolation of the product being produced.

_____ **7.** Mutations account for the greatest amount of genetic diversity while gene transfers contribute only a small amount of change.

Multiple Choice

Select best possible answer

_____ 8. Plasmids:
a. cannot replicate on their own.
b. are small circular pieces of DNA.
c. represent the DNA that is incorporated into the host chromosome.
d. are transmitted during transformation.
e. None of the above is true.

_____ 9. *Richia coli:*
a. can transfer a great variety of genes.
b. can undergo generalized transduction.
c. can sometimes transduce specific genes.
d. always kills the host cell.
e. All of the above are true.

_____ 10. An Hfr:
a. transfers only a few genes at a time.
b. possesses the F′ pilus.
c. transfers large numbers of genes in a linear cycle.
d. mates only with F′ cells.
e. none of the above are true.

_____ 11. Protoplast fusion:
a. can only occur between strains of the same species.
b. can only occur with prokaryotic cells.
c. has only been attempted experimentally.
d. can be used to obtain the best genetic characteristics of two organisms.
e. None of the above are true.

Concept Question

12. A research laboratory isolated and purified a new interleukin type molecule from macrophage. This new protein seems to greatly increase the activity of cytotoxic T-cells. How could this laboratory produce enough of this molecule to study?

Matching

Select the answer from the right-hand side that corresponds to the term or phrase on the left-hand side of the page. An answer may be used more than once. In some cases, more than one answer may be required.

Topic: Genetic Transfer Mechanisms

_____ 13. Determined DNA structure
_____ 14. Involves virus transfer of DNA
_____ 15. Discovered transformation
_____ 16. Identified chemical composition of transforming principles
_____ 17. Discovered bacterial conjugation
_____ 18. Discovered bacterial transduction
_____ 19. Involves the transfer of naked DNA
_____ 20. Discovered bacteriocins

a. Frederick Griffith
b. Joshua Lederberg and Norton Zinder
c. André Gratia
d. Oswald Avery, Colin MacCleod and Maclyn McCarty
e. transformation
f. transduction
g. James Watson and Francis Crick
h. Joshus Lederberg

Topic: Plasmids and Genetic Engineering

_____ 21. Joining of genes from two different operons

_____ 22. Simplest type of transposable element

_____ 23. Growth-inhibiting proteins

_____ 24. Combining two organisms without cell walls to allow genetic information mixing

_____ 25. Inducing plasmids to reproduce rapidly within a cell

_____ 26. Manipulation of DNA outside of cells

_____ 27. Fusion of a bone marrow cancer cell with an antibody-producing white blood cell

a. bacteriocins

b. protoplast fusion

c. recombinant DNA technology

d. insertion sequence

e. genetic fusion

f. gene amplification

g. hybridoma creation

Fill-Ins

Provide the correct term or phrase for the following. Spelling counts.

28. The movement of genetic information between organisms is known as ________________ ________________.

29. The first mechanism of gene transfer, which was discovered by Griffith, was ________________.

30. A special protein that is released into a medium and appears to facilitate the uptake of DNA is called a ________________ ________________.

31. Viruses that specifically infect bacteria are called ________________.

32. A virulent phage is capable of causing infection and, eventually, the destruction and death of a bacterial cell in a process known as a ________________ cycle.

33. A virus that does not actually kill its host cell but instead incorporates into it is a ________________ phage.

34. If phage DNA is incorporated into a host bacterium's DNA, the phage DNA is called a ________________

35. In ________________ ________________, only specific genes are transduced by a bacteriophage.

36. In ________________ ________________, a fragment of DNA from the degraded chromosome of an infected bacterial cell is accidentally incorporated into a new phage particle during viral replication and thereby transferred to another bacterial cell.

37. ________________ differs from transformation and transduction in two ways: (1) it requires contact between donor and recipient cells, and (2) it transfers much larger quantities of DNA.

38. Extrachromosomal circular pieces of DNA are called ________________.

39. The protein bridge which is used in conjugation is called an ________________ ________________.

40. A microbial strain that can induce a very large number of genetic recombinations is a ________________ ________________ recombination strain.

41. Plasmids that commonly carry genes for resistance to antibiotics are called ________________.

42. If a genetic sequence moves from one location to another, the sequence is called a ________________ ________________.

43. Genetic units that can change their location within the chromosome and that contain both the genes responsible for transposition and one or two other genes are called ________________.

44. Bacteriocinogens produced by several strains of *E. coli* are called ________________.

45. ________________ ________________ is a technique of genetic engineering that allows the transposition of genes from one location on a chromosome to another location.

46. The fusion of two bacteria without cell walls is called ________________ ________________.

47. The addition of plasmids to microorganisms to increase the yield of useful substances is known as ________________ ________________.

48. DNA that contains information from two different species is ________________ DNA.

49. If recombinant DNA genes integrate permanently into egg or sperm cells, allowing the genes to be transferred the offspring, the resulting organism is said to be ________________.

50. The process by which plasmids, or in some cases bacteriophages, are induced to reproduce within cells at a rapid rate is known as ________________

51. One important application of hybridomas was the production of large quantities of highly specific antibodies called ________________ antibodies.

Labeling

Questions 52–57: Select the best possible label for each step in Griffith's experiment with pneumococcal infections in mice: live, nonvirulent, rough pneumococci, without capsule; live, virulent, smooth pneumococci, with capsule; live smooth pneumococci (with capsule) plus live rough pneumococci (without capsule) isolated from dead mouse; heat-killed smooth pneumococci, with capsule; live, nonvirulent, rough pneumococci, no capsule; heat-killed smooth pneumococci with capsule. *Place your answers on the spaces provided on page 99.*

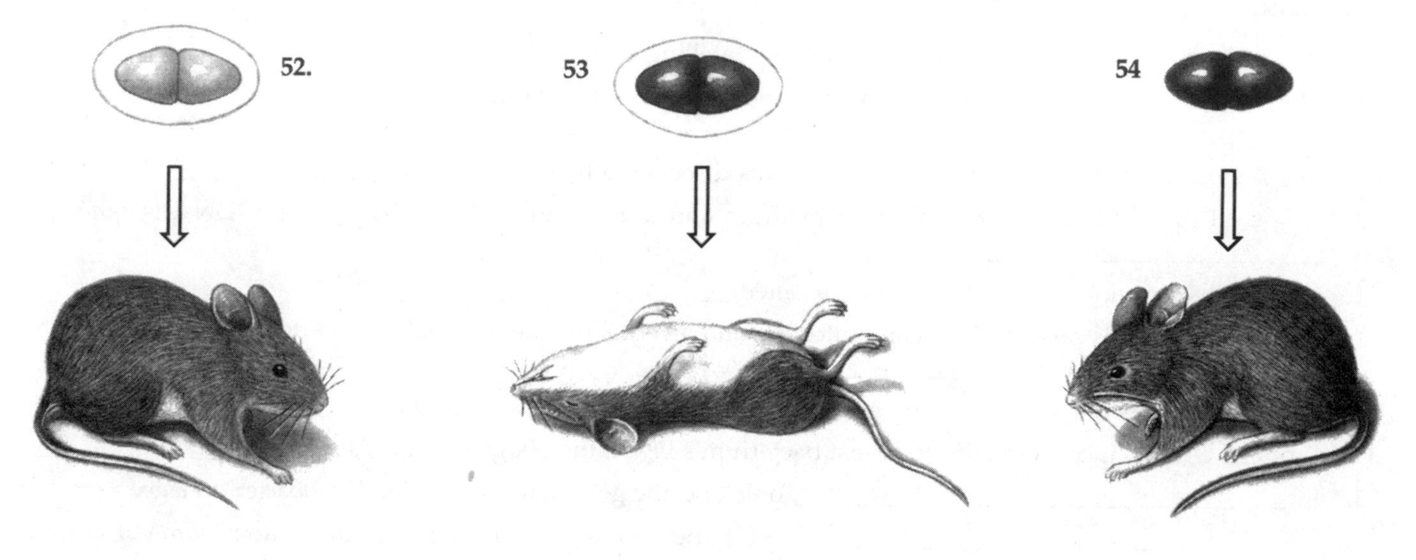

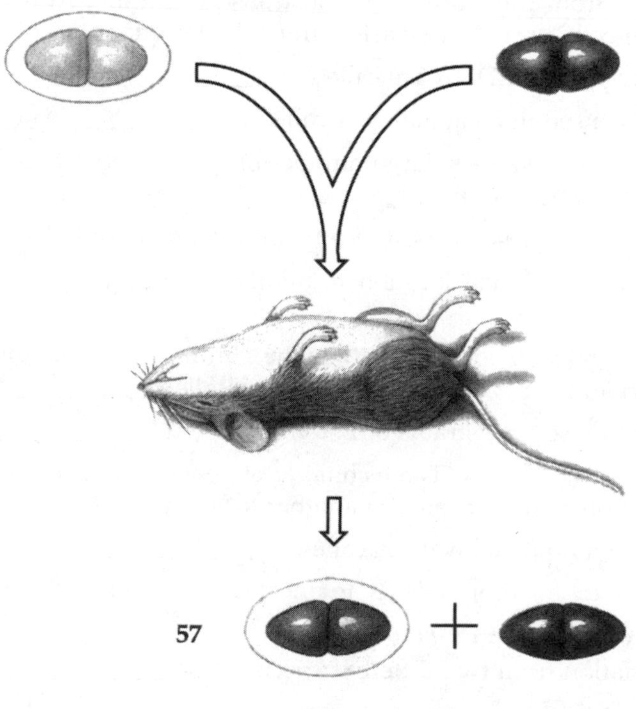

52. ____________________

53. ____________________

54. ____________________

55. ____________________

56. ____________________

57. ____________________

Questions 58–65: Select the best possible label for each step of specialized transduction by lambda phage *E. coli*: (a) induction of the lytic cycle; (b) bacterial cell is lysed and phages are released; (c) phage DNA incorporating some bacterial genes; (d) phage replicates; (e) host cell chromosome acquires both phage DNA and genes from previous host; (f) phage infects new host cell; (g) prophage is incorrectly excised from bacterial chromosome; (h) phage incorporated into chromosome.

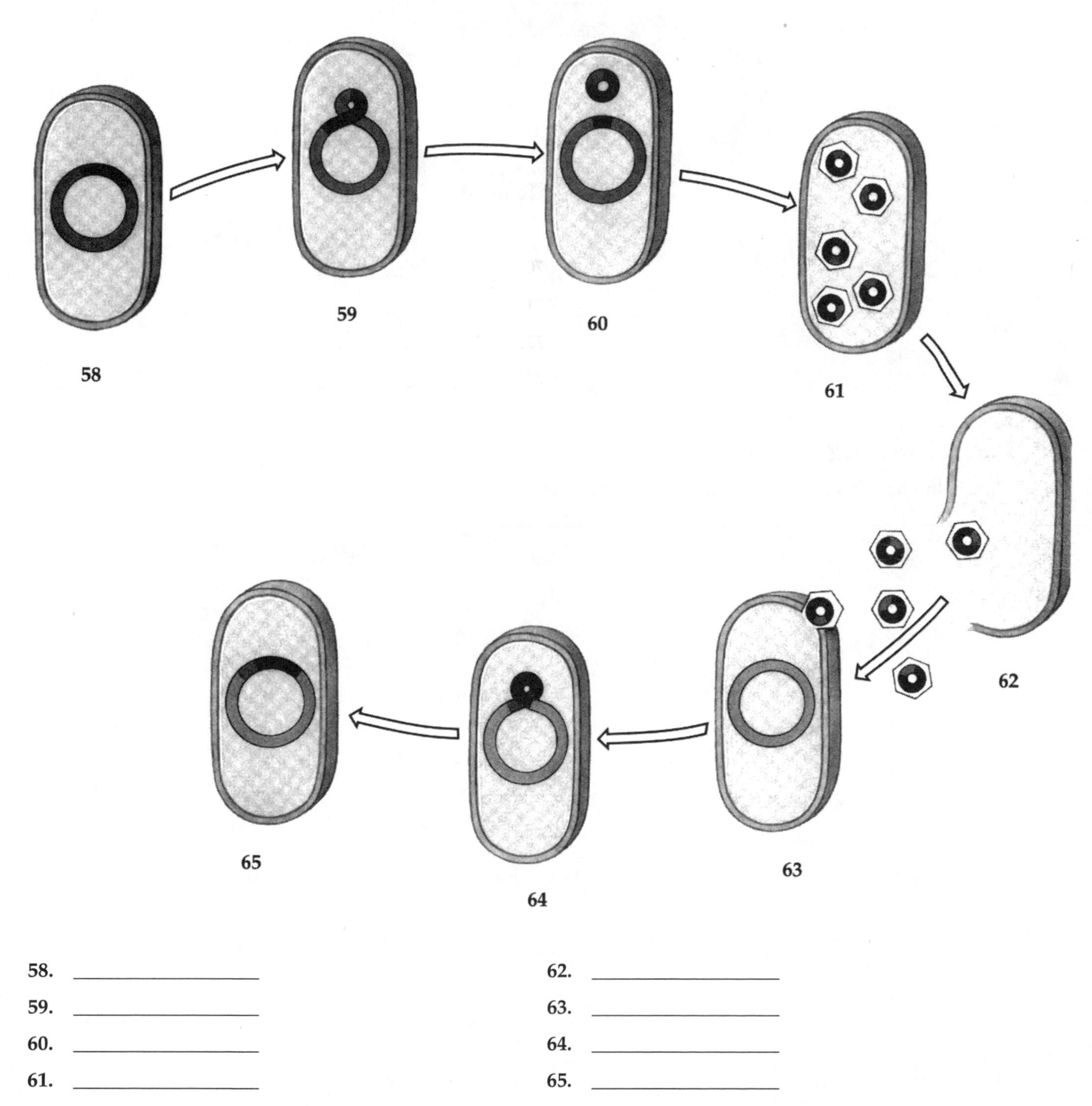

58. ____________________

59. ____________________

60. ____________________

61. ____________________

62. ____________________

63. ____________________

64. ____________________

65. ____________________

Questions 66–73: Select the best possible label for each element in a typical transposon: repressor; plasmid; inverted repeat terminal; transposase; inverted repeat terminal (inserted sequence); transposon genes; binding site for repressor protein; ampicillin resistance.

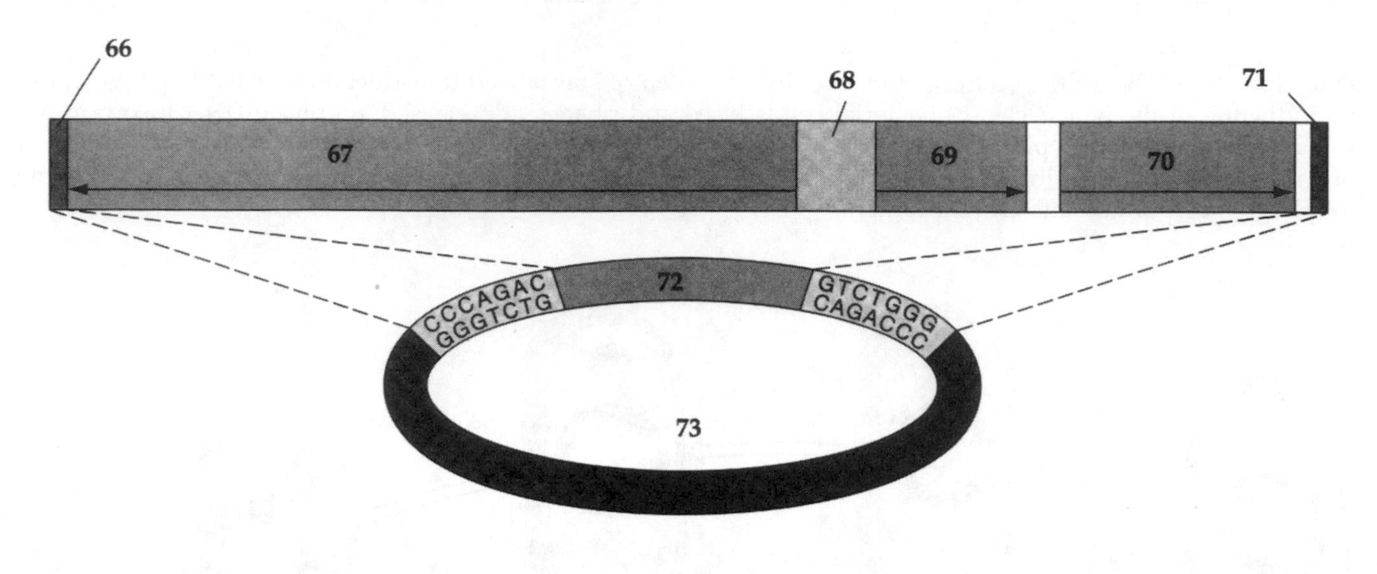

66. ____________________

67. ____________________

68. ____________________

69. ____________________

70. ____________________

71. ____________________

72. ____________________

73. ____________________

Critical Thinking

74. What is the nature and significance of gene transfer?

75. What is the mechanism and significance of transformation?

76. What are the mechanisms and significance of transduction?

77. What is the significance of conjugation?

78. What are the characteristics and actions of plasmids?

79. How are the following techniques of genetic engineering used: (a) genetic fusion, (b) protoplast fusion, (c) gene amplification, (d) recombinant DNA, and (e) hybridomas?

80. Why are scientists concerned about uses of recombinant DNA?

ANSWERS

True/False

1. **False** Transformation can occur with a living or dead donor cell. With a dead cell, the wall is usually disrupted and the DNA can be automatically released. With a live donor, small sections of DNA may be released especially during the log phase of reproduction.
2. **True** The incorporation of prophages into bacterial chromosomes illustrates the close evolutionary relationship between the virus DNA and the host DNA. In order for incorporation to occur, there must be similar base sequences.
3. **True** High frequency of recombination strains can allow the transfer of the entire host chromosome into the recipient whereas other mechanisms can only transfer short segments.
4. **False** Hybridomas are normally formed by the fusion of two eukaryotic cells, one a normal cell and the other a cancer cell. Bacterial protoplast fusion can occur if the cell walls are removed, protoplasts formed, and then the cells fused.
5. **True** Some prophages can incorporate into the host DNA without replication and destruction of the bacterial cell. However, these phages can be induced by a number of means to reenter the lytic cycle.
6. **False** A major problem encountered in industry when using recombinant DNA technology is to isolate the products after they are made. Although the product is made in large quantity it must still be purified.
7. **False** Mutations account for some genetic diversity; however, gene transfer provides the greatest increases in the genetic diversity of organisms.

Multiple Choice

8. b; 9. c; 10. c; 11. d.

Concept Question

12. Since the molecule is a protein, recombinant DNA technology could be one possible way. This would involve removal of DNA from the macrophage, cutting it into small segments, and incorporating the desired DNA pieces into a replicon such as a plasmid. The plasmid could then be cloned, reinserted into a bacterium, and the cells stimulated to divide and produce the desired protein in a sizeable quantity.

Matching

13. g; 14. f; 15. a; 16. d; 17. h; 18. b; 19. e; 20. c; 21. e; 22. d; 23. a; 24. b; 25. f; 26. c; 27. g.

Fill-Ins

28. gene transfer; 29. transformation; 30. competence factor; 31. bacteriophages; 32. lytic; 33. temperate; 34. protoplast; 35. specialized transduction; 36. generalized transduction; 37. conjugation; 38. plasmids; 39. F pilus; 40. high frequency; 41. R factors; 42. transposable element; 43. transposons; 44. colicins; 45. genetic fusion; 46. protoplast fusion; 47. gene amplification; 48. recombinant; 49. transgenic; 50. gene amplification; 51. monoclonal.

Labeling

52. heat-killed smooth pneumococci, with capsule; 53. live, virulent, smooth pneumococci, with capsule; 54. live, nonvirulent, rough pneumococci, no capsule; 55. heat-killed smooth pneumococci, with capsule; 56. live, nonvirulent, rough pneumococci, without capsule; 57. live smooth pneumococci (with capsule) plus live rough pneumococci (without capsule) isolated from dead mouse; 58. a; 59. g; 60. c; 61. d; 62. b; 63. f; 64. h; 65. e. ; 66. inverted repeat terminal; 67. transposase; 68. binding site for repressor protein; 69. repressor; 70. ampicillin resistance; 71. inverted repeat terminal (insertion sequence); 72. transposon genes; 73. plasmid.

Critical Thinking

74. **What is the nature and significance of gene transfer?**
Gene transfer refers to the movement of genetic information between organisms and occurs in bacteria by transformation, transduction, and conjugation. This type of transfer is significant because it increases genetic

diversity in a population and therefore insures that some of the members of the population will always survive environmental changes. These mechanisms were not thought to occur with bacteria until the discoveries by Griffith and others.

75. What is the mechanism and significance of transformation?

Bacterial transformation was discovered in 1928 by Griffith, who showed that live, nonlethal rough pneumococci and heat-killed previously lethal, smooth pneumococci could produce live, lethal pneumococci when introduced together into a mouse. This discovery was especially important because it demonstrated an ability of bacteria to change genetically. Following this discovery others, including Avery and Watson and Crick, demonstrated that DNA was responsible for this genetic change and that the genetic material was coded in the DNA molecules.

Transformation involves the release of DNA fragments from donor cells and their uptake by recipient cells during specific stages in their growth cycle. Competence factors are required for the DNA uptake so that endonucleases can cut the DNA into units and single strands of DNA can be incorporated into the recipient cell's DNA.

This mechanism is significant because it represents an important method of genetic transfer, it can be used in genetic manipulation, and it can be used to create recombinant DNA.

76. What are the mechanisms and significance of transduction?

A second mechanism of genetic transfer is transduction. This involves the transfer of DNA from a donor cell to a recipient via a bacteriophage. When it was discovered that phages could be either virulent or temperate, the likelihood of this mechanism occurring became apparent. Transduction, therefore, involves two basic processes: the lytic cycle and the lysogenic cycle.

The lytic cycle proceeds as follows: (1) phage adsorbs to receptor site on bacterial cell wall, penetrates it, and inserts its DNA; (2) phage DNA inserts itself into bacterial chromosome; (3) phage DNA directs the cell's metabolism to produce viral components—proteins and copies of phage DNA; (4) empty phage heads and pieces of DNA are created within the bacterial cytoplasm; (5) heads are packed with DNA; (6) collars, sheaths, and base plates are attached to heads, with tail fibers being added last; (7) bacterial cell bursts (lysis), releasing completed infective phages.

The lysogenic cycle proceeds as follows: (1) phage adsorbs to receptor site on bacterial cell wall, penetrates it, and inserts its DNA; (2) phage DNA inserts itself into bacterial chromosome; (3) phage is replicated within the bacterial DNA prior to binary fission; (4) binary fission occurs; (5) each resulting cell has the phage DNA incorporated.

First, a virulent phage must infect and destroy a host cell's DNA and then cause lysis. In the process the virulent phage may pick up a few genes of the host cell and incorporate them into its genome. The second step involves the virulent phage infecting a new host (recipient) and instead of lysing the cell, it persists in the cell and becomes a temperate phage. The temperate phage, now called a prophage, can be incorporated with the recipient's DNA or it can exist as a plasmid. Bacterial cells containing the prophage are considered lysogenized since the virus could reenter the lytic phase.

Specialized transduction involves the transfer of only those host genes adjacent to the phage. Generalized transduction involves the transfer of any DNA fragment since the phage exists as a plasmid.

Transduction is significant because it represents another mechanism of genetic transfer and because it demonstrates that prophages and bacterial cells must have had a close evolutionary relationship. Furthermore, it suggests a viral origin of cancer and it provides a mechanism for studying gene linkage.

77. What is the significance of conjugation?

Conjugation involves transfer of genetic material from a donor cell to a recipient cell via a conjugation bridge called a pilus. Donors are called "males" or F+ cells and recipients are called "females" or F− cells. F+ cells can transfer extrachromosomal units called plasmids. If the plasmid becomes integrated into the host chromosome, the F+ cell, now called an Hfr, can transfer a large number of host genes. If the plasmid is incorporated and subsequently separated from the host chromosome, the cell becomes an F′ cell and can transfer host genes that are attached to the plasmid.

This mechanism is significant because it represents another method of genetic transfer and it increases the prospects of genetic diversity in bacteria. It helps provide another means of mapping bacterial chromosomes and indicates additional evolutionary stages of development.

78. What are the characteristics and actions of plasmids?

Plasmids are circular, self-replicating, double-stranded extrachromosomal units of DNA that carry nonessential but often useful pieces of genetic information. Some plasmids carry antibiotic resistant genes called

R factors, some provide information to produce bacteriocins which can inhibit the growth of other bacteria, some are virulence plasmids that cause disease symptoms, and some can cause tumors in plants. Most, if not all, plasmids carry genetic information to produce the pili necessary for transfer of the plasmids.

79. How are the following techniques of genetic engineering used: (a) genetic fusion, (b) protoplast fusion, (c) gene amplification, (d) recombinant DNA, and (e) hybridomas?

Genetic engineering is the manipulation of genetic material to alter the characteristics of an organism. In order to accomplish this, several techniques have been developed. Genetic fusion involves the alteration of genes within a single species of organism. This is accomplished by deleting small sections or relocating certain sections to create different characteristics. Protoplast fusion involves the combination of two, cell wall-less organisms with the resultant fusion of their DNA. This allows the mixing of favorable characteristics of two different organisms. Gene amplification involves the addition or multiplication of plasmids in organisms to increase the yield of products the cell can produce. This has increased the antibiotic production of certain organisms. Recombinant DNA technology involves the insertion of foreign DNA into the DNA of bacteria and usually involves plasmids. This technology has provided a mechanism to produce very large quantities of substances such as human insulin, interferon, and interleukins made only by eukaryotic cells. Hybridomas are genetic recombinations usually involving eukaryotic cells. This technique has allowed the production of large quantities of highly specific antibodies and other molecules.

80. Why are scientists concerned about uses of recombinant DNA?

Recombinant DNA technology has carried with it advantages and disadvantages. The advantages included the ability to alter genes of any organism to produce large quantities of proteins that were very difficult to produce. This was contrasted with the prospect that the technology might create a virulent uncontrollable pathogen. Fortunately, this prospect seems very remote at the present time.

Chapter 9 An Introduction to Taxonomy: The Bacteria

INTRODUCTION

At first, taxonomy may seem like a dull and boring subject. Now I have to memorize rules and names; what a dreary business. Of course, if one is ill from a bacterial infection, we expect our health care providers to know exactly what is making us sick. In such a circumstance, a name may have meaning. Oddly enough, this subject is more abstract than one might think. In the world of bacteria it is especially difficult. When we move the concept of Kingdoms, we consider an even more abstract concept.

The author of this guide has a friend, Dr. Katherine Baker, who is an environmental microbiologist. In 1989, a large oil tanker, the Exxon Valdez spilled large amounts of petroleum into the pristine waters of Alaska. The government hired Dr. Baker to help remediate the spill. She applied fertilizers on coastal areas to promote the growth of bacteria that would break down the oil into harmless chemicals. I asked her exactly what bacterium was helping to clean up the spill. Dr. Baker reminded me that no one knows the name of most environmental bacteria and in this case it does not matter. The fact is, we can identify and name very few bacteria in nature.

What exactly is a species or kind of bacterium? To the great Charles Darwin, the species concept was arbitrary. The modern concept of a species was developed using birds not bacteria as a model. Ernst Mayr defined a species as an interbreeding population. This implies one can observe sex. In the case of birds this is relatively easy to do. Obviously, Bacteria are not like this. They do not engage in sex like birds and cannot be easily observed. Just what is a species of bacterium or fungus or protozoan? In this chapter we will see how modern biologists approach this issue.

When we move to the concept of the kingdom, it does not get easier. When the author of this manual was in college, I learned that there were only two kingdoms. When you were in high school, you were likely to have learned that there are five kingdoms. In this chapter you will learn how some biologists believe that all life is divided into three domains. What is a student to believe? Which of these are correct? No wonder students can be confused. The answer is that all of these ideas are both true and false. The act of placing large numbers of organisms into groups is an act of concept formation. We make connections in our mind that we feel are useful. The meaning of these concepts can be measured against one's values. That is, is the goal to create a system with practical value? Is the purpose of the system to create a concept that most accurately reflects evolution? In any event, the numbers of kingdoms or domains is in fact arbitrary. There are 'lumpers' and 'splitters'. In the end, the author of this introduction thinks that the great Charles Darwin was likely to be as correct as anyone.

So approach this chapter with a little more interest than you would have if you'd only read its title. Taxonomy isn't as boring as you might think it will be—and who knows, you may make your mark in science someday just because of the things you learn about in this chapter.

STUDY OUTLINE

I. Taxonomy: the Science of Classification

A. Linnaeus, the father of taxonomy

The eighteenth-century Swedish botanist Carolus Linnaeus is credited with founding the science of *taxonomy,* since his *binomial nomenclature* system is still used today.

B. Using a taxonomic key

Biologists often use a *dichotomous taxonomic key* to identify organisms using paired (either-or) statements to describe characteristics.

C. Problems in taxonomy

Scientists would like to classify organisms according to their *phylogenetic,* or evolutionary, relationships, and so taxonomy must change with evolutionary changes and new knowledge. It is far more important to have a taxonomic system that reflects our current knowledge than to have a system that never changes.

D. Developments since Linnaeus's time

In 1866, Ernst H. Haeckel created a third kingdom, the Protistia. Changes in the understanding and classification of the bacteria arose as tools to study bacteria were developed. Cells have been recognized as prokaryotic or eukaryotic, leading to the placement of bacteria in a separate kingdom of anucleate organisms. In 1956, Margulis and Copeland proposed a four-kingdom system of classification. R.H. Whittaker's taxonomic system, created in 1969, led to the five-kingdom classification system of today.

II. The Five-Kingdom Classification System

A. General features

It is important to remember that organisms may be classified in very diverse taxonomic groups even though their cells have many similarities in structure and function. In the five-kingdom classification system, prokaryotes are all placed in the kingdom Monera. Most unicellular eukaryotes are placed in the kingdom Protista. Fungi are in their own separate kingdom.

B. Kingdom Monera

This kingdom consists of all prokaryotic organisms, including eubacteria ("true bacteria"), the cyanobacteria, and the archaeobacteria. Members of the Monera kingdom are prokaryotic and are found alone (unicellular) or grouped. They usually possess cell walls, obtain nutrients through absorption or photosynthesis or chemosynthesis, and reproduce asexually, usually by fission.

C. Kingdom Protista

Members of the Protista kingdom are eukaryotic and are unicellular or occasionally multicellular. They sometimes possess cell walls, obtain nutrients through ingestion or absorption or photosynthesis, and usually reproduce asexually, although some reproduce sexually.

D. Kingdom Fungi

The kingdom Fungi includes mostly multicellular and some unicellular eukaryotic organisms whose cells are surrounded by cell walls. Nutrients are obtained through absorption and reproduction is both sexual and asexual.

E. Kingdom Plantae

The kingdom Plantae includes multicellular eukaryotic organisms whose cells are all surrounded by cell walls. Nutrients are obtained through absorption and photosynthesis and reproduction is both sexual and asexual.

F. Kingdom Animalia

Members of the Animalia kingdom are muticellular eukaryotic organisms. Their cells are not surrounded by cell walls and nutrients are obtained through ingestion and, occasionally, absorption. Reproduction is primarily sexual.

III. The Three-Domain Classification System

A. General features

In the late 1970s, studies by Woese, Fox, and others proposed a new classification system based on the existence of three types of cells: urkaryotes, prokaryote, and eukaryotes.

B. The evolution of prokaryotic organisms

Stromalites are fossilized photosynthetic prokaryotes that appear as masses of cells or microbial mats. Evidence from studies of archaeobacteria and the most ancient stromalites convinced many scientists, but not all scientists, that three branches of the tree of life formed during the Age of Microorganisms and that each branch gave rise to distinctly different groups of organisms. In 1990, Woese suggested that the category *domain* be placed above the level of kingdom, and, in 1998, he proposed theories of how the three domains may have arisen.

C. The Bacteria
All Bacteria are unicellular prokaryotes. This domain includes the eubacteria ("true bacteria").

D. The Archaea
All Archaea are unicellular prokaryotes that possess a cell wall made of materials other than peptidoglycan.

E. The Eukarya
All Eukarya are eukaryotic cells, meaning they have a true nucleus.

IV. Classification of Viruses

Viruses are submicroscopic, parasitic, acellular microorganisms composed of a nucleic acid (DNA or RNA) core inside a protein coat. They are not assigned to a kingdom because they display only a few characteristics of living organisms.

V. The Search for Evolutionary Relationships

A. Special methods needed for prokaryotes
Prokaryotes are not classified according to morphology, genetic features, and evolutionary relationships. They are classified according to metabolic reactions, genetic relatedness, and other specialized properties.

B. Numerical taxonomy
Numerical taxonomy is a comparison of organisms based on a quantitative assessment of a large number of their characteristics. Computers are being used in the assessment process.

C. Genetic homology
Genetic homologies are similarities of DNA base sequences among organisms. These homologies can be determined by: (1) directly determining the base composition of the DNA (which determines the total amount of each nucleotide present without giving any indication of the sequence of these bases); (2) sequencing the bases in DNA or RNA by using probes; (3) using DNA hybridization (in which the double strands of DNA of two organisms are split apart and the split strands from the two organisms are allowed to combine; and (4) using protein profiles and amino acid sequences.

D. Other techniques
Other techniques used for studying evolutionary relatedness include: (1) determining the properties of ribosomes, since ribosomes have evolved very slowly; (2) comparing immunological reactions, since monoclonal antibodies only bind to specific proteins and can show that similar proteins exist between two different organisms; and (3) using phage typing, since bacteriophages attack only related bacteria.

E. The significance of findings
The main significance of methods of determining evolutionary relationships is that these methods can be used to group closely-related organisms and to separate them from less closely-related ones.

VI. Bacterial Taxonomy

A. Criteria for classifying bacteria
Bacteria are classified according to criteria such as: morphology, staining, growth, nutrition, physiology, biochemistry, genetics, serology, phage typing, sequence of bases in rRNA, and protein profiles. By using these classification criteria, we can identify an organism as belonging to a particular genus and species. But assigning bacterial genera to higher taxanomic levels is difficult because of the incompleteness of the fossil record and the limited information that can be gleaned from what fossils have been found.

B. The History and Significance of Bergey's Manual
The first edition of *Bergey's Manual of Determinative Bacteriology* was published in 1923. Since then, eight editions, an abridged version, and several supplements have been published. *Bergey's Manual of Systemic Bacteriology*, a four-volume manual with a much broader scope than previous manuals, has also been published; the first volume was published in 1984, the second in 1986, and the final two in 1989. The ninth edition of *Bergey's Manual of Determinative Bacteriology* was published in 1994, and, in 1995, became available on CD-ROM. Bergey's Manual is organized into sections, rather than families and orders, with each section devoted to a group of similar organisms. The manual has become an internationally recognized reference for bacterial taxonomy and has served as a reliable standby for medical workers.

C. Problems associated with bacterial taxonomy
   Despite tremendous effort, no complete classification of bacteria from kingdom to species has been established, since too little is known about evolutionary relationships to establish clearly-defined taxonomic classes and orders for many bacteria. Those bacteriologists looking from the top level down can propose plausible divisions of the kingdom Monera (Prokaryotae). Those looking from the bottom up can establish strains, species, and genera and can sometimes assign bacteria to higher-level groups. Section assignments will continue to be used until discrepancies can be resolved and taxonomic levels between genera and divisions can be more precisely determined.

D. Bacterial nomenclature
   Bacterial nomenclature refers to the naming of species according to internationally agreed-upon rules, which are subject to change as new information is obtained.

E. Bacterial by section in Bergey's Manual
   There are 33 sections of prokaryotes described in Bergey's Manual. Some contain only a few organisms, others contain hundreds; some contain organisms of medical significance, others are of interest mainly to ecologists, taxonomists, or researchers.

F. Bacterial taxonomy and you
   You can use the endpapers, which contain information from Bergey's Manual and are located inside the front and back covers of your textbook, to find out about any given organism or any given disease.

VII. The Discovery of New Organisms

There are still new worlds to discover, with new creatures in them. For example, recent discoveries have been made inside submarine hot vents, inside volcanoes, and in deep oil wells. Many more microbes have yet to be discovered.

SELF-TESTS

Continue with this section only after you have read Chapter 9 of your textbook. Write your answers in the appropriate space provided. Correct answers to all questions can be found at the end of the Self-Test section. A score of 80 percent or better is good. If your score is less than 65, reread the chapter.

True/False

Mark T for True, F for False

_____ **1.** The animal kingdom is the only kingdom that lacks any organisms that relate to microbiology.

_____ **2.** With proper taxonomy, the genus and species names are always italicized and capitalized.

_____ **3.** A dichotomous key always requires paired statements.

_____ **4.** Viruses are currently classified according to their nucleic acids, their capsid, and the presence or absence of envelopes.

_____ **5.** Quite often, the genus name of an organism is used to honor the scientist credited with its discovery.

_____ **6.** In DNA hybridization, a high degree of similarity exists when the DNA of two organisms have long identical sequences of bases.

_____ **7.** The ATCC represents a branch of the CDC whose task is to identify new diseases.

Multiple Choice

Select the best possible answer.

_____ **8.** Viruses are:
- **a.** considered to be prokaryotes.
- **b.** not classified by any classification scheme.
- **c.** considered to be acellular particles.
- **d.** placed in the kingdom Monera.
- **e.** None of the above is true.

_____ **9.** The scientist credited with the development of the binomial scheme is:
a. Linnaeus.
b. Semmelweiz.
c. Leeuwenhoek.
d. Aristotle.

_____ **10.** The scientist most responsible for establishing the five-kingdom classification system currently in use is:
a. Linnaeus.
b. Bergey.
c. Whittaker.
d. Margulis.

_____ **11.** A characteristic associated *only* with the kingdom Prokaryotae is the:
a. cell membrane.
b. unicellularity.
c. photosynthesis.
d. lack of a true nucleus.

_____ **12.** Which of the following kingdoms are characterized by members that are eukaryotic and mostly all unicellular?
a. Monera.
b. Protista.
c. Fungi.
d. Plantae.
e. Animalia.

Matching

Select the answer from the right-hand side that corresponds to the term or phrase on the left-hand side of the page. An answer may be used more than once. In some cases, more than one answer may be required.

Topic: Contributions to Classification

_____ **13.** E. H. Haeckel
_____ **14.** R. H. Whittaker
_____ **15.** C. Linnaeus
_____ **16.** C. Woese
_____ **17.** C. Woese and E. Fox

a. binomial system
b. urkaryote
c. taxonomic category, the domain
d. five-kingdom system
e. created Protista kingdom

Topic: Eukaryotic and Prokaryotic Organization

_____ **18.** Mitochondria
_____ **19.** Nucleoli
_____ **20.** Chloroplasts
_____ **21.** Nuclear envelope
_____ **22.** Nucleoid
_____ **23.** Ribosomes
_____ **24.** Flagella
_____ **25.** Chromosomes
_____ **26.** Both DNA and RNA present
_____ **27.** Either DNA or RNA, never both

a. associated with prokaryotes
b. associated with eukaryotes
c. associated with both prokaryotes and eukaryotes
d. associated with neither prokaryotes nor eukaryotes

Fill-Ins

Provide the correct term or phrase for the following. Spelling counts.

28. The science of classification is known as ________________.

29. The "Father of Taxonomy" is ________________ ________________.

30. The use of a two-name system for each organism is called ________________ ________________.

31. The "first" name of an organism that is never capitalized is the ________________ name.

32. A taxonomic key that uses paired statements describing characteristics is a ________________ key.

33. The kingdom that includes all the prokaryotes is ________________.

34. The "true bacteria" are the ________________.

35. Prokaryotic, blue-green algae are also known as ________________.

36. Primitive bacteria that lack peptidoglycan and are found only in very extreme environments are classified as ________________.

37. Eukaryotic organisms that are all unicellular are placed in the ________________ kingdom.

38. A kingdom of organisms that obtains nutrients solely by absorption is the ________________ kingdom.

39. ________________ are fossilized photosynthetic prokaryotes that appear as masses of cells or microbial mats.

40. In 1990, Carl Woese proposed that a new taxonomic category be erected above the level of kingdom, known as the ________________.

41. The ________________ are strictly anaerobic organisms that have been isolated from waterlogged soils, lake sediments, marshes, marine sediments, and the gastrointestinal tracts of animals, including humans.

42. Acellular entities that are not classified in any kingdom are called ________________.

43. Comparing organisms based on a large number of characteristics and grouping them according to the percent of shared characteristics is called ________________ taxonomy.

44. Taxonomic studies can be based on DNA similarities, or ________________ ________________.

45. A DNA fragment that has sequences complementary to those being sought could be called a DNA ________________.

46. A taxonomic method in which studies are made of the degree of annealing between strands of DNA from test organisms is called the ________________ ________________ method.

47. A taxonomic method that studies the similarities between cellular proteins is the ________________ ________________ method.

48. A taxonomic method in which bacterial strains are compared, based on the type of bacteriophage that can infect them, is called the ________________ ________________ method.

49. A bacterial ________________ consists of the descendants of a single isolation in pure culture.

50. The taxonomic "Bible" of the prokaryotes is ________________ Manual.

51. Instead of using the categories "family" or "order," Bergey's Manual divides bacteria into ________________.

52. ________________ ________________ refers to the naming of species according to internationally agreed-upon rules.

Labeling

Questions 53–61: Select the best possible label for each of the classification categories corresponding to the classification of a human: specific epithet, order, division/phylum, kingdom, genus, class, subphylum, family.

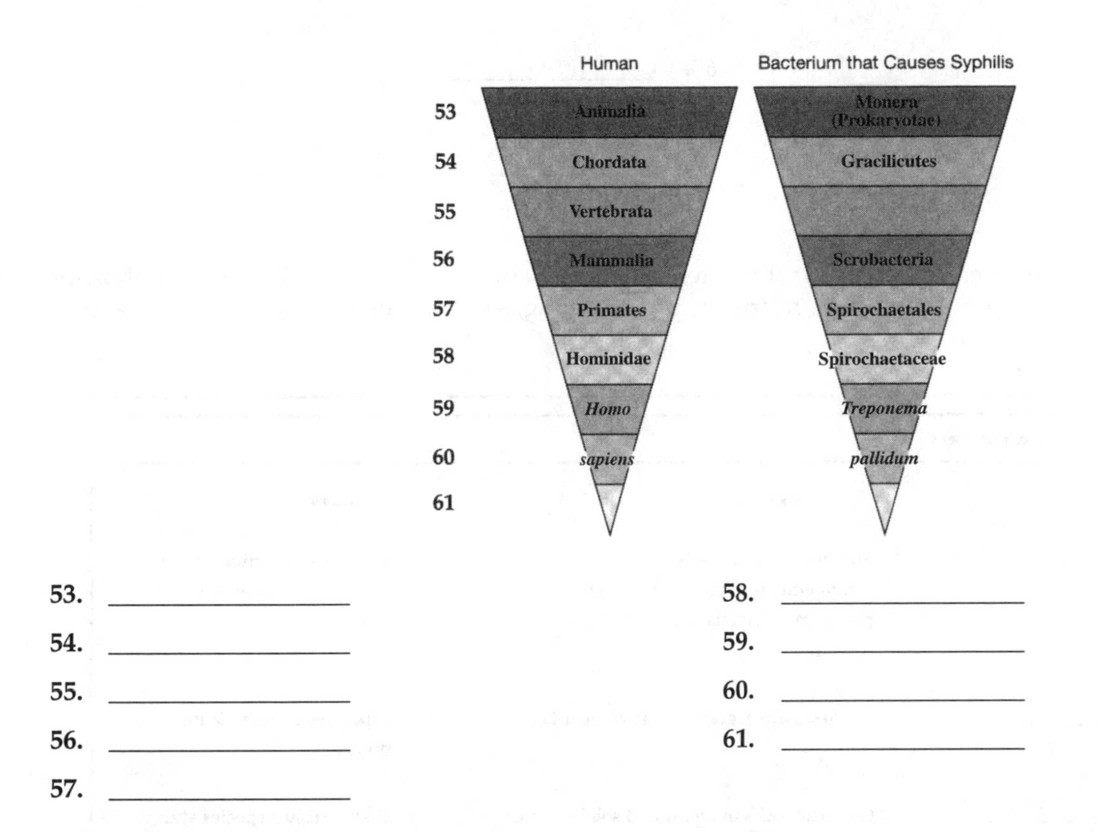

53. ____________________

54. ____________________

55. ____________________

56. ____________________

57. ____________________

58. ____________________

59. ____________________

60. ____________________

61. ____________________

Questions 61–67: Select the best possible label for each step in DNA hybridization: (a) incubate together and cool to allow annealing (joining); (b) if organisms are closely related, a high percentage of annealing occurs; (c) DNA from organism 1; (d) heat to "melt" (separate) DNA into single strands; (e) if organisms are unrelated, very little annealing occurs; (f) DNA from organism 2; (g) if organisms are identical, get 100% annealing. *Place your answers in the spaces provided on page 114.*

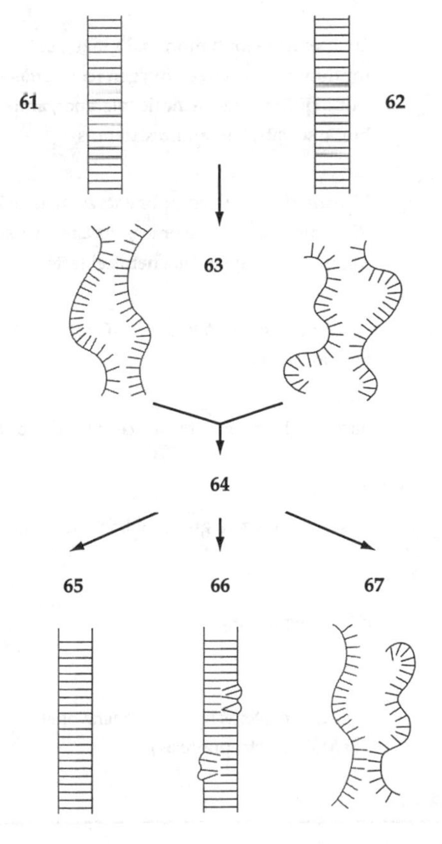

61. ____________________

62. ____________________

63. ____________________

64. ____________________

65. ____________________

66. ____________________

67. ____________________

Questions 68–78: Select the best possible label for the criteria used to classify bacteria: physiology, morphology, genetics, biochemistry, staining, protein profiles, nutrition, phage typing, serology, growth, sequence of bases in tRNA.

CRITERIA FOR CLASSIFYING BACTERIA

CRITERIA	EXAMPLES	USES
68. ____________________	Size and shape of cells; arrangements in pairs; or filaments; presence of flagella, pili, endospores, capsules	Primary distinction of genera and sometimes species
69. ____________________	Gram-positive, gram-negative, acid-fast	Separates eubacteria into divisions
70. ____________________	Characteristics in liquid and solid cultures colony morphology, development of pigment	Distinguishes species and genera
71. ____________________	Autotrophic, heterotrophic, fermentative with different products, energy sources, carbon sources, nitrogen sources, needs for special nutrients	Distinguish species, genera, and higher groups
72. ____________________	Temperature (optimum and range), pH (optimum and range), oxygen requirements, salt requirements, osmotic tolerance, antibiotic sensitivities and resistances	Distinguish species, genera, and higher groups
73. ____________________	Nature of cellular components such as cell wall, RNA molecules, ribosomes, storage inclusions, pigments, antigens; biochemical tests	Distinguish species, genera, and higher groups
74. ____________________	Percentage of DNA bases (G+C ratio); DNA hybridization	Determine relatedness within genera and families
75. ____________________	Slide agglutination, fluorescent-labeled antibodies	Distinguish strains and some species
76. ____________________	Susceptibility to a group of bacteriophages	Identification and distinguishing of strains
77. ____________________	rRNA sequencing	Determine relatedness among all living things
78. ____________________	Separate proteins by two-dimensional PHAGE (electrophoresis)	Distinguish strains

Critical Thinking

79. How are microorganisms named?

80. What did Linnaeus contribute to taxonomy?

81. How is a dichotomous taxonomic key used to identify organisms?

82. What are some problems and developments in taxonomy since Linnaeus?

83. What are the main characteristics of the kingdoms in the five-kingdom system of taxonomy?

84. How are viruses classified?

85. What special methods are needed for determining evolutionary relationships among prokaryotes?

86. What criteria are used for classifying bacteria?

87. What are the history and significance of *Bergey's Manual of Systematic Bacteriology*?

88. What problems are associated with bacterial taxonomy?

ANSWERS

True/False

1. **False** Several groups of animals are of interest to microbiologists since some act as parasites such as the tapeworms and roundworms which live on or in other animals and some act as carriers of microbial disease agents such as certain insects.
2. **False** The genus name is always capitalized and underlined or italicized and the species name is only underlined or italicized.
3. **True** A dichotomous key always uses paired statements to present an "either or" choice such that only one statement can be true.
4. **True** Viruses, since they are acellular particles, require a separate classification system. Currently the scheme uses the three parameters mentioned.
5. **True** Many scientists including Pasteur, Rickets, Lister, Yersin, Neisser and Escherich all are honored in this manner.
6. **True** When the DNA of two organisms are split apart and then allowed to reanneal, the amount of relatedness can be determined. The greater the amount of base pairing by the strands the more they are related.
7. **False** ATCC refers to American Type Culture Collection which is based in Rockville, Maryland. The company maintains thousands of lyophilized cultures to ensure their exact characteristics.

Multiple Choice

8. c; 9. a; 10. c; 11. d; 12. b.

Matching

13. e; 14. d; 15. a; 16. c; 17. b; 18. b; 19. b; 20. b; 21. b; 22. a; 23. c; 24. c; 25. c; 26. c; 27. d.

Fill-Ins

28. taxonomy; 29. Linnaeus; 30. binomial nomenclature; 31. species; 32. dichotomous; 33. Monera; 34. eubacteria; 35. cyanobacteria; 36. archaeobacteria; 37. Protista; 38. Fungi; 39. stromatolites; 40. domain; 41. methanogens; 42. viruses; 43. numerical; 44. genetic homologies; 45. probe; 46. DNA hybridization; 47. protein profile; 48. phage typing; 49. strain; 50. Bergey's; 51. sections; 52. bacterial nomenclature.

Labeling

53. kingdom; 54. division/phylum; 55. subphylum; 56. class; 57. order; 58. family; 59. genus; 60. specific epithet; 61. c; 62. f; 63. d; 64. a; 65. g; 66. b; 67. e.; 68. morphology; 69. staining; 70. growth; 71. nutrition; 72. physiology; 73. biochemistry; 74. genetics; 75. serology; 76. phage typing; 77. sequence of bases in rNRA; 78. protein profiles.

Critical Thinking

79. **How are microorganisms named?**
Organisms are named according to several characteristics such as where they are found, who discovered them, and what diseases they cause. In science however, it is essential to have accurate and standardized names. The science of classification is called taxonomy. This system insures that we name organisms based on fundamental concepts of unity and diversity among living organisms.

80. **What did Linnaeus contribute to taxonomy?**
Linnaeus is credited with a system of classification called binomial nomenclature, which specifically identifies each living organism and provides them with a last or genus name, and a first or species name. He also established the basic hierarchy of taxonomy of families, orders, classes, phyla, and kingdoms.

81. **How is a dichotomous taxonomic key used to identify organisms?**
Taxonomic keys are used to identify organisms according to physical and/or biochemical characteristics. A dichotomous key uses a series of paired statements that are presented as either/or choices. By select-

ing the appropriate statement, an organism can be classified rather quickly. If the key is sufficiently detailed and if enough information is known about the organism, it can be identified down to the genus and species.

82. What are some problems and developments in taxonomy since Linnaeus?

Several problems have developed since Linnaeus. One major problem has been that evolutionary changes occur at rapid rates and our knowledge of the evolutionary history of organisms is incomplete. Therefore any taxonomic system must be flexible to accommodate the new discoveries that are made. Another problem has been in deciding what constitutes a kingdom and what constitutes a species.

Major changes in taxonomy since Linnaeus have been in the formulation of five kingdoms rather than the two he proposed. Also, a descriptive taxonomic scheme edited by Bergey was developed to accommodate the increasing variety of bacteria. This Manual of Determinative Bacteriology has also undergone several revisions as a result of new evolutionary evidence.

83. What are the main characteristics of the kingdoms in the five-kingdom system of taxonomy?

The five kingdoms proposed by Whittaker are Monera, Protista, Fungi, Plantae, and Animalia. Each kingdom has different distinguishing characteristics. First, Monera—prokaryotic; unicellular, although cells are sometimes grouped; nutrition by absorption, but in some forms by photosynthesis or chemosynthesis; reproduction asexual, usually by fission. Second, Protista—eukaryotic; unicellular, but sometimes cells are grouped; nutrition varies among phyla and can be by ingestion, photosynthesis, or absorption; reproduction asexual, and, in some forms, both sexual and asexual. Third, Fungi—eukaryotic; unicellular or multicellular; nutrition by absorption; reproduction usually both sexual and asexual and often involves a complex life cycle. Fourth, Plantae—eukaryotic; multicellular; nutrition by photosynthesis. Fifth, Animalia—eukaryotic; multicellular; nutrition by ingestion, but in some parasites by absorption; reproduction primarily sexual.

84. How are viruses classified?

Viruses are acellular particles that consist of nucleic acids and a protein coat. They are not assigned to any kingdom and there is much debate over whether they are even "alive." Currently they are classified on the basis of their nucleic acid, their protein coat, and the presence of an envelope.

85. What special methods are needed for determining evolutionary relationships among prokaryotes?

Prokaryotic organisms have few morphological characteristics and have left only a sparse fossil record. Consequently, special methods are needed for determining their relationships. These methods include numerical taxonomy, genetic homology, and other techniques.

With numerical taxonomy, organisms are compared on a large number of characteristics such as Gram staining, morphology, types of enzymes, and oxygen requirements. Genetic homology involves comparisons based on the cell's DNA. Various techniques have been developed to accomplish this. These include nucleotide base comparisons involving G-C ratios, DNA hybridization involving the matching of strands of DNA between two organisms, DNA and RNA sequencing, and protein profiles involving a study of the amino acid sequences of certain proteins.

Other techniques include studies of the ribosomes of organisms, immunological reactions, and phage typing.

86. What criteria are used for classifying bacteria?

Because many bacteria have similar shapes and sizes, several criteria are necessary for classification of the bacteria. These include cell morphology, staining properties (especially Gram stain), growth, nutrition, physiology, biochemistry, and genetics. These criteria can be used to classify bacteria into a genus and, in some cases, into species and even strains.

87. What are the history and significance of Bergey's Manual?

Bergey's Manual of Determinative Bacteriology is the taxonomic "Bible" for microbiology. It was first published in 1923 with David H. Bergey as the chairperson of the editorial board. It has been revised numerous times, the ninth edition being the latest. The manual contains the names and descriptions of organisms and diagnostic keys and tables for identifying each organism.

88. What problems are associated with bacterial taxonomy?

Ever since the first manual of determinative bacteriology was written, microbiologists have been unable to agree on exactly how the members of the kingdom Monera (Prokaryotae) should be divided. Because of recent advances in taxonomic technology, many changes have been made resulting in an extremely wide diversity of opinions. Those that are looking from the top down, that is, from the kingdom level down can assign the bacteria to several divisions but have difficulty with classes and orders. Those that are looking from the bottom up, that is, from the genus level up have established individual groups or sections but have difficulty with many of the organisms within the sections. Recent data have created even more discrepancies in assignments. There is even more trouble when considering orders and classes. Hopefully, with additional information that should be obtained soon, some of the questions will be answered.

Chapter 10 Viruses

INTRODUCTION

In 1969, Dr. John Frame from Columbia University was in Nigeria and noted a new and deadly virus. Dr. Frame returned to Columbia-Presbyterian Hospital with a missionary nurse who was infected with the newly discovered Lassa virus. The virus is so infective that one drop of patient fluid on unbroken human skin could infect. Dr. Jorde Cassals from Yale University came to New York to help with the problem and became infected by working on the case. At the time, it was believed that 80% infected would die. Through extraordinary means both the patient and Dr. Cassals would live. The emergence of this virus from the urine of rodents in Africa to people stimulated the Centers for Disease Control in Atlanta to create special laboratories to contain deadly viruses. Since this time new viruses are emerging on a regular basis and threaten each of us.

By 1980, we learned of the emergence of the HIV virus that causes AIDS. West Nile virus emerged in the United States just a few years ago. More recently, SARS virus emerged from China and quickly spread to North America. As this it being written the first cases of human-to-human transmission of avian flu virus may have been detected.

Viruses are like the mythical Zombies of the horror movies, for both are not alive. How do you kill the undead? Antibiotics will not work on what is not alive. The need to understand, prevent, and treat viral infections has never been greater. Consider, three thousand people died in the 9/11 attacks but more than ten times that number died of influenza flu last year in the United States.

STUDY OUTLINE

I. **General Characteristics of Viruses**
   A. What are viruses?
      Viruses are submicroscopic, parasitic, acellular microorganisms composed of a nucleic acid (DNA or RNA) core inside a protein coat. They are obligate intracellular parasites (they can only replicate inside living cells).
   B. Components of viruses
      Typical viral components include a nucleic acid core and a surrounding protein coat called a *capsid*. Some viruses also have a lipid bilayer membrane called an *envelope* surrounding their capsids.

C. Shapes and sizes
Most viruses are too small to be seen with a light microscope, but they range in size from less than 30 nm in diameter to 240 nm by 300 nm. Most viruses also have a specific shape that is determined by their capsomeres or envelopes. These shapes include *helical, icosahedral, complex, bullet-shaped,* or *spherical.*

D. Host ranges and specificity of viruses
The *host range* of a virus refers to the spectrum of hosts that a virus can infect. *Viral specificity* refers to the specific kinds of cells a virus can infect.

II. Classification of Viruses

A. Basic methods
Virus families are often distinguished on the basis of: (1) their nucleic acid type (DNA or RNA, positive sense or negative sense); (2) their capsid symmetry (shape); (3) whether or not they have an envelope; and (4) their size.

B. RNA viruses
The different families of RNA viruses are distinguished from one another by their nucleic acid content, their capsid shape, and the presence or absence of an envelope. Important groups of RNA viruses include picornaviruses, togaviruses, flaviviruses, retroviruses, paramyxoviruses, rhabdoviruses, orthomyxoviruses, filoviruses, bunyaviruses, and reoviruses.

C. DNA viruses
The animal DNA viruses are grouped into families according to their DNA organization; the dsDNA viruses are separated into their families on the basis of the shape of their DNA (linear or circular), their capsid shape, and the presence or absence of an envelope. Important groups of DNA viruses include adenoviruses, herpesviruses, poxviruses, papovaviruses, hepadnaviruses, and parvoviruses.

D. Emerging Viruses
Viruses that was once very localized or found in other species and crossed species to humans are called emerging viruses. Increases in human population, increases in the numbers of people living in cities, the use of land that was previously uninhabited and other factors are causing viruses to emerge, and infect large numbers of people.

III. Viral Replication

A. General characteristics of replication
In general, viruses go through five steps in their replication cycles: (1) adsorption, (2) penetration, (3) synthesis, (4) maturation, and (5) release.

B. Replication of bacteriophages
Bacteriophages (or *phages)* are viruses that infect bacterial cells. Replication of certain bacteriophages, known as *T-even phages*, includes: (1) adsorption, which is a chemical attraction that occurs between host cells and phages; (2) penetration, which involves the bacterial cell wall-weakening *lysozyme;* (3) synthesis, which results in the formation of viral DNA and viral proteins by the host cell; (4) maturation, which involves the assembly viral components to form mature, infective phages; and (5) release, in which phages escape from the host cell. This process can be described by a *replication curve.*

C. Lysogeny
Lysogeny refers to the ability of temperate bacteriophages to persist in a bacterium by the integration of the viral DNA (the *prophage*) into the host chromosome (in a process called *lysogenic conversion*), without the replication of new viruses or cell lysis.

D. Replication of animal viruses
Replication of animal viruses differs from the replication of bacteriophages in the following ways: (1) animal viruses attach to plasma membrane proteins, instead of to cell wall proteins; (2) animal viruses enter the host cell through endocytosis or fusion, instead of through the injection of their nucleic acids through host cell walls; (3) animal viruses need to be uncoated through the enzymatic digestion of viral proteins; (4) animal viruses are synthesized in both the cytoplasm and the nucleus, instead of just in the cytoplasm; (5) animal viruses can be released through budding, instead of just through host cell lysis; (6) animal viruses establish chronic infections through latency, chronic infections and cancer, instead of only through lysogeny.

IV. Culturing of Animal Viruses

A. Development of culturing methods
Biologists found that proteolytic enzymes could free animal cells from the surrounding tissues without injuring the freed cells; this led to the development of *monolayers,* which are suspensions of cells that attach to plastic or glass surfaces as sheets, one cell layer thick. Biologists then discovered that monolay-

ers could be *subcultured*, or transferred from an existing culture to new containers with fresh nutrient media.

B. Types of cell cultures

Three basic types of cell cultures are widely used in clinical and research virology: (1) primary cell cultures, (2) dipoid fibroblast strains, and (3) continuous cell lines. The cells in these cultures show different visible *cytopathic effects* that can allow the infecting virus to be identified.

V. Viruses and Teratogenesis

A. Teratogenesis

Teratogenesis is the induction of defects during embryonic development.

B. Viral groups responsible

Certain viruses—including cytomegalovirus, herpes simplex virus types 1 and 2, and rubella virus—can cause teratogenesis when they are transmitted across the placenta and infect the fetus.

VI. Viruslike Agents: Viroids and Prions

A. Viroids

Viroids are infectious RNA particles smaller than viruses. They differ from viruses in six ways: (1) they consist of single circular RNA molecules of low molecular weight; (2) they exist inside cells as particles of RNA without capsids or envelopes; (3) they do not require helper viruses; (4) their RNA does not produce proteins; (5) their RNA is always copied in the host cell nucleus; (6) they are identified in infected tissues only with special techniques.

B. Prions

Prions are exceedingly small proteinaceous infectious particles. They are resistant to inactivation by heating them to 90 °C, they are not sensitive to virus-damaging radiation, they are not destroyed by DNA- or RNA-digesting enzymes, they are sensitive to protein denaturing agents, and they have direct pairing of amino acids.

VII. Viruses and Cancer

A. Human cancer viruses

Known human cancer viruses include the Epstein-Barr virus, several human papillomaviruses, hepatitis B virus, and HTLV-1.

B. How cancer viruses cause cancer

DNA tumor viruses cause *neoplastic transformation*, or the uncontrollable division of infected cells. RNA tumor viruses transcribe their (+) sense RNA into DNA and then integrate this DNA as a provirus into the chromosome. The provirus then codes for proteins that transform host cells into neoplastic ones.

C. Oncogenes

The proteins produced by tumor viruses that cause uncontrolled host cell division come from segments of DNA called *oncogenes*. These oncogenes cause neoplasm and also contain information for synthesizing viral proteins needed for viral replication. A *proto-oncogene* is a normal gene that can act as an oncogene when under the control of a virus.

SELF-TESTS

Continue with this section only after you have read Chapter 10 of your textbook. Write your answers in the appropriate space provided. Correct answers to all questions can be found at the end of the Self-Test section. A score of 80 percent or better is good. If your score is less than 65, reread the chapter.

True/False

Mark T for True, F for False

_____ **1.** Some of the complex viruses have both DNA and RNA in their genome.

_____ **2.** Following the release of viral particles from the cell, the cell usually, but not always, dies.

_____ **3.** The diseases of botulism and diphtheria are actually viral related since a temperate phage provides the bacterium with the ability to make a toxin.

_____ **4.** The most common childhood viral diseases are those found in the arenavirus group.

_____ **5.** A tiny viral group that includes the viruses that cause polio is the picornaviruses.

Multiple Choice

Select best possible answer.

______ **6.** Viral groups known to have members that cause tumors include all the following except:
a. herpesviruses.
b. reoviruses.
c. papovaviruses.
d. adenoviruses.

______ **7.** Viruses with single-strand, negative-sense RNA:
a. cannot replicate inside cells.
b. must replicate inside the nucleus of cells.
c. must first make a positive sense RNA strand
d. do not require a transcriptase to make viral components.

______ **8.** Viroids differ from viruses in all the following ways except:
a. viroids lack capsids.
b. viroids possess RNA and infectious protein units.
c. viroids possess nucleic acids of low molecular weight.
d. viroids are hard to identify in tissue.

______ **9.** Viral groups that cause childhood infections include all the following except:
a. picornaviruses.
b. herpesviruses.
c. rhabdoviruses.
d. reoviruses.
e. paramyxoviruses.

______ **10.** Members of the herpesviridae include all the following except:
a. varicella.
b. zoster.
c. cytomegalloviruses.
d. rubellaviruses.

______ **11.** In the penetration step during replication of bacteriophages:
a. Tail fibers contract to push the internal tube through the wall.
b. The entire phage unit penetrates through the cell wall.
c. Lysozyme in the phage tail weakens the bacterial cell wall.
d. Cell wall receptor sites act to bind the tail core proteins, which osmotically open the wall.

Matching

Select the answer from the right-hand side that corresponds to the term or phrase on the left-hand side of the page. An answer may be used more than once. In some cases, more than one answer may be required.

Topic: RNA Viruses

______ **12.** Small, enveloped, (1) sense RNA; multiplies in host cytoplasm
______ **13.** Medium-sized, enveloped, (2) sense RNA; helical nucleocapsid
______ **14.** Small, naked, polyhedral, (1) sense RNA
______ **15.** Medium-sized, enveloped, (1) sense RNA, vary from spherical to helical
______ **16.** Enveloped, polyhedral, (1) sense RNA, transmitted by ticks
______ **17.** Enveloped, filamentous, single (2) sense RNA
______ **18.** Enveloped, two (1) sense RNA strands; reverse transcriptase
______ **19.** Enveloped, (2) sense RNA; three-segment genome
______ **20.** Enveloped, (2) sense RNA, two-segment genome

a. Picornaviridae
b. Flaviviridae
c. Retroviridae
d. Togaviridae
e. Paramyxoviridae
f. Orthomyxoviridae
g. Filoviridae
h. Bunyaviridae
i. Arenaviridae

Topic: RNA Virus Families

______	21. Rhinoviruses	a. Picornaviridae
______	22. HIV	b. Togaviridae
______	23. Enteroviruses	c. Retroviridae
______	24. Equine encephalitis virus	d. Paramyxoviridae
______	25. Measles virus	e. Orthomyxoviridae
______	26. Hepatoviruses	f. Flaviviridae
______	27. Influenza viruses	g. Rhabdoviridae
______	28. Hantavirus	h. Bunyaviridae
______	29. Flaviviruses	i. Filoviridae
______	30. Rabies virus	j. Reoviridae
______	31. Ebola virus	
______	32. Rotaviruses	

Topic: DNA Viruses

______	33. Fifth disease	a. Parvoviridae
______	34. Cowpox	b. Hepadnaviridae
______	35. Hepatitis B	c. Papovaviridae
______	36. Warts	d. Poxviridae
______	37. Chickenpox	e. Herpesviridae
______	38. Shingles	f. Adenoviridae
______	39. Respiratory infections	
______	40. Cervical cancer	

Topic: DNA Viruses

______	41. Small, naked, polyhedral dsDNA	a. Adenoviridae
______	42. Naked, medium-sized, linear dsDNA	b. Herpesviridae
______	43. Large, enveloped, linear dsDNA	c. Papovaviridae
______	44. Largest and most complex of all the viruses; enveloped, linear dsDNA	d. Poxviridae
______	45. Small, naked, linear ssDNA	e. Parvoviridae

Topic: Structure and Function of Viruses

______	46. Capsid	a. can be either single- or double-stranded
______	47. Nucleocapsid	b. a subunit of the capsid
______	48. Virion	c. contains the viral genetic material
______	49. Genome	d. a mature virus particle
______	50. Envelope	e. viral outer covering obtained from host membrane material
______	51. Spike	f. used for attachment to host cells
______	52. RNA	g. a 20-sided figure
______	53. Icosahedron	h. composed of capsomeres

Topic: Replication Cycle of Bacteriophages

_____ 54. Initial release of lysozyme
_____ 55. Uncoating of viral capsid
_____ 56. Capsomere production
_____ 57. Attachment to bacterial cell wall receptor sites
_____ 58. Injection of viral nucleic acid

a. adsorption
b. penetration
c. new viral DNA replication
d. burst size
e. mature virus release
f. all of these
g. none of the above

Fill-Ins

Provide the correct term or phrase for the following. Spelling counts.

59. Since viruses can only grow inside living cells, they are called _______________ _______________ parasites.
60. The viral nucleic acid is covered by a protein coat called a _______________.
61. Some viruses have a lipid bilayer around them called an _______________.
62. Viruses that exhibit a many-sided box-like capsid are called _______________viruses.
63. Viruses are usually measured in metric units called _______________.
64. The different kinds of organisms a virus can infect are called its _______________ _______________.
65. Viruses are primarily classified on the basis of their _______________ _______________.
66. The nucleic acid strand that encodes the information for making proteins needed by a virus is called the _______________ _______________ nucleic acid.
67. An enzyme that certain RNA viruses possess to copy RNA into DNA is called _______________ _______________.
68. The first step in the viral replication process is _______________.
69. Virus particles are assembled inside the host cell during the _______________ phase.
70. The time from adsorption to release of phage particles is the _______________ time.
71. During the _______________ phase, phages disappear from the culture, due to absorption and/or penetration into the host cell.
72. Clear areas on a bacterial lawn caused by lytic viruses are called _______________.
73. The stimulation that causes a temperate phage to become virulent is called _______________.
74. Once an animal virus enters the host cell's cytoplasm, the viral genome must be separated from its protein coat through a process called _______________.
75. Enveloped viruses obtain their coating from the _______________ _______________ of the host cell.
76. Viral cell cultures that come directly from the animal and are not subcultured are called _______________ _______________ cultures.
77. When viruses cause a visible effect on their host cells, the effect is called the _______________ effect.
78. An agent that can induce defects duringembryonic development is known as a _______________.
79. A series of blood tests often referred to as the _______________ _______________ is sometimes used to identify teratogenic diseases in pregnant women and newborn infants.
80. Viruses that lack capsids, contain RNA, and infect tissue may be _______________.
81. Viral-like agents that many only contain infectious protein particles are called _______________.

82. When cells cannot stop dividing, the result is a ________________, or a localized accumulation of cells (a tumor).

83. DNA tumor viruses cause one major cytopathic effect: ________________ ________________, or the uncontrollable division of infected cells.

84. The proteins produced by tumor viruses that cause uncontrolled host cell division come from segments of DNA called ________________.

85. A ________________-________________ is a normal gene that, when under the control of a virus, can cause uncontrolled cell division.

Labeling

Questions 86–89: Select the best possible label for each component of the herpesvirus (an animal virus): envelope, nucleic acid (DNA) core, capsid composed of many capsomeres, spikes.

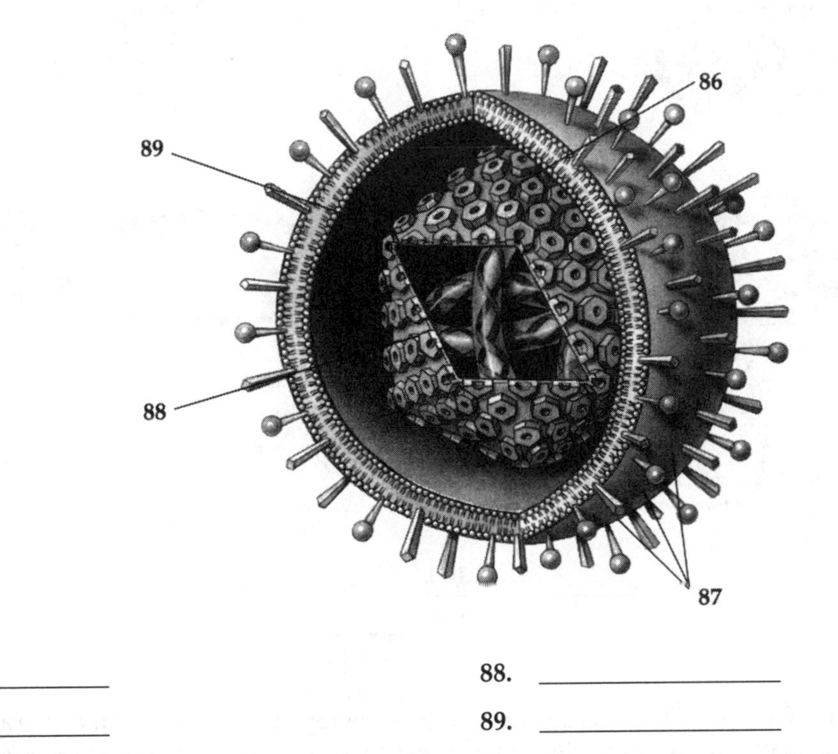

86. ________________

87. ________________

88. ________________

89. ________________

Questions 90–96: Select the best possible label for each component of the T4 bacteriophage: head, pin, tail fibers, collar, tail, tail sheath, plate. *Place your answers in the spaces provided on page 130.*

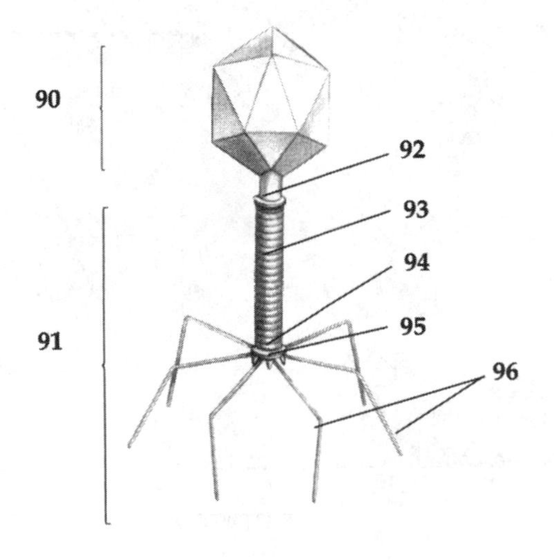

90. ________________

91. ________________

92. ________________

93. ________________

94. ________________

95. ________________

96. ________________

Questions 97–101: Select the best possible label for each step in the replication of a virulent bacteriophage: release, maturation, adsorption, penetration, synthesis.

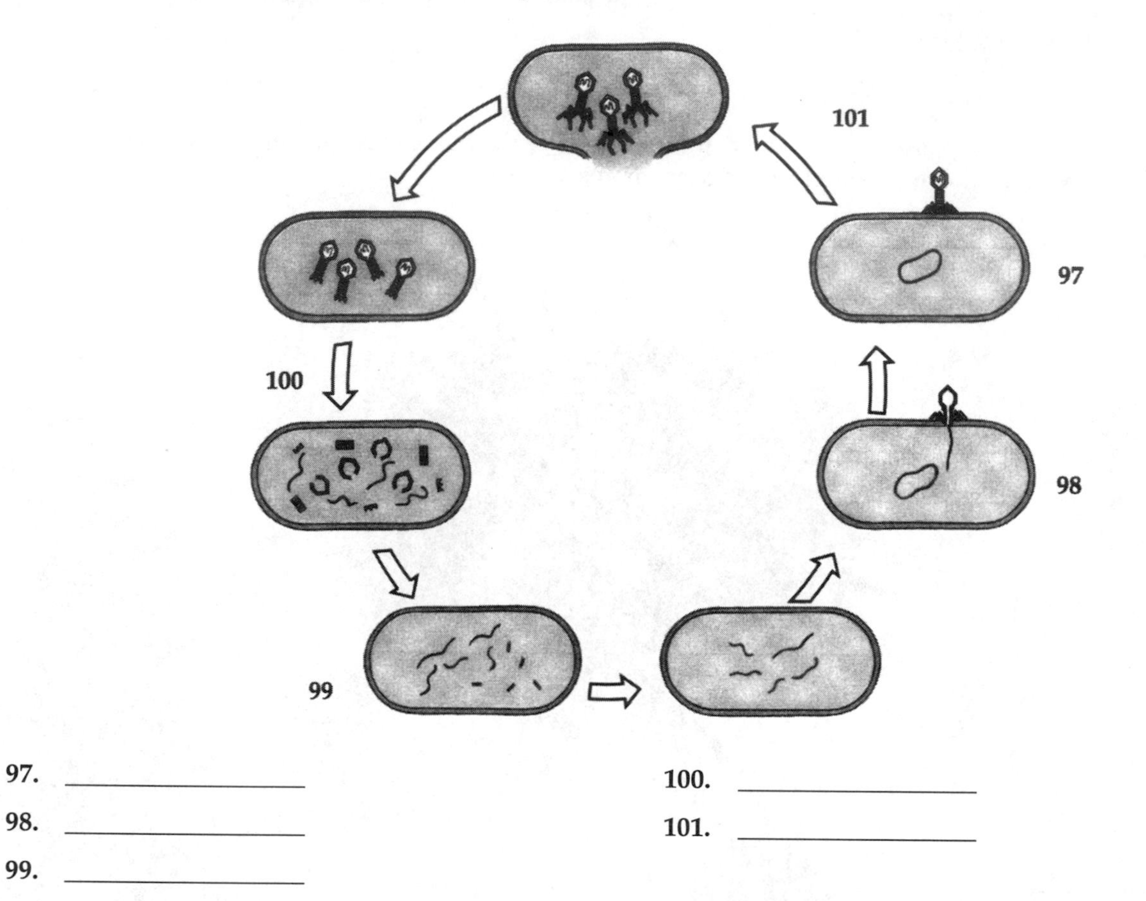

97. ________________

98. ________________

99. ________________

100. ________________

101. ________________

Questions 102–106: Select the best possible label for each step in the replication curve of a bacteriophage: eclipse period, viral yield, total virus, latent period, extracellular virus. *Place your answers in the spaces provided on page 131.*

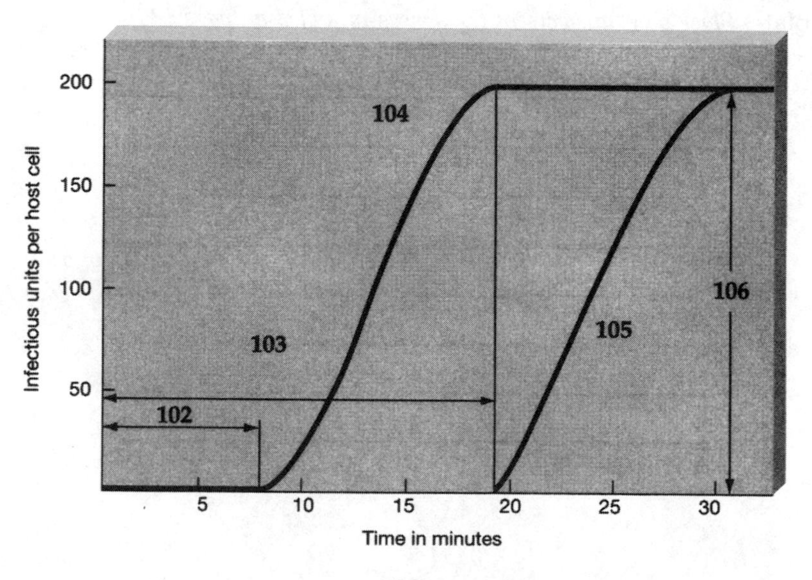

102. ____________________

103. ____________________

104. ____________________

105. ____________________

106. ____________________

Critical Thinking

107. What are the general properties of viruses?

108. How are viruses classified?

109. What are the properties of viroids and prions?

110. How do viruses replicate in general?

111. How do lytic and temperate bacteriophages replicate?

112. How do animal viruses replicate?

113. How were methods developed to culture animal viruses?

114. What types of viral cultures are currently in use?

115. What is a teratogen and how do viruses act as teratogens?

ANSWERS

True/False

1. **False** All viruses possess either DNA or RNA but not both.
2. **True** Viruses that are released by budding through the cell membrane may or may not kill the cell. Viruses that are released by cell lysis will kill the cell.
3. **True** Certain bacteria require the presence of temperate phages with the genetic information that codes for the production of a toxin. Without the phages the organisms do not cause diseases.
4. **False** The arenavirus group contains members that cause hemorrhagic fevers. The picornaviruses and paramyxoviruses contain members that cause the common cold, measles and mumps.
5. **True** The picornaviridae are small (27-30 nm) icosahedral viruses that include the enteroviruses (polio and echo) and rhinoviruses (common cold).

Multiple Choice

6. b; 7. c; 8. b; 9. c; 10. d; 11. c.

Matching

12. d; 13. e; 14. a; 15. f; 16. b; 17. g; 18. c; 19. h; 20. i; 21. a; 22. c; 23. a; 24. b; 25. d; 26. a; 27. e; 28. h; 29. f; 30. g; 31. i; 32. j; 33. a; 34. d; 35. b; 36. c; 37. e; 38. e; 39. f; 40. c; 41. c; 42. a; 43. b; 44. d; 45. e; 46. h; 47. c; 48. d; 49. c; 50. e; 51. f; 52. a, c; 53. g; 54. a; 55. g; 56. g; 57. a; 58. b.

Fill-Ins

59. obligate intracellular; 60. capsid; 61. envelope; 62. icosahedral; 63. nanometers; 64. host range; 65. nucleic acid; 66. positive sense; 67. reverse transcriptase; 68. adsorption; 69. maturation; 70. burst; 71. eclipse; 72. plaques;

73. induction; 74. uncoating; 75. cell membrane; 76. primary cell; 77. cytopathic; 78. teratogen; 79. TORCH series; 80. viroids; 81. prions; 82. neoplasm; 83. neoplastic transformation; 84. oncogenes; 85. proto-oncogene.

Labeling

86. envelope; 87. spikes; 88. nucleic acid (DNA) core; 89. capsid composed of many capsomeres; 90. head; 91. tail; 92. collar; 93. tail sheath; 94. plate; 95. pin; 96. tail fibers; 97. adsorption; 98. penetration; 99. synthesis; 100. maturation; 101. release; 102. eclipse period; 103. latent period; 104. total virus; 105. extracellular virus; 106. viral yield.

Critical Thinking

107. What are the general properties of viruses?

Viruses are acellular, infectious particles that are too small to be seen with a light microscope. They consist of a nucleic acid, either DNA or RNA, but not both, and a protein coat. Some viruses also have an envelope that they acquired from cells they previously infected. Viruses range in size from 5 to 300 nm and have an icosahedral, helical, or complex shape.

Viruses also have a very specific host range in that they generally only infect one kind of organism. They also have a specificity in that most can infect only one kind of cell.

108. How are viruses classified?

Viruses are classified on the basis of their type of nucleic acid, their structure, presence of an envelope, chemical and physical characteristics, method of replication, and host range.

RNA viruses are divided into five classes, and DNA viruses are divided into three classes. Virtually all DNA and RNA classes contain medically important human viruses.

109. What are the properties of viroids and prions?

Viroids are infectious particles smaller than viruses that lack a protein coat, consist of RNA, and are self-replicating. They appear to infect only plant tissues but are not well understood.

Prions are proteinaceous infectious particles believed to cause certain slow viral diseases such as Creutzfeldt-Jacob disease and kuru. Their exact origin, role, method of replication, and disease-causing potential are not clearly known.

110. How do viruses replicate in general?

Viral replication involves five basic steps: (1) adsorption or attachment to the host cell; (2) penetration, which refers to the entry of the virus into the cell; (3) synthesis, which refers to the making of the nucleic acids and protein components; (4) maturation, which refers to the assembly of the components; and (5) release, which is the departure of the intact virus from the cell.

111. How do lytic and temperate bacteriophages replicate?

Bacteriophages can either lyse the cell and undergo the lytic cycle or they can reside within the cell and establish a lysogenic cycle.

Lytic phages, which include the T-even phages, follow the classic five steps of replication. They possess recognition factors that allow them to attach or adsorb to a specific cell surface. Penetration is accomplished by means of an enzyme to weaken the wall and a syringe tube-like structure to inject the nucleic acid. The virus then takes over the cell's machinery and directs the construction or synthesis of all the necessary viral parts. Maturation occurs with the assembly of all the viral components in a very orderly fashion. With the completion of the intact viruses, an enzyme is produced and lysis of the cell occurs followed by release of the phages. A complete cycle takes only about 30 minutes with the release of hundreds of viruses! Lytic bacteriophage viruses grown on a bacterial lawn will form clear areas called plaques, which are filled with viruses.

Lysogenic or temperate phages do not immediately kill the cell but generally establish a long-term relationship. They can either exist as a prophage or revert to a lytic phage in the future. This relationship is important because the virus can provide extra genes for the bacterial cell such as the ability to produce specific toxins or enzymes.

112. How do animal viruses replicate?

Animal viruses follow the same basic steps found with bacteriophages but generally have a greater variety of mechanisms. Adsorption involves receptors on the viral surface, which allows attachment to the cell membrane. Penetration follows by a process such as pinocytosis. Synthesis and maturation differ in DNA and RNA viruses.

DNA viruses have a traditional mechanism with the DNA being transcribed into mRNA, which then directs the protein synthetic mechanisms. RNA viruses exhibit a great variety of mechanisms depending on whether the virus is single-stranded or double stranded and negative or positive sense RNA. In some cases, the RNA virus requires a reverse transcriptase to form DNA, which is then used to direct the synthesis of components. In some cases, the viral DNA is incorporated into the cell's DNA. Viruses are assembled in the cell and released by direct lysis or by budding through the membrane.

113. How were methods developed to culture animal viruses?
The development of cultural methods proceeded very slowly because of difficulty in developing tissue cultures and keeping them free of bacterial contamination. Antibiotics and special enzymes aided in the development of useful culture systems.

114. What types of viral cultures are currently in use?
Several cell culture systems are in use today and include primary cell cultures, diploid fibroblast strains, and continuous cell lines. The primary cell cultures come directly from animals, cannot be subcultured, and normally do not last long. They can be used to support the growth of a wide variety of viruses. Diploid fibroblast strains are the most widely used because they can be maintained for several years. They are often used in making vaccines. Continuous cell lines are usually derived from cancer cells and can grow almost indefinitely. They are commonly used in research.

115. What is a teratogen and how do viruses act as teratogens?
Teratogens are agents that can induce embryonic defects. Viruses can act as teratogens by crossing the placenta and infecting embryonic cells. The greatest damage seems to occur in the early developmental stages. Virtually any body organ or tissue can be affected. Rubellaviruses, cytomegaloviruses, and herpesviruses seem to be the most common viral agents.

Chapter 11 Eukaryotic Microorganisms and Parasites

INTRODUCTION

There are more worm or helminthic infections than there are people on Earth. That means that if you are not currently infected by a species of worm, there is a person somewhere infected with at least two different worm infections! It is truly a wormy world. Even in the United States, worm infections are common. Consider, autopsy data reveals one American in twenty has trichinosis or pork worm.

Each person is like a world. Each part of our body is like a distinct habitat. Instead of rivers, forests, and oceans we have organs like the liver, intestine, and skeletal muscles. Just as organisms in nature are adapted to different habitats, different parasites are adapted to different organs. Thus, there are more parasites than there are free-living species. In additions to worms, there are protist, fungal, and arthropod parasites.

Some parasites are tiny and infect huge numbers of people. If you wanted to know how many people had malaria (caused by the protozoan, *Plasmodium*) before World War II, all you had to do was to take the world's population and divide by four. After the war some progress occurred but even today in Africa one child dies of malaria every thirty seconds. In the poor countries of Africa, what limited resources are available are diverted to HIV infections and so malaria is getting less attention. Sadly, for the protist and helminthic infections of the Third World, there are few effective medicines or vaccines.

Even in urban America, we are not without parasites and not all parasites are tiny. Once, Dr. Michael Katz from Columbia University was at Kennedy airport in New York City and he heard a mother cry out in panic. Dr. Katz was the Chair of Tropical Medicine at Columbia and in charge of pediatric infectious diseases for the hospital complex. Dr. Katz saw a child with an *Ascaris* worm emerging from the child's nostril. The good doctor told the mother that she should not be alarmed because this happens all the time. This of course was a lie. Dr. Katz proceeded to pull out a worm that was almost a foot in length.

This chapter will introduce you to world that few of us have thought about. It is a world of forgotten people and forgotten diseases. It is also a world of amazing relationships between parasite and host. A world that is the rule for most animals on Earth that spend most days of their lives infected with multiple organisms.

STUDY OUTLINE

I. Principles of Parasitology

A. General characteristics

A *parasite* is an organism that lives at the expense of another organism, called the *host*. Parasites that cause disease are called *pathogens*. *Parasitology* is the study of parasites.

B. Significance of parasitism

There are more parasitic infections than there are living humans. One-fourth of the 60 million people dying each year die of parasitic infections or their complications. Parasites play an important role in the worldwide economy.

C. Parasites in relation to their hosts

Parasites can be divided into *ectoparasites,* which live on the surface of other organisms, and *endoparasites,* which live within the bodies of other organisms. A *biological vector* is a vector that a parasite inhabits as it is going through during part of its life cycle, while a *mechanical vector* is a vector in which the parasite does not go through any part of its life cycle during transit. Hosts are classified as *definitive hosts* if they harbor a parasite while they reproduce sexually; they are said to be *intermediate hosts* if they harbor the parasite during some other developmental stages. *Reservoir hosts* are infected organisms that make parasites available for transmission to other hosts. *Host specificity* refers to the range of different hosts in which a parasite can mature.

II. Protists

A. Characteristics of protists

Protists are unicellular (though sometimes colonial), eukaryotic organisms. They vary in diameter from 5 μm to 5 mm.

B. Importance of protists

Protists are a key part of food chains. Protists can be economically beneficial—helping to determine the age of rocks or to determine the presence of oil—or detrimental—producing toxins that cause disease or death. Other protists cause the *eutrophication* of bodies of water, leading to the death of plants and animals. Protists can also cause parasitic diseases.

C. Classification of protists

There are plantlike protists (including *euglenoids, diatoms,* and *dinoflagellates*), funguslike protists (including *water molds, plasmodial slime molds,* and *cellular slime molds*), and animal-like protists (including *mastigophorans, sarcodines, apicomplexans,* and *ciliates.*

III. Fungi

A. Characteristics of fungi

The kingdom Fungi includes nonphotosynthetic, eukaryotic organisms that absorb nutrients from their environment. The body of a fungus is called a *thallus* and consists of a *mycelium.* The cell walls of most fungi contain *chitin.* Many fungi reproduce both sexually and asexually, but a few have only asexual production.

B. Importance of fungi

In ecosystems, fungi are important decomposers. In the health sciences, they are important as facultative parasites and producers of antibiotics.

C. Classification of fungi

Fungi are classified according to the nature of the sexual stage in their life cycles. The four phylum discussed in the book are: (1) Zygomycota, or bread molds; (2) Ascomycota, or sac fungi; (3) Basidiomycota, or club fungi; and (4) Deuteromycota, or Fungi Imperfecti.

IV. Helminths

A. Characteristics of helminths

Helminths, or worms, are bilaterally symmetrical, have head and tail ends, and have three distinct tissue layers (ectoderm, mesoderm, and endoderm). Flatworms and roundworms are examples of helminths.

B. Parasitic helminths

The parasitic helminths discussed in the book include flukes, tapeworms, adult roundworms of the intestine, and roundworm larvae.

V. Arthropods

A. Characteristics of arthropods

Arthropods constitute the largest group of living organisms. Arthropods are characterized by jointed chitinous exoskeletons, segmented bodies, jointed appendages associated with some or all of the segments, true coeloms, small brains and an extensive network of nerves, different structures that extract oxygen from air or from aquatic environments, and a distinctive sex.

B. Classification of arthropods

The book discusses three subgroups (classes) of arthropods that are important either as parasites or as disease vectors: the arachnids, insects, and crustaceans.

SELF-TESTS

Continue with this section only after you have read Chapter 11 of your textbook. Write your answers in the appropriate space provided. Correct answers to all questions can be found at the end of the Self-Test section. A score of 80 percent or better is good. If your score is less than 65, reread the chapter.

True/False

Mark T for True, F for False

_____ **1.** Plasmogamy is a process that is associated with the sexual reproduction of certain types of fungi.

_____ **2.** Fungi are generally classified on the basis of their type of asexual reproduction.

_____ **3.** Parasites that can effectively kill their host in a short period of time are those that have maintained the longest relationships with their hosts.

_____ **4.** All animal protists are parasitic.

_____ **5.** Flatworms lack a coelom and have a simple digestive tract with a single opening.

_____ **6.** Most roundworms that parasitize humans live most of their life cycle in the circulatory system.

Multiple Choice

Select the best possible answer.

_____ **7.** Conidia are associated with:
a. Club fungi.
b. Plant-like protists.
c. Sporozoites.
d. Ascomycetes.

_____ **8.** Nematodes are characterized by:
a. having a pseudocoelom and separate sexes.
b. possessing a scolex.
c. forming woronin bodies.
d. having life cycles involving snails.

_____ **9.** Hydatid cysts:
a. are formed by the sac fungi.
b. contain tapeworm larvae.
c. can be called ectoparasites.
d. usually exhibit dimorphism.

_____ **10.** Arachnids:
a. include the crabs and crayfish.
b. have six legs.
c. are not known to transmit disease.
d. are vectors of Rocky Mountain spotted fever and scrub typhus.

Matching

Select the answer from the right-hand side that corresponds to the term or phrase on the left-hand side of the page. An answer may be used more than once. In some cases, more than one answer may be required.

Topic: Parasites in Relation to Their Hosts

_____	**11.** Endoparasite	**a.** found on host's exterior
_____	**12.** Obligate parasite	**b.** found in host's tissues
_____	**13.** Hyperparasitism	**c.** remains in a host after penetration
_____	**14.** Ectoparasite	**d.** feeds on and then leaves host

_____ **15.** Permanent parasite

_____ **16.** Temporary parasite

_____ **17.** Vector

e. a parasite having parasites

f. must spend some of its life cycle on or in a host

g. agent of transmission for infectious dieases

Topic: Protists

_____ **18.** Water molds

_____ **19.** Euglenoids

_____ **20.** Photosynthesis

_____ **21.** Diatoms

_____ **22.** Fruiting bodies

_____ **23.** Slime molds

_____ **24.** Dinoflagellates

_____ **25.** Pseudoplasmodium

_____ **26.** Sporozoites

_____ **27.** Trophozoites

a. plantlike protists

b. funguslike protists

c. animal-like protists

Topic: Animal-like Protists and Infectious Diseases (Part 1)

_____ **28.** *Leishmania*

_____ **29.** *Plasmodium*

_____ **30.** *Trichomonas*

_____ **31.** *Entamoeba*

_____ **32.** *Giardia*

_____ **33.** *Toxoplasma*

_____ **34.** *Balantidium*

a. Mastigophora

b. Sarcodina

c. Apicomplexa

d. Ciliata

Topic: Animal-like Protists and Infectious Diseases (Part 2)

_____ **35.** *Trichomonas*

_____ **36.** *Entamoeba*

_____ **37.** *Giardia*

_____ **38.** *Plasmodium*

_____ **39.** *Leishmania*

_____ **40.** Trypanosomes

_____ **41.** *Toxoplasma*

_____ **42.** *Balantidium*

_____ **43.** *Paramecium*

_____ **44.** *Trichonympha*

a. diarrhea

b. amoebic dysenteries

c. malaria

d. vaginal inflammation

e. African sleeping sickness

f. skin lesions

g. dysentery

h. no disease

i. blindness in adults

Topic: Characteristics of Fungi

_____ **45.** Reproductive unit

_____ **46.** Structural unit

_____ **47.** Cell wall polysaccharide

a. hypha

b. spore

c. chitin

______ **48.** Fungus body — **d.** mycelium

______ **49.** Cross-wall — **e.** septum

______ **50.** Mass of hyphae — **f.** thallus

Topic: Classification of Fungi

______ **51.** Sac fungi		**a.** Zygomycota
______ **52.** *Neurospora*		**b.** Ascomycota
______ **53.** *Amanita*		**c.** Basidiomycota
______ **54.** Ascus formation		**d.** Deuteromycota
______ **55.** Fungi Imperfecti		
______ **56.** *Rhizopus*		
______ **57.** Zygospores		
______ **58.** Mushrooms		
______ **59.** No sexual stages		
______ **60.** Basidia		
______ **61.** *Claviceps purpurea*		

Topic: Parasitic Helminths

______ **62.** *Paragonimus westermani*		**a.** liver fluke
______ **63.** *Schistosoma* species		**b.** hydatid worm
______ **64.** *Diphyllobothrium*		**c.** lung fluke
______ **65.** *Clonorchis sinensis*		**d.** cysticercus
______ **66.** bladder worm		**e.** fish tapeworm
______ **67.** *Echinococcus*		**f.** blood fluke
______ **68.** *Taenia saginata*		**g.** beef tapeworm
______ **69.** *Enterobius vermicularis*		**h.** pinworm
______ **70.** *Trichinella spiralis*		**i.** hookworm
______ **71.** *Necator americanus*		**j.** pork roundworm
______ **72.** *Dracunculus medinensis*		**k.** guinea worm
______ **73.** *Wuchereria bancrofti*		**l.** elephantiasis

Topic: Arthropods and Disease Transmission

______ **74.** Tick paralysis		**a.** *Ixodes*
______ **75.** Lyme disease		**b.** *Dermacentor andersoni*
______ **76.** Relapsing fever		**c.** *Ornithodorus*
______ **77.** Colorado tick fever		**d.** body louse
______ **78.** Typhus		**e.** *Pulex irritans*
______ **79.** Plague		**f.** tsetse fly
______ **80.** Trench fever		**g.** *Culex tarsalis*

_____ **81.** Rocky Mountain spotted fever
_____ **82.** African sleeping sickness
_____ **83.** Malaria
_____ **84.** Chagas' disease
_____ **85.** Yellow fever
_____ **86.** Dengue fever
_____ **87.** Western equine encephalitis

h. *Aedes aegypti*
i. reduviid bug
j. *Anopheles*

Topic: Causative Agents of Disease

_____ **88.** Plague
_____ **89.** Lyme disease
_____ **90.** Tularemia
_____ **91.** Q fever
_____ **92.** Salmonellosis
_____ **93.** Rocky Mountain spotted fever
_____ **94.** Relapsing fever
_____ **95.** Chagas' disease
_____ **96.** Yellow fever
_____ **97.** Kala azar
_____ **98.** Dengue fever

a. *Salmonella species*
b. *Yersinia pestis*
c. *Francisella tularensis*
d. *Borrelia* species
e. *Borrelia burgdorferi*
f. *Coxiella burnetii*
g. *Rickettsia rickettsii*
h. togaviruses
i. *Trypanosoma cruzi*
j. *Leishmania* species

Topic: Protist Structure and Function

_____ **99.** Trichocysts
_____ **100.** Cilium
_____ **101.** Tests
_____ **102.** Pellicle
_____ **103.** Cyst
_____ **104.** Gametes
_____ **105.** Tentacles

a. organelle of locomotion
b. provides protection
c. involved in food capture
d. reproduction
e. none of these

Fill-Ins

Provide the correct term or phrase for the following. Spelling counts.

106. An organism that lives at the expense of another organism is a _________________.

107. Ticks and lice, since they live on the surface of a host, act as _________________.

108. Parasites that normally are free-living but can obtain nutrients from a host are _________________ parasites.

109. Vectors that are required as part of the life cycle of the parasite are _________________ vectors.

110. If a host harbors a parasite while it reproduces sexually, it is a _________________ host.

111. Worms are _________________ if both male and female organs are found together.

112. The red tide is caused by plant-like protists called ________________.

113. Fungus-like protists include organisms called ________________ molds.

114. Fungi that act as decomposers of organic matter are commonly called ________________ fungi.

115. Ciliates and amoebas are protists that are ________________ - ________________.

116. Protozoans that move by pseudopodia are grouped as ________________.

117. The presence of a complex-like cycle including merozoites and trophozoites is found with the ________________.

118. Fungi are studied in the specialty called ________________.

119. A loosely-organized mass of thread-like structures that forms the thallus of most fungi is called the ________________.

120. Cross walls found in the hyphae of most fungi are called ________________.

121. Fungi with single septal pores often have organelles that act in protection called ________________ ________________.

122. In the reproductive cycle of ascomycetes, the cytoplasmic fusion of the male and female strains is called ________________.

123. When fungi cause diseases in humans, the diseases are known as ________________.

124. The ability of a fungus to exhibit different structural forms when it changes habitats is called ________________.

125. The bread molds, including *Rhizopus*, are characterized because they form ________________.

126. The sac fungi are so called because of the presence of sac-like ________________.

127. The majority of the mushrooms and toadstools are found with the club fungi or ________________.

128. The "imperfect" fungi are so called because they lack a ________________ ________________.

129. Free-swimming larval forms of flukes that penetrate snails or other mollusks are called ________________.

130. Flukes can give rise to aquatic free-swimming larva called ________________.

131. Parasitic worms that look much like a measuring tape are called ________________.

132. The head end of a tapeworm that is equipped with hooks and suckers is the ________________.

133. Body units of tapeworms that contain reproductive organs are called ________________.

134. Tapeworm larva that form into sac-like bladder worms in human tissues are called ________________.

135. Worms that enter the human body as immature larvae, usually by means of arthropod vectors, are ________________.

136. ________________ constitute the largest group of living organisms.

137. ________________ have two body regions—a cephalothorax and an abdomen, four pairs of legs, and mouthparts that are used in capturing and tearing apart prey.

138. ________________ are generally aquatic arthropods that typically have a pair of appendages associated with each segment, including mouthparts, claws, walking legs, and appendages that aid in swimming or in copulation.

Labeling

Questions 139–141: Select the best possible group name corresponding to the characteristics and examples of protists given in the chart below: animal-like protists, plantlike protists, funguslike protists.

PROPERTIES OF PROTISTS

GROUP	CHARACTERISTICS	EXAMPLES
139.	Have chloroplasts; live in moist, sunny environments	Euglenoids, diatoms, dinoflagellates
140.	Most are saprophytes; may be unicellular or multicellular	Water molds, plasmodial and cellular slime molds
141.	Heterotrophs; most are unicellular, most are free-living, but some are commensals or parasites	Mastigophorans, sarcodines, apicomplexans, and ciliates

139. ________________

140. ________________

141. ________________

Questions 142–144: Select the best possible group name corresponding to the identifying characteristics and examples of three classes of arthropods given in the chart below: insects, arachnids, crustaceans.

PROPERTIES OF THREE CLASSES OF ARTHROPODS

GROUP	IDENTIFYING CHARACTERISTIC	EXAMPLES
142.	Have eight legs	Spiders, scorpions, ticks, mites
143.	Have six legs	Lice, fleas, flies, mosquitoes, true bugs
144.	A pair of appendages on each body segment	Crabs, crayfish, copepods

142. ________________

143. ________________

144. ________________

Critical Thinking

145. What is a parasite, and what are the principles of parasitism?

146. What are protists, and why are they important?

147. How do groups of protists differ?

148. What are the fungi, and why are they important?

149. How do groups of fungi differ?

150. What are parasitic helminths, and why are they important?

151. How do groups of parasitic helminths differ?

152. What are the characteristics of parasitic and vector arthropods?

ANSWERS

True/False

1. **True** Plasmogamy is the process whereby haploid gametes of fungi can unite and allow their cytoplasm to fuse, ultimately forming a dikaryotic or diploid cell.
2. **False** Fungi are actually classified on the basis of their sexual spore formation. Asexual reproduction is a very common method of propagation.
3. **False** An effective, efficient parasite ensures continued survival by not killing its host. A parasite that has only a brief experience with a host will often kill it quickly because it lacks an appropriate tolerance.
4. **False** Most animal protists are free-living and are often found in fresh or salt water. Only a few are actually parasites.
5. **True** Flatworms are primitive worms usually no more than 1 mm thick with lengths of over 30 feet. Roundworms have a pseudocoelom and a complete digestive system.
6. **True** Most roundworm infestations in the United States affect the gastrointestinal tract. Therefore the majority of their life cycle is spent in the digestive tract.

Multiple Choice

7. d; 8. a; 9. b; 10. d.

Matching

11. b; 12. f; 13. e; 14. a; 15. c; 16. d; 17. g; 18. b; 19. a; 20. a; 21. a; 22. b; 23. b; 24. a; 25. b; 26. c; 27. c; 28. a; 29. c; 30. a; 31. b; 32. a; 33. c; 34. d; 35. d; 36. b; 37. a; 38. c; 39. f; 40. e; 41. i; 42. g; 43. h; 44. h; 45. b; 46. a; 47. c; 48. f; 49. e; 50. d; 51. b; 52. b; 53. c; 54. b; 55. d; 56. a; 57. a; 58. c; 59. d; 60. c; 61. c; 62. c; 63. f; 64. e; 65. a; 66. d; 67. b; 68. g; 69. h; 70. j; 71. i; 72. k; 73. l; 74. b; 75. a; 76. c; 77. b; 78. d; 79. e; 80. d; 81. b; 82. f; 83. j; 84. i; 85. h; 86. h; 87. g; 88. b; 89. e; 90. c; 91. f; 92. a; 93. g; 94. d; 95. i; 96. h; 97. j; 98. h; 99. b; 100. a; 101. b; 102. b; 103. b, d; 104. c; 105. c.

Fill-Ins

106. parasite; 107. ectoparasites; 108. facultative; 109. biological; 110. definitive; 111. hermaphroditic; 112. dinoflagellates; 113. slime; 114. saprophytic; 115. animal-like; 116. sarcodinas; 117. sporozoites; 118. mycology; 119. mycelium; 120. septa; 121. Woronin body; 122. plasmogamy; 123. mycoses; 124. dimorphism; 125. zygospores; 126. asci; 127. Basidiomycota; 128. sexual stage; 129. miracidia; 130. cercariae; 131. tapeworms; 132. scolex; 133. proglottids; 134. cysticerci; 135. microfilaria; 136. arthropods; 137. arachnids; 138. crustaceans.

Labeling

139. plantlike protists; 140. funguslike protists; 141. animal-like protists; 142. arachnids; 143. insects; 144. crustaceans.

Critical Thinking

145. What is a parasite, and what are the principles of parasitism?

A parasite is an organism that lives at the expense of another organism called the host. The study of parasites, which include members of the protozoa, helminths, and arthropods, is parasitology. Parasites are responsible for a major portion of human disease and death and play an important role in the worldwide human economy.

They appear to live on or in their hosts and many are obligate parasites having to spend some of their life cycle with the host. Some parasites are permanent residents others only temporary or accidental. They can reproduce sexually in their definitive hosts and spend other life stages in intermediate hosts. They can also be transmitted from a reservoir host via a vector to a human host. Most importantly, a good parasite is one that has become well adapted to its host and causes little damage.

146. What are protists, and why are they important?

The protists are members of the kingdom Protista, are eukaryotic, microscopic, and mostly unicellular. They can be self-supporting or autotrophic or heterotrophic where they feed on other organisms. Some are parasitic.

Many are important in the food chains as producers or decomposers and many are very important economically. Some are responsible for forming calcified deposits used in building materials while others multiply so rapidly that they form thick impenetrable layers in lakes, causing many useful organisms such as fish to die.

147. How do groups of protists differ?

Protists may be grouped in several ways. The simplest way is to group them according to the kingdom of macroscopic organisms they most resemble. For example, plantlike protists (like euglenoids, diatoms, and dinoflagellates) have chloroplasts and live in moist, sunny environments. Funguslike protists (like true slime molds and cellular slime molds) are usually saprophytes and may be unicellular or multicellular. Animal-like protists (like mastigophorans, sarcodinas, sporozoans, and ciliates) are heterotrophs, usually unicellular, and usually free-living, although some are commensals or parasites.

148. What are the fungi, and why are they important?

The fungi are members of the kingdom Fungi, are eukaryotic, saprophytic (live off of dead organic material), and mostly multicellular except for the yeasts which are unicellular. They form threads called hyphae that assemble into a mat called a mycelium. Molds typically form an unorganized mycelium while mushrooms have a very organized mycelial structure. Most reproduce both sexually and asexually with their sexual stage being used for classification.

Fungi are very important as decomposers in the environment but also as parasites on plants, animals, and humans. Some fungi are economically important as foods and as producers of enzymes and antibiotics.

149. How do groups of fungi differ?

Fungi are usually classified according to the nature of their sexual stage in their life cycle. Unfortunately, some fungi do not exhibit a sexual stage and others are difficult to match their sexual and asexual stages properly. Therefore, they will be grouped into the water molds (Oomycota), bread molds (Zygomycota), sac fungi (Ascomycota), club fungi (Basidiomycota), and imperfect fungi (Deuteromycota).

150. What are parasitic helminths, and why are they important?

The helminths, or worms, are members of the Animal kingdom, are eukaryotic, multicellular, bilaterally symmetrical, have head and tail ends, and are differentiated into tissue layers. Most of these worms are harmless, free-living, environmental or aquatic organisms. A few have established a parasitic relationship with man, animals, and plants.

151. How do groups of parasitic helminths differ?

The helminths are commonly grouped into the flatworms and roundworms or nematodes. The flatworms include the tapeworms which look much like a flattened measuring tape and the flukes, which look like a flat leaf. They all lack a coelom, have a simple digestive tract with only one opening, and are mostly hermaphroditic, that is, both sexes within the same worm. The worms commonly possess hooks and suckers at the head end, which are used for attachment.

The nematodes have a pseudocoelom, separate sexes, and a cylindrical body with a tough outer covering or cuticle. They include hookworms, pinworms, and other parasitic worms of the intestinal tract, tissues, and lymphatics.

152. What are the characteristics of parasitic and vector arthropods?

The arthropods constitute the largest group of living organisms, are members of the Animal kingdom, have jointed, chitinous exoskeletons, segmented bodies, and jointed appendages. They include the arachnids, the insects, and the crustacea. Most are free-living in the soil, on vegetation, and in fresh and salt water but some also serve as intermediate hosts for other human parasites or act directly as the parasite.

Chapter 12 Sterilization and Disinfection

INTRODUCTION

As you may recall, it was necessary to develop the ability to sterilize a broth filled with bacteria in order to establish the germ theory. To prove that bacteria can cause disease one must first show that bacteria reproduce and do not arise by spontaneous generation. Pasteur believed that one could kill all microbes by boiling a solution for two hours. He was wrong. Ferdinand Cohn showed that some bacteria can form heat resistant endospores and can survive boiling. This eventually led to the development of the autoclave commonly used today in hospitals and other laboratories. This chapter will in part explain how the autoclave can heat water to 121 degrees Celsius and sterilize solutions. At least, we believed that until 2003, when we learned that there are bacteria that live in deep-sea vents in the ocean that can survive and reproduce under the conditions of the autoclave. These usually heat resistant bacteria may be closely related to the first cells to have evolved on Earth. They are so tough; they could survive travel through space. Fortunately, bacteria that infect people are not like that and we do not have to invent new technology to sterilize fluids in hospitals . . . yet. The ability to kill and control microbes is still evolving.

In our own lives, we often have to be concerned more about the control of microbes. We wash our hands as often as possible during the flu and cold season. Soap and water will not kill many microbes, but reducing their number may also reduce our chances of becoming ill.

In our kitchen, any product that we place on a surface is likely to wind up in our mouth. If we have children at home, any cleaner we place on the floor will enter their bodies. Cleaning surfaces with toxic chemicals may not be in our interest. A good weapon in our kitchen in the war on bacteria is our microwave. If we place our sponges in the microwave at the end of the day for a minute, we can kill bacteria that may do us harm without the dangers of eating dangerous chemicals.

This chapter will outline the many ways and tools we have to protect ourselves from microbes. Different circumstance will demand different approaches.

STUDY OUTLINE

I. Principles of Sterilization and Disinfection

A. Definitions

Disinfectants are agents that are typically applied to inanimate objects. *Antiseptics* are agents applied to living tissue. *Sterilization* is the killing or removal of all microorganisms in a material or on an object.

Disinfection is the reduction of the number of pathogenic organisms on objects or in materials so that they pose no threat of disease.

B. Control of microbial growth

A definite proportion of the organisms treated with antimicrobial agents will die in a given time interval. The fewer organisms present, the shorter the time needed to achieve sterility. Microorganisms differ in their susceptibility to antimicrobial agents.

II. Chemical Antimicrobial Agents

A. Potency of chemical agents

Time, temperature, pH, and concentration affect the *potency*, or effectiveness, of a chemical antimicrobial agent.

B. Evaluation of effectiveness of chemical agents

In order to evaluate the effectiveness of chemical agents, the following are sometimes used: the phenol coefficient, the filter paper method, and the use-dilution test.

C. Disinfectant selection

An ideal disinfectant should: (1) be fast-acting; (2) be effective against all types of infectious agents; (3) easily penetrate the material to be disinfected; (4) be easy to prepare and stable; (5) be inexpensive and easy to obtain and use; and (6) not have an unpleasant odor.

D. Mechanisms of action of chemical agents

Chemical antimicrobial agents kill microorganisms by participating in one or more chemical reactions that damage cell components. Agents can thus be classified by whether they: (1) affect proteins; (2) affect membranes; (3) affect other cell components; or (4) affect viruses.

E. Specific chemical antimicrobial agents

Some specific chemical antimicrobial agents are: soaps and detergents, acids and alkalis, heavy metals, halogens, alcohols, phenols, oxidizing agents, akylating agents, and dyes.

III. Physical Antimicrobial Agents

A. Principles and applications of heat killing

Heat is a preferred agent of sterilization for all materials not damaged by it. The *thermal death point, thermal death time,* and *decimal reduction time (DRT* or *D value)* are all measurements that have practical significance in industry as well as in the laboratory.

B. Dry heat, moist heat, and pasteurization

Dry heat does most of its damage by oxidizing molecules. *Moist heat* destroys microorganisms mainly be denaturing proteins and disrupting membrane lipids, and *pasteurization* kills pathogens that might be present in milk, other dairy products, and beer.

C. Refrigeration, freezing, drying, and freeze-drying

Many fresh foods can be prevented from spoiling by keeping them at ordinary refrigerator temperature. Freezing significantly slows the rate of chemical reactions so that microorganisms do not cause foods to spoil. Drying can be used to preserve foods because the absence of water inhibits the action of enzymes. Freeze-drying can be used to preserve cultures of microorganisms for long periods of time.

D. Radiation

Four types of radiation can be used to control microorganisms and to preserve foods: ultraviolet light, ionizing radiation, microwave radiation, and strong visible light (under special circumstances).

E. Sonic and ultrasonic waves

Sonic, or sound, waves can destroy bacteria if the waves are of sufficient intensity.

F. Filtration

Sterilization by *filtration* requires that materials be passed through filters with exceedingly small pores.

G. Osmotic pressure

High concentrations of salt, sugar, or other substances create a hyperosmotic medium that draws water from microorganisms by osmosis. *Plasmolysis*, or loss of water, severely interferes with cell function and eventually leads to cell death.

SELF-TESTS

Continue with this section only after you have read Chapter 12 of your textbook. Write your answers in the appropriate space provided. Correct answers to all questions can be found at the end of the Self-Test section. A score of 80 percent or better is good. If your score is less than 65, reread the chapter.

True/False

Mark T for True, F for False

______ **1.** Chemical disinfectants are usually adversely affected by organic material.

______ **2.** If a label on a disinfectant indicates it is bactericidal, it means it can kill any microbe and should be an effective sterilizing agent.

______ **3.** If 20% of the organisms in an antimicrobial agent die in the first minute, then the rest should die within the next four minutes.

______ **4.** It is virtually impossible to find a perfect disinfectant.

______ **5.** Effective hand washing with bar soap can eventually sterilize the hands since most soaps kill bacteria on contact.

______ **6.** Pasteurization does not achieve sterility but does kill pathogens present in the product.

Multiple Choice

Select the best possible answer

______ **7.** Select the most incorrect statement relative to sterilization principles.
- **a.** A definite proportion of organisms will die in a given time interval.
- **b.** Microorganisms tend to be similar in their susceptibility to antimicrobial agents.
- **c.** The smaller the number of organisms present, the shorter the time needed to achieve sterility.
- **d.** If an agent cannot kill endospores, it cannot be a sterilizing agent.
- **e.** All of the above are correct.

______ **8.** Select the most appropriate setting for sterilization.
- **a.** Dry heat oven at 130°C for 1 hour.
- **b** Moist steam at 10 pounds of pressure for 15 minutes.
- **c** Moist steam at 121°C for 20 minutes.
- **d.** Ethylene oxide at 20°C for 25 minutes
- **e.** Freezing at −10°C for 2 days.

______ **9.** Microwave ovens:
- **a.** cook by means of UV radiation
- **b.** cannot be used for sterilization because of uneven distribution of high-frequency vibrations.
- **c.** can easily kill most bacteria and parasitic cysts.
- **d.** actually cook by pasteurization.

______ **10.** The phenol coefficient test:
- **a.** can be applied to all disinfectants.
- **b.** uses *Escherichia coli* and *Bacillus subtilis* as indicator organisms.
- **c.** compares the effectiveness of diluted phenol samples to dilutions of test samples.
- **d.** is no longer used for testing purposes.

Matching

Select the answer from the right-hand side that corresponds to the term or phrase on the left-hand side of the page. An answer may be used more than once. In some cases, more than one answer may be required.

Topic: General Properties of Disinfection and Sterilization (Part I)

______ **11.**	Active form of one type of halogen disinfectant	**a.** disinfection
______ **12.**	Method of testing new agents	**b.** sterilization
______ **13.**	Effectiveness decreased by various agents	**c.** crystal violet
______ **14.**	Chlorine	**d.** mercury
______ **15.**	Alkylating agent	**e.** detergent

_____ 16. Complete killing		f. phenol coefficient
_____ 17. Heavy metal		g. quats
_____ 18. Pathogen removal		h. halogen
_____ 19. Cationic or anionic		i. high tissue toxicity
_____ 20. Cell wall synthesis inhibitor		j. hypochlorite

Topic: General Properties of Disinfection and Sterilization (Part 2)

_____ 21. Ultraviolet radiation	a. pasteurization
_____ 22. Relative killing measure	b. filtration
_____ 23. Flash method	c. proteins
_____ 24. Cellulose acetate	d. DNA
_____ 25. Photoreactivation	e. autoclave
_____ 26. Major heat target	f. *Bacillus* species
_____ 27. Ionizing radiation	g. D-value
_____ 28. Sterilization monitoring	h. moist heat
_____ 29. Surgical masks	
_____ 30. High pressures	

Fill-Ins

Provide the correct term or phrase for the following. Spelling counts.

31. A chemical agent used to destroy most microbes on inanimate objects is a _______________.
32. The killing or removal of all microorganisms in a material or on an object is called _______________.
33. Many chemical agents are _______________ , or killing, at high concentrations and _______________, or cell growth-inhibiting, at low concentrations.
34. Disinfectants that are compared to phenol under the same conditions are employing a test called the _______________ _______________ _______________.
35. When paper disks are soaked in selected disinfectants and placed on a plate seeded with a test organism, the test is called the _______________ _______________ _______________.
36. Disinfectants that reduce the surface tension to break up grease particles are called _______________.
37. Solutions that assist in the penetration of the disinfectant are called _______________ _______________.
38. An agent that can kill viruses is called a _______________.
39. Cationic agents that can be neutralized by soaps are _______________ _______________ compounds.
40. Pennies, nickels, and old silver dimes can act as inhibitory agents because they release tiny quantities of _______________ _______________.
41. Chemicals dissolved in alcohols are called _______________.
42. Iodine combined with organic molecules used in skin preparation agents, especially in surgery, is _______________.
43. _______________ and its derivatives, called _______________, disrupt cell membranes, denature proteins, and inactivate enzymes.
44. Hydrogen peroxide acts as an _______________ agent to disrupt disulfide bonds.
45. A gas that is commonly used to sterilize rubber and plastic tubing is _______________ _______________.

46. The time required to kill all bacteria in a particular culture at a specified temperature is the _______________ _______________ _______________.

47. The length of time needed to kill 90% of the organisms in a given population at a specified temperature is the _______________ _______________.

48. _______________ heat probably does most of its damage by oxidizing molecules, while _______________ heat destroys microorganisms mainly by denaturing proteins and disrupting membrane lipids.

49. A device that sterilizes objects by moist steam under pressure is the _______________.

50. The temperature required for sterilization in a typical autoclave is _______________.

51. A quality control device that can determine whether an autoclave has actually sterilized a load of material contains _______________.

52. A technique used to destroy pathogens in food products such as milk and wine is _______________.

53. A pasteurization technique called _______________ _______________ raises the temperature of liquids from 74°C to 148°C and back in less than 5 seconds.

54. The process used in freeze-drying bacterial cultures and in making instant coffee is known as _______________.

55. A unit of radiation energy absorbed per gram of tissue is a _______________ unit.

56. Disruption of cells by sound waves is called _______________.

57. A special type of filter called a _______________ - _______________ _______________ filter is used in surgical rooms and burn units that require complete microbial control of the air.

58. _______________, or loss of water due to osmotic pressures, severely interferes with cell function and eventually leads to cell death.

Labeling

Questions 59–69: Select the best possible term that corresponds to each definition in the table below: bacteriostatic agent, disinfection, viricide, sanitizer, bactericide, disinfectant, sporocide, germicide, sterilization, antiseptic, fungicide. *Place your answers in the spaces provided on page 157.*

Terms Related to Sterilization and Disinfection

Term	Definition
59	The killing or removal of all microorganisms in a material or an object.
60	The reduction of the number of pathogenic microorganisms to the point where they pose no danger of disease.
61	A chemical agent that can safely be used externally on living tissue to destroy microorganisms or to inhibit their growth.
62	A chemical agent used on inanimate objects to destroy microorganisms. Most disinfectants do not kill spores.
63	A chemical agent typically used on food-handling equipment and eating utensils to reduce bacterial numbers so as to meet public health standards. Sanitization may simply refer to thorough washing with only soap or detergent.
64	An agent that inhibits the growth of bacteria.
65	An agent capable of killing microbes rapidly; some such agents effectively kill certain microorganisms but only inhibit the growth of others.
66	An agent that kills bacteria. Most such agents do not kill spores.
67	An agent that inactivates viruses.
68	An agent that kills fungi.
69	An agent that kills bacterial endospores or fungal spores.

59. ____________

60. ____________

61. ____________

62. ____________

63. ____________

64. ____________

65. ____________

66. ____________

67. ____________

68. ____________

69. ____________

Questions 70–80: Select the best possible chemical antimicrobial agent corresponding to the actions and uses listed in the chart below: soaps and detergents, phenols, heavy metals, alkalis, dyes, surfactants, oxidizing agents, alkylating agents, alcohols, acids, halogens.

PROPERTIES OF CHEMICAL ANTIMICROBIAL AGENTS

AGENT	ACTION	USE
70	Lower surface tension, make microbes accessible to other agents	Hand washing, laundering, sanitizing kitchen and dairy equipment
71	Dissolve lipids, disrupt membranes, denature proteins, and inactivate enzymes in high concentrations; act as wetting agents in low concentrations	Cationic detergents are used to sanitize utensils, anionic detergents to launder clothes and clean household objects; quaternary ammonium compounds are sometimes used as antiseptics on skin.
72	Lower pH and denature proteins	Food preservation
73	Raise pH and denature proteins	Found in soaps
74	Denature proteins	Silver nitrate is used to prevent gonococcal infections, mercury compounds to disinfect skin and inanimate objects, copper to inhibit algal growth, and selenium to inhibit fungal growth.
75	Oxidize cell components in absence of organic matter	Chlorine is used to kill pathogens in water and to disinfect utensils; iodine compounds are used as skin antiseptics.
76	Denature proteins when mixed with water	Isopropyl alcohol is used to disinfect skin; ethylene glycol and propylene glycol can be used in aerosols.
77	Disrupt membranes, denature proteins, and inactivate enzymes; not impaired by organic matter	Phenol is used to disinfect surfaces and destroy discarded cultures; amylphenol destroys vegetative organisms and inactivates viruses on skin and inanimate objects; chlorhexidine gluconate is especially effective as a surgical scrub.
78	Disrupt disulfide bonds	Hydrogen peroxide is used to clean puncture wounds, potassium permanganate to disinfect instruments.
79	Disrupt structure of proteins and nucleic acids	Formaldehyde is used to inactivate viruses without destroying antigenic properties, glutaraldehyde to sterilize equipment, betapropiolactone to destroy hepatitis viruses, and ethylene oxide to sterilize inanimate objects that would be harmed by high temperatures.
80	May interfere with replication or block cell wall synthesis	Acridine is used to clean wounds, crystal violet to treat some protozoan and fungal infections.

70. ____________

71. ____________

72. ____________

73. ____________

74. ____________

75. ____________

76. ____________

77. ____________

78. ____________

79. ____________

80. ____________

Questions 81–94: Select the best possible physical antimicrobial agent corresponding to the actions and uses listed in the chart below: osmotic pressure, filtration, drying, strong visible light, dry heat, moist heat, ultraviolet light, freezing, ionizing radiation, pasteurization, refrigeration, microwave radiation, sonic and ultrasonic waves, freeze-drying.

PROPERTIES OF PHYSICAL ANTIMICROBIAL AGENTS

AGENT	ACTION	USE
81	Denatures proteins	Oven heat used to sterilize glassware and metal objects; open flame used to incinerate microorganisms.
82	Denatures proteins	Autoclaving sterilizes media, bandages, and many kinds of hospital and laboratory equipment not damaged by heat and moisture; pressure cooking sterilizes canned foods.
83	Denatures proteins	Kills pathogens in milk, dairy products, and beer.
84	Slows th rate of enzyme-controlled reactions	Used to keep fresh food for a few days; does not kill most microorganisms.
85	Greatly slows the rate of enzyme-controlled reactions	Used to keep fresh foods for several months; does not kill most microorganisms; used with glycerol to preserve microorganisms.
86	Inhibits enzymes	Used to preserve some fruits and vegetables; sometimes used with smoke to preserve sausages and fish.
87	Dehydration inhibits enzymes	Used to manufacture some instant coffees; used to preserve microorganisms for years.
88	Denatures proteins and nucleic acids	Used to reduce the number of microorganisms in air in operating rooms, animal rooms, and where cultures are transferred.
89	Denatures proteins and nucleic acids	Used to sterilize plastics and pharmaceutical products and to preserve foods.
90	Absorbs water molecules, then releases microwave energy to surroundings as heat	Cannot be used reliably to destroy microbes except in special media-sterilizing equipment.
91	Oxidation of light-sensitive materials	Can be used with dyes to destroy bacteria and viruses; may help sanitize clothing.
92	Causes cavitation	Not a practical means of killing microorganisms but useful in fractionating and studying cell components.
93	Mechanically removes microbes	Used to sterilize media, pharmaceutical products, and vitamins, in manufacturing vaccines, and in sampling microbes in air and water.
94	Removes water from microbes	Used to prevent spoilage of foods such as pickles and jellies.

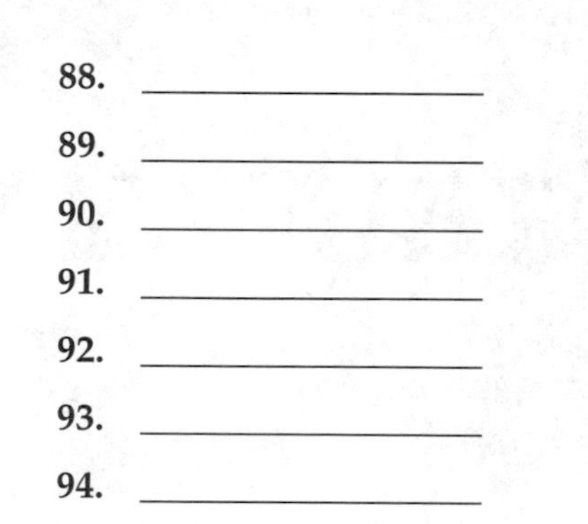

81. ____________________
82. ____________________
83. ____________________
84. ____________________
85. ____________________
86. ____________________
87. ____________________
88. ____________________
89. ____________________
90. ____________________
91. ____________________
92. ____________________
93. ____________________
94. ____________________

Critical Thinking

95. How do sterilization and disinfection differ, and what terms are used to describe these processes?

96. What important principles apply to the processes of sterilization and disinfection?

97. What factors affect the potency of antimicrobial chemical agents?

98. How is the effectiveness of an antimicrobial chemical agent assessed?

99. By what mechanisms do antimicrobial chemical agents act?

100. What are the properties of commonly-used antimicrobial chemical agents?

101. How are dry heat, moist heat, and pasteurization used to control microorganisms?

102. How are refrigeration, freezing, drying, and freeze-drying used to control and to preserve microorganisms?

103. How are radiation, sonic and ultrasonic waves, filtration, and osmotic pressure used to control microorganisms?

ANSWERS

True/False

1. **True** Most chemical disinfectants are unable to penetrate oily or organic material. The addition of alcohol, which can dissolve oils, often facilitates its action.
2. **False** A bactericidal agent is one that kills bacteria but often cannot kill bacterial endospores or viruses.
3. **False** The death rate of organisms follows a logarithmic scale, therefore, if 20% die in the first minute, 20% of the remaining organisms die in the next minute and so forth. At this rate, it may require 10 to 15 minutes to achieve sterility.
4. **True** An ideal disinfectant should be fast acting, effective, penetrating, easily prepared, inexpensive, and safe for human use. No disinfectant currently available can accomplish all of these.
5. **False** Most hand soaps do not kill organisms but only act to remove surface bacteria, oily substances, and dirt. Because of the pores and crevices in the skin, it is impossible to sterilize the skin.
6. **True** Pasteurization is effective in killing organisms such as *Salmonella, Mycobacteria, Listeria, Q-fever, rickettsia,* and *Brucella.*

Multiple Choice

7. b; 8. c; 9. b; 10. c.

Matching

11. j; 12. f; 13. g; 14. h; 15. i; 16. b; 17. d; 18. a; 19. e; 20. c; 21. d; 22. g; 23. a; 24. b; 25. d; 26. c; 27. d; 28. f; 29. b; 30. e.

Fill-Ins

31. disinfectant; 32. sterilization; 33. bactericidal, bacteriostatic; 34. phenol coefficient test; 35. filter paper method; 36. surfactants; 37. wetting agents; 38. viricide; 39. quaternary ammonium; 40. heavy metals; 41. tinctures; 42. iodophors; 43. phenol, phenolics; 44. oxidizing; 45. ethylene oxide; 46. thermal death time; 47. D value; 48. dry, moist; 49. autoclave; 50. 121°C; 51. endospores; 52. pasteurization; 53. ultrahigh temperature; 54. lyophilization; 55. rad; 56. sonication; 57. high-efficiency particulate air; 58. plasmolysis.

Labeling

59. sterilization; 60. disinfection; 61. antiseptic; 62. disinfectant; 63. sanitizer; 64. bacteriostatic agent; 65. germicide; 66. bactericide; 67. viricide; 68. fungicide; 69. sporocide; 70. soaps and detergents; 71. surfactants; 72. acids; 73. alkalis; 74. heavy metals; 75. halogens; 76. alcohols; 77. phenols; 78. oxidizing agents; 79. alkylating agents; 80. dyes; 81. dry heat; 82. moist heat; 83. pasteurization; 84. refrigeration; 85. freezing; 86. drying; 87. freeze-drying; 88. ultraviolet light; 89. ionizing radiation; 90. microwave radiation; 91. strong visible light; 92. sonic and ultrasonic waves; 93. filtration; 94. osmotic pressure.

Critical Thinking

95. **How do sterilization and disinfection differ, and what terms are used to describe these processes?**
Sterilization refers to the absolute killing or removal of *all* organisms in any material or on any object. There is no such thing as "almost" sterile. Disinfection refers to the reduction in numbers of pathogenic organisms on objects or in materials so that the organisms no longer pose a disease threat. Generally, disinfectants can not be used on the human body. Antisepsis refers to those agents that can be safely used on the body surfaces. Other important terms include: (1) sanitizer—a chemical agent typically used on food-handling equipment and eating utensils to reduce bacterial numbers so as to meet public health standards; (2) bacteriostatic agent—an agent that inhibits the growth of bacteria; (3) germicide—an agent capable of killing microbes rapidly; (4) bactericide—an agent that kills bacteria; (5) viricide—an agent that inactivates viruses; (6) fungicide—an agent that kills fungi; and (7) sporocide—an agent that kills bacterial endospores or fungal spores

96. **What important principles apply to the processes of sterilization and disinfection?**
Microorganisms follow the same laws regarding death rates as those declining in numbers from natural causes. Therefore a definite proportion of organisms will die in a given time interval and will follow a logarithmic death rate. The smaller the number of organisms that are present, the less time will be needed to achieve sterility.

    Based on these principles, thermal death points and thermal death times have been devised for virtually all types of chemicals and equipment. It should be remembered that microbes will differ greatly in their susceptibility to antimicrobial agents, especially if they produce endospores.

97. **What factors affect the potency of antimicrobial chemical agents?**
Various factors affect the activity of antimicrobial chemical agents. These include the time used for exposure to the agent, the temperature of the agent, the pH, and the concentration of the agent. Potency increases with the length of time the organisms are exposed to the agent; to an increase in temperature; to an acidic or alkaline pH; and, to an extent, to an increase in the concentration of the agent. It is always extremely important to *read the label* to know exactly how to prepare the agent for the greatest effectiveness.

98. **How is the effectiveness of an antimicrobial chemical agent assessed?**
Effective evaluation of antimicrobial chemical agents is extremely difficult. No completely satisfactory method is currently available. What works in laboratory conditions may not work in clinical conditions. The most common evaluation technique is the phenol coefficient, which can only be used for those disinfectants similar to phenol. The test involves the determination of ratios between phenol and the test disinfectant. Other tests include the filter paper method of evaluating chemical agents using small filter paper disks soaked with different chemical agents, and the use-dilution test where bacteria are added to tubes containing different dilutions of the agent.

99. **By what mechanisms do antimicrobial chemical agents act?**
Antimicrobial chemical agents kill microbes by participating in one or more chemical reactions that damage the proteins, cell membranes, or other cell components.

Reactions that affect cellular proteins include hydrolysis, oxidation, and attachment of atoms or chemical groups to protein molecules. The reaction causes the protein to be denatured and nonfunctional. Agents include hydrogen peroxide, boric acid, halogens, heavy metals, and formaldehyde.

Reactions that affect cell membranes are surfactants that reduce the surface tension and dissolve lipids and protein denaturing agents such as those mentioned above. Surfactant agents include soaps and detergents, alcohols, phenols, and quaternary ammonium compounds.

Reactions of other chemical agents damage nucleic acids and energy-capturing systems. Damage to nucleic acids is an important means of inactivating viruses. Examples include alkylating agents such as ethylene oxide and nitrous acid, dyes, detergents, and alcohols.

100. What are the properties of commonly used antimicrobial chemical agents?

(1) Soaps and detergents lower surface tension and make microbes accessible to other agents. (2) Surfactants dissolve lipids, disrupt cell membranes, denature proteins, and inactivate enzymes in high concentrations. They also act as wetting agents in low concentrations. (3) Acids lower pH and denature proteins. (4) Alkalines raise pH and denature proteins. (5) Heavy metals denature proteins. (6) Halogens oxidize cell components in the absence of organic matter. (7) Alcohols denature proteins when they are mixed with water. (8) Phenols disrupt cell membranes, denature proteins, and inactivate enzymes. They are not impaired by the presence of organic matter. (9) Oxidizing agents disrupt disulfide bonds. (10) Alkylating agents disrupt the structure of proteins and nucleic acids. (11) Dyes may interfere with replication or block cell wall synthesis.

101. How are dry heat, moist heat, and pasteurization used to control microorganisms?

Heat destroys microorganisms by denaturing proteins, melting lipids, and when open flame is used, by incineration. Dry heat can be used to sterilize metal objects, oils, powders, and glassware but no liquid-containing material. A temperature of 160°C is used for 2 hours or longer.

Moist heat, especially when placed under pressure as with an autoclave, is the most effective method for sterilization. A temperature of 121°C is used for 15 minutes or longer. Heat-sensitive materials cannot be autoclaved. Quality control is essential when using autoclaves. Devices include heat-sensitive autoclave tape, wax pellets, and vials and strips containing endospores. The latter is the only device that can detect sterility.

Pasteurization does not achieve sterility. It does kill important pathogens such as those that cause tuberculosis, salmonellosis, brucellosis, Q-fever, and listeriosis.

102. How are refrigeration, freezing, drying, and freeze-drying used to control and to preserve microorganisms?

All of these mechanisms can be used to retard the growth of microorganisms. Lyophilization, which is drying in the frozen state, can be used for long-term preservation of live cultures of microorganisms.

103. How are radiation, sonic and ultrasonic waves, filtration, and osmotic pressure used to control microorganisms?

Radiation affects the DNA of microbes by damaging the nucleotides. Ultraviolet light, ionizing radiation, microwave radiation, and strong visible light can be used to control microorganisms and preserve foods.

Sonic and ultrasonic waves can kill microorganisms but they are used mostly for disruption of cells called sonication.

Filtration as a physical separation mechanism can be used to sterilize heat-sensitive substances, separate viruses, and collect microorganisms from air and water samples. Filters can be made of glass, asbestos, diatomaceous earth, and porcelain. The most common type used today is the membrane filter which is made of nitrocellulose.

Osmotic pressure can be created using high concentrations of sugar or salt that will plasmolyze the cells. It can be used to prevent the growth of microorganisms in highly sweetened or salted foods.

Chapter 13 Antimicrobial Therapy

INTRODUCTION

The antimicrobial medicines and their limitations are discussed in this chapter. Antibiotics can protect us from only some bacteria but not from viruses. Bacterial pneumonia is the sixth most common cause of death. Bacteria are increasingly becoming resistant to many antibiotics and death from influenza killed 36,000 American in the same year the attacks on the Trade Towers and Washington killed 3,000. Our need to develop better drugs to fight and prevent infectious disease is clear. In 2004, about 5 million people died of HIV-AIDS and the fate of 40 million living people who are infected remains unclear.

Bioterror in the United States with anthrax spores showed us that American can be killed despite antibiotic therapy. In an uncertain world, we all need to know as much as possible about the use and misuse of antimicrobial agents.

STUDY OUTLINE

I. Antimicrobial Chemotherapy

A. General features

Today, the word *chemotherapy* refers to the use of chemical substances to treat various aspects of disease.

B. Terms

A *chemotherapeutic agent*, or a *drug*, is any chemical substance used in medical practice. *Antimicrobial agents* are members of a special group of chemotherapeutic agents used to treat diseases caused by microbes. *Antibiosis* means "against life," while an *antibiotic* is "a chemical substance produced by microorganisms which has the capacity to inhibit the growth of bacteria and even destroy bacteria and other microorganisms in dilute solution." *Synthetic drugs* are agents synthesized in the laboratory. *Semisynthetic drugs* are antimicrobial agents made partly by laboratory synthesis and partly by microorganisms.

II. History of Chemotherapy

A. Early history

Throughout history, humans have often attempted to alleviate suffering by treating disease by taking concoctions of plant substances. Because of this, traditional healers of primitive societies, especially in the tropics, are quite knowledgeable about the medicinal properties of plants. Pharmaceutical companies are now attempting to learn from these healers and make written records of their treatments, as well as test the plants that they use.

B. Erlich's contributions

In Western civilization, the first systematic attempt to find specific chemical substances to treat infectious disease was made by Paul Erlich. He developed many concepts important to the science of chemotherapy.

C. Development of sulfa drugs

Sulfa drugs, or *sulfonamides*, were developed to attack a variety of pathogens. However, they do not attack all pathogens and sometimes cause kidney damage and allergies.

D. Fleming's discoveries

Alexander Fleming identified *penicillin* as an inhibitory agent. Because of him, penicillin was produced in mass during World War II, saving the lives of many people.

III. General Properties of Antimicrobial Agents

A. Selective toxicity

For internal use, an antimicrobial drug must have *selective toxicity*. This means that it must harm the microbes without causing significant damage to the host. The *toxic dosage level* is the dosage that causes host damage. The *therapeutic dosage level* is the dosage that successfully eliminates the pathogen if the level is maintained over a period of time. The *chemotherapeutic index* is the maximum tolerable dose per kilogram of body weight, divided by the minimum dose per kilogram of body weight that will cure the disease.

B. Spectrum of activity

The range of different microbes against which an antimicrobial agent acts is called its *spectrum of activity*. Agents that are effective against a great number of microorganisms from a wide range of taxonomic groups have a *broad spectrum* of activity, while those that are effective against a small number of microorganisms or a single taxonomic group have a *narrow spectrum* of activity.

C. Modes of action

The five different modes of action of antimicrobials discussed in the textbook are: (1) inhibition of cell wall synthesis; (2) disruption of cell membrane function; (3) inhibition of protein synthesis; (4) inhibition of nucleic acid synthesis, and (5) action as antimetabolites.

D. Kinds of side effects

The side effects of antimicrobial agents on infected persons (hosts) fall into three general categories: (1) toxicity; (2) allergy; and (3) disruption of normal microflora.

E. Resistance of microorganisms

Resistance of a microorganism to an antibiotic means that a microorganism formerly susceptible to the action of the antibiotic is no longer affected by it. It is important to understand how resistance is acquired; what the mechanisms of resistance are; how first-line, second-line, and third-line drugs can be used; what cross-resistance is; and how scientists can try to limit drug resistance.

IV. Determination of Microbial Sensitivities to Antimicrobial Agents

A. Disk diffusion method

Microbial sensitivity to antimicrobial agents can be determined using the *disk diffusion method*, or the *Kirby-Bauer method*, in which antibiotic disks are placed on an inoculated petri dish, incubated, and observed for inhibition of growth.

B. Dilution method

Microbial sensitivity to antibiotics can also be determined using the *dilution method*, in which organisms are incubated in a series of tubes containing known quantities of a chemotherapeutic agent.

C. Serum killing power

Another method of determining the effectiveness of a chemotherapeutic agent is to measure its *serum killing power* by adding a bacterial suspension to the serum of a patient who is receiving an antibiotic and incubating the solution.

D. Automated methods

Automated methods are now available to identify pathogenic organisms and to determine which antimicrobial agents will effectively combat them. These methods make laboratory identification of organisms and their sensitivities to antimicrobials more efficient and less expensive.

V. Attributes of an Ideal Antimicrobial Agent

The characteristics of an ideal antimicrobial agent are: (1) solubility in body fluids; (2) selective toxicity; (3) toxicity not easily altered; (4) nonallergenic; (5) stability—maintenance of a constant, therapeutic concentration in blood and tissue fluids; (6) resistance by microorganisms not easily acquired; (7) long shelf life; (8) reasonable cost.

VI. Antibacterial Agents

A. Sources of antibiotics

The book chooses to categorize antibacterial agents by their modes of action. Another way of grouping antibiotics is by the microorganisms that produce them: fungi, streptomycetes, actinomycetes, or bacillus, for example.

B. Inhibitors of cell wall synthesis

Antibiotics that work by inhibiting the synthesis of microbial cell walls include penicillins, cephalosporins, carbapenems, bacitracin, and vancomycin.

C. Disrupters of cell membranes

Antibiotics that work by disrupting microbial cell membranes include polymyxins and tyrocidins.

D. Inhibitors of protein synthesis

Antibiotics that work by inhibiting microbial protein synthesis include aminoglycosides, tetracyclines, chloramphenicol, macrolides, and lincosamides.

E. Inhibitors of nucleic acid synthesis

Antibiotics that work by inhibiting microbial nucleic acid synthesis include rifampin and quinolones.

F. Antimetabolites and other antibacterial agents

Other antibacterial agents include sulfonamides (sulfa drugs), isoniazid, ethambutol, and nitrofurans.

VII. Other Antimicrobial Agents

A. Antifungal agents

The *antifungal agents* include imidazoles and triazoles, polyenes (amphotericin B and nystatin), griseofulvin, flucytosine, tolnaftate (Tinactin), and terbinafine (Lamisil).

B. Antiviral agents

Antiviral agents currently in use include purine and pyrimidine analogues, amantadine, inteferons, and immunoenhancers.

C. Antiprotozoan agents

The *antiprotozoan agents* include quinine, chloroquine and primaquine, metronidazole, pyrimethamine, suramin sodium, nifurtimox, arsenic and antimony compounds, and pentamidine isethionate.

D. Antihelminthic agents

Antihelminthic agents currently in use include niclosamide, mebendazole, piperazine (Antepar), and ivermectin.

E. Special problems with resistant hospital infections

As soon as antibacterial agents became available, resistant organisms began to appear. Many organisms are now resistant to several different antibiotics, and new resistant strains are constantly being encountered, especially among hospitalized patients. Treatment of resistant infections is very difficult, so care should be taken to prevent infections caused by antibiotic-resistant strains of microorganisms.

SELF-TESTS

Continue with this section only after you have read Chapter 13 of your textbook. Write your answers in the appropriate space provided. Correct answers to all questions can be found at the end of the Self-Test section. A score of 80 percent or better is good. If your score is less than 65, reread the chapter.

True/False

Mark T for True, F for False

______ **1.** Antibiotics such as penicillin that can inhibit the formation of the cell wall have a bacteriocidal effect.

______ **2.** The use of antibiotics in animal food supplements has greatly reduced the number of infections transmitted in poultry, especially chickens and turkeys.

______ **3.** A major problem with overuse of clindamycin is that superinfections of Clostridium difficile can occur.

______ **4.** Certain cold viruses seem to be controlled best by antibiotics such as the aminoglycosides that affect their protein synthetic processes.

_____ 5. Once a patient feels better following a bacterial infection, antibiotic therapy should be halted to reduce the problems of overuse of antibiotics.

_____ 6. Tetracyclines can actually stain teeth if given prior to the eruption of the teeth.

Multiple Choice

Select the best possible answer.

_____ 7. Select an attribute of an antibiotic that would not be beneficial.
- **a.** Toxicity not easily altered.
- **b.** Insolubility in body fluids.
- **c.** Nonallergenic.
- **d.** Maintenance of constant, therapeutic concentration in blood and tissues.
- **e.** All of the above are beneficial.

_____ 8. The minimal inhibitory concentration test:
- **a.** uses disks soaked in antibiotics.
- **b.** allows measurement of zones of inhibition.
- **c.** can help determine the appropriate dilution of an antibiotic for use in therapy.
- **d.** involves the use of the patient's serum.

_____ 9. With respect to antibiotic therapy:
- **a.** Antibiotics should always be used to treat viral infections.
- **b.** Superinfections can be controlled using large doses of tetracyclines.
- **c.** Because of the development of resistant strains, some drugs have gone to second and third generations of derivatives.
- **d.** Drugs that affect protein synthesis will have the greatest bactericidal effect.

_____ 10. Antiprotozoan agents:
- **a.** include metronidazole and chloroquine.
- **b.** All are designed to affect the cell wall.
- **c.** are laboratory-synthesized penicillin derivatives.
- **d.** can also be used to treat viral agents.

_____ 11. Which of the following is not true with respect to antibiotic resistance in bacteria?
- **a.** Chromosome resistance usually occurs against a single type of antibiotic.
- **b.** Extrachromosomal resistance usually involves R factors.
- **c.** Resistance can occur by altering the receptors to which antimicrobial agents can bind.
- **d.** Resistance can occur by altering a metabolic pathway that bypasses a reaction inhibited by an antimicrobial agent.
- **e.** All of the above are true.

Matching

Select the answer from the right-hand side that corresponds to the term or phrase on the left-hand side of the page. An answer may be used more than once. In some cases, more than one answer may be required.

Topic: Principles of Chemotherapy

_____ 12. Effective against several gram-negative and gram-positive bacteria

_____ 13. Bactericidal drug

_____ 14. Effective against a single group of microbes

_____ 15. Static effect

- **a.** inhibits growth
- **b.** causes death
- **c.** broad spectrum
- **d.** narrow spectrum
- **e.** none of the above

Topic: specific mechanisms of antimicrobial drugs

_______ 16. Interferon

______ 17. Quinolones

______ 18. Rifampin

______ 19. Penicillin

______ 20. Carbapenems

______ 21. Aminoglycosides

______ 22. Tetracyclines

______ 23. Cephalosporins

______ 24. Tyrocidins

______ 25. Ethambutol

______ 26. Polymyxin

______ 27. Sulfonamides

______ 28. Erythromycin

______ 29. Streptomycin

______ 30. Isoniazid

a. inhibit(s) some aspects of protein synthesis

b. inhibit(s) cell wall formation

c. damage(s) cell membranes

d. inhibit(s) nucleic acid synthesis

e. function as antimetabolites

f. none of the above

Topic: Antifungal, Antiviral, Antiprotozoan, and Antihelminthic Drugs

______ 31. Acyclovir

______ 32. Griseofulvin

______ 33. Amphotericin B

______ 34. Nystatin

______ 35. Amantadine

______ 36. Piperazine

______ 37. Ivermectin

______ 38. Quinine

______ 39. Chloroquine

______ 40. Metronidazole

______ 41. Mebendazole

a. antifungal

b. antiprotozoan

c. antiviral

d. antihelminthic

e. none of the above

Topic: Specific Mechanisms of Antimicrobial Drugs

______ 42. Cephalosporin

______ 43. PABA

______ 44. Sulfonamides

______ 45. Penicillin

______ 46. Ampicillin

______ 47. Streptomycin

______ 48. Chloramphenicol

______ 49. Tetracycline

a. interfere(s) with the synthesis of folic acid

b. interfere(s) with cell wall formation

c. inactivated by -lactamase

d. cause(s) misreading of the genetic code

e. block(s) transfer of activated amino acid from transfer RNA to growing polypeptide chain

f. increase(s) permeability of membranes

g. none of the above

_____ 50. Polymyxin B

_____ 51. Adenine dinucleotide

_____ 52. Isonicotininc hydrazide

Fill-Ins

Provide the correct term or phrase for the following. Spelling counts.

53. The use of chemicals to kill pathogenic organisms is known as ________________.

54. A chemotherapeutic agent is also referred to as a ________________.

55. A chemical substance produced by one microbe that is inhibitory or lethal to another is an ________________.

56. Chemical agents produced in the laboratory that are used in therapy are ________________ drugs.

57. Alexander Fleming is credited with discovering the first antibiotic, known as ________________.

58. The ________________ ________________ level is the drug level that causes host damage.

59. The ________________ ________________ level is the drug level that successfully eliminates the pathogenic organism if the level is maintained over a period of time.

60. The maximum tolerable dose of an antibiotic per kilogram body weight divided by the minimum dose per kilogram body weight that will cure the disease is called its ________________ ________________.

61. Antimicrobial agents that can attack a wide variety of microbes can be said to have a ________________ ________________ of activity.

62. Penicillin contains a chemical structure called the ________________ - ________________ ________________ that acts as the active portion of the molecule.

63. ________________ are substances that affect the utilization of metabolites and therefore prevent a cell from carrying out necessary metabolic reactions.

64. Antimetabolites that imitate the normal molecule required for their metabolism are probably affecting it by ________________ ________________.

65. An ________________ is a condition in which the body's immune system responds to a foreign substance, usually a protein.

66. Some antimicrobial agents that allow invasion by replacement flora tend to cause ________________.

67. ________________ of a micro-organism to an antibiotic means that a microorganism formerly susceptible to the action of an antibiotic is no longer affected by it.

68. Extrachromosomal pieces of DNA that can carry resistance genes and can be transmitted from one cell to another are called ________________ ________________.

69. ________________ - ________________ is resistance to two or more similar antimicrobial agents via a common mechanism.

70. If two separate antibiotics cannot inhibit an organism but can exert an effect together, the effect is known as ________________.

71. Some drugs are less effective when used in combination than when used alone; this is known as ________________.

72. A method that involves filter paper disks soaked in an antibiotic to determine the antibiotic's sensitivity is the ________________ - ________________ test.

73. Measurable clear areas around disks used in the determination of sensitivities to antibiotics in the Kirby-Bauer test are called ________________ of ________________.

74. In the dilution method, the lowest concentration of the chemotherapeutic agent that yields no growth following a second inoculation is the MBC, or the ________________ ________________ ________________.

75. Cephalosporins are derived from species of ________________.

76. Purine and pyrimidine analogs appear to be effective in controlling infections caused by ________________.

77. A substance acquired from the bark of the chinchona tree that was used to treat malaria is ________________.

78. Black hairy tongue sometimes occurs as a result of the antiprotozoan drug ________________.

Labeling

Questions 79–83: Select the correct antimicrobial effect caused by the following antimicrobial examples: inhibition of protein synthesis; inhibition of nucleic acid synthesis; disruption of cell membrane function; inhibition of cell wall synthesis; action as antimetabolites.

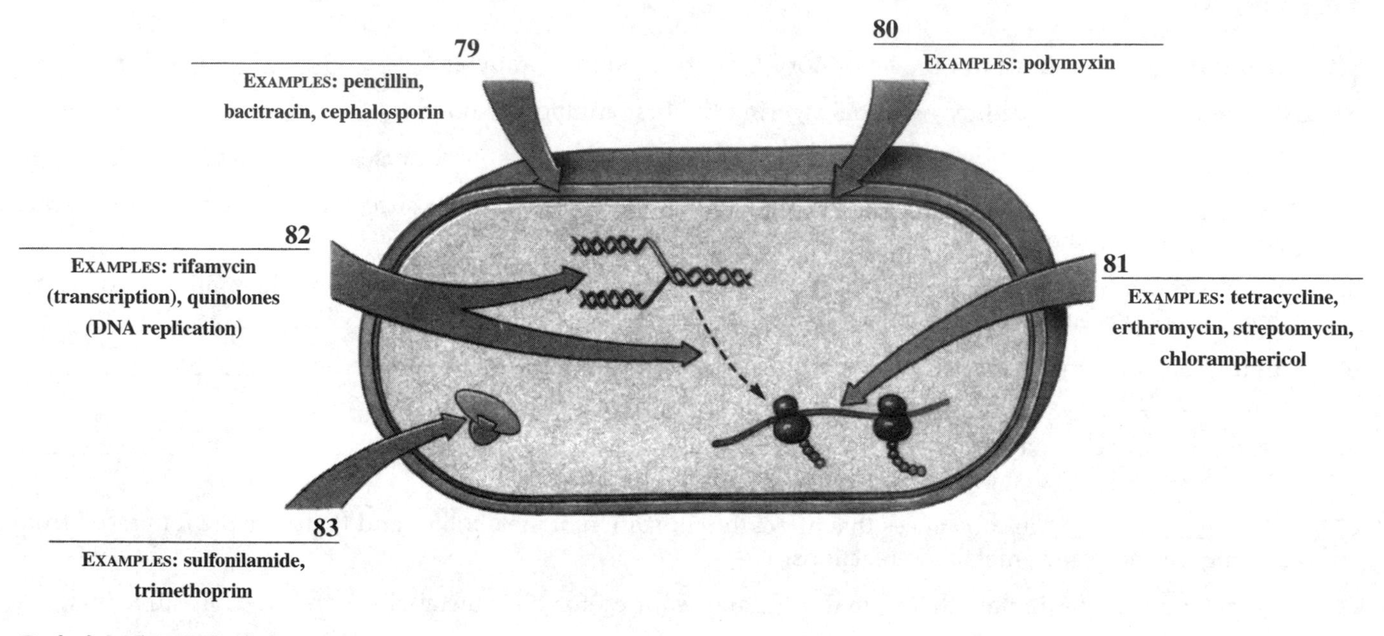

Critical Thinking

84. What terms are used to discuss chemotherapy and antibiotics, and what do they mean?

85. How have chemotherapeutic agents been developed?

86. How do the terms selective toxicity, spectrum of activity, and modes of action apply to antimicrobial agents?

87. What kinds of side effects are associated with antimicrobial agents?

88. What is resistance to antibiotics, and how do microorganisms acquire it?

89. How are sensitivities of microbes to chemotherapeutic agents determined?

90. What are the attributes of an ideal antimicrobial agent?

91. What are the properties, uses, and side effects of antibacterial agents?

92. What are the properties, uses, and side effects of antifungal agents, antiviral agents, antiprotozoan agents, and antihelminthic agents?

93. How do resistant hospital infections arise, and how can they be treated and prevented?

ANSWERS

True/False

1. **True** Antibiotics that inhibit the cell wall synthesis selectively damage bacterial cells. Since many bacteria have a very high internal osmotic pressure, the loss of a wall will cause the cells to burst when subjected to a low osmotic condition.

2. **False** Antibiotics used in animal feeds have greatly increased the resistant strains of many bacteria such as Salmonella and Escherichia coli. The low dosage levels in the products provide the opportunity for resistance to develop.

3. **True** Clindamycin is effective against Bacteroides and other anaerobes but not Clostridium difficile. Toxins produced by this anaerobe can cause severe and sometimes fatal colitis.

4. **False** Since viruses reproduce only inside living cells, they do not respond to any antibiotics and only respond to just a few chemosynthetic agents.

5. **False** It is very important to finish all antibiotic medications to insure that all microbes are destroyed. Failure to do so may allow a few persisters to survive as well as a few resisters, which could then establish a resistant population.

6. **True** Tetracyclines, while effective against a wide variety of organisms, have a major side effect, which is permanent staining of teeth. The drug should be avoided by pregnant mothers and very young children.

Multiple Choice

7. b; 8. c; 9. c; 10. a; 11. e.

Matching

12. c; 13. b; 14. d; 15. a; 16. a; 17. d; 18. d; 19. b; 20. b; 21. a; 22. a; 23. b; 24. c; 25. e; 26. c; 27. e; 28. a; 29. a; 30. e; 31. c; 32. a; 33. a; 34. a; 35. c; 36. d; 37. d; 38. b; 39. b; 40. b; 41. d; 42. b; 43. g; 44. a; 45. b and c; 46. b and c; 47. d; 48. e; 49. e; 50. f; 51. g; 52. g.

Fill-Ins

53. chemotherapy; 54. drug; 55. antibiotic; 56. synthetic; 57. penicillin; 58. toxic dosage; 59. therapeutic dosage; 60. chemotherapeutic index; 61. broad spectrum; 62. beta-lactam ring; 63. antimetabolites; 64. molecular mimicry; 65. allergy; 66. superinfections; 67. resistance; 68. R plasmids; 69. cross-resistance; 70. synergism; 71. antagonism; 72. Kirby-Bauer; 73. zones of inhibition; 74. minimum bactericidal concentration; 75. fungi; 76. viruses; 77. quinine; 78. metronidazole.

Labeling

79. inhibition of cell wall synthesis; 80. disruption of cell membrane function; 81. inhibition of protein synthesis; 82. inhibition of nucleic acid synthesis; 83. action as antimetabolites.

Critical Thinking

84. What terms are used to discuss chemotherapy and antibiotics, and what do they mean?
Chemotherapy refers to the use of any chemical agent in the practice of medicine. Chemotherapeutic agents are any chemicals used in medical practice while antimicrobial agents are chemicals used to treat diseases caused by microbes. A chemosynthetic agent is one that is made in the laboratory while an antibiotic agent is one that is produced by other microorganisms. Many of the drugs available today are semisynthetic in that they are partly made by microorganisms and partly by laboratory synthesis.

85. How have chemotherapeutic agents been developed?
Chemotherapeutic agents were used for thousands of years and most were extracts from plants and herbs. The first systematic attempt to find chemotherapeutic agents was begun by Erlich. His work helped in the discovery of agents such as sulfa drugs. Fleming was the first to discover the antibiotics with his work on penicillin.

86. How do the terms selective toxicity, spectrum of activity, and modes of action apply to antimicrobial agents?
Antimicrobial agents share certain common properties. These include selective toxicity, spectrum of activity, and mode of action. Selective toxicity refers to the property of antimicrobial agents that allows them to exert greater toxic effects on microbes than on the host. Some drugs such as penicillin have a very wide range between the level sufficient to affect the pathogen and the level of toxicity to the host. Others such as those that contain arsenic must be monitored very carefully because of their narrow range.

Spectrum of activity refers to the variety of microbes sensitive to the agent. Agents such as ampicillin and gentamicin that affect a large number of organisms are called broad spectrum. Agents such as erythromycin and penicillin that affect only a few are called narrow spectrum.

Mode of action refers to the mechanism by which the antibiotic exerts its effect. Some affect only the cell wall, some the cell membrane and some the cellular machinery or its genetic mechanisms.

87. What kinds of side effects are associated with antimicrobial agents?
The side effects caused by antimicrobial agents include excessive toxicity, allergy, and disruption of normal flora bacteria. Many agents are not licensed for use because of their high toxicity to the patient even though they are very effective antimicrobial agents. Some drugs cause allergic reactions which represent a reaction to the agent as a foreign substance. Many drugs attack not only the pathogen but the normal flora that in many cases help prevent secondary infections. Elimination of these bacteria can cause superinfections with new pathogens.

88. What is resistance to antibiotics, and how do microorganisms acquire it?
Resistance means that a microorganism formerly susceptible to an antibiotic is no longer affected by it. Nongenetic resistance occurs when microbes hide from the effects of the drug or undergo a temporary change that allows them to resist it.

Genetic resistance occurs when microbes are able to acquire new genetic units such as plasmids, which carry resistant genes or when the microbes can alter their genes to avoid damage by the antibiotic. Chromosomal resistance is due to a mutation in the microbial DNA while extrachromosomal resistance is due to the acquisition of plasmids. Five mechanisms of resistance have been identified and include alteration of receptors, alteration of cell membrane permeability, development of new enzymes, alteration of existing enzymes, and alteration of metabolic pathways.

Drug resistance can be minimized by continuing the treatment until all sensitive pathogens are eliminated, combining the effectiveness of two antibiotics that do not have cross-resistance and by using antibiotics only when absolutely necessary.

89. How are sensitivities of microbes to chemotherapeutic agents determined?

Microbes vary greatly in their sensitivities to different chemotherapeutic agents, and their susceptibilities can change over time. By exposing the microbes to various agents in the laboratory, a complete pattern of sensitivity or resistance can be determined. Several methods are available and include disk diffusion (Kirby-Bauer) method, dilution method, and the automated method.

The disk-diffusion method involves placing antibiotic-soaked paper disks on plates seeded with the test organism. Following incubation, zone diameters indicating the degree of sensitivity can be measured.

The dilution method involves the addition of an organism into tubes or walls containing nutrient media and varying concentrations of the antibiotic. A minimum inhibitory concentration of the agent can be determined and represents the lowest concentration in which no growth of the organism is observed.

Automated methods allow rapid identification of microorganisms along with the determination of its sensitivities to various antimicrobial agents.

90. What are the attributes of an ideal antimicrobial agent?

An ideal antimicrobial agent is soluble in body fluids, selectively toxic, and nonallergenic; can be maintained at a constant therapeutic concentration in blood and body fluids; is unlikely to elicit resistance; has a long shelf life; and is reasonable in cost.

91. What are the properties, uses, and side effects of antibacterial agents?

Antibacterial agents such as penicillin and cephalosporins inhibit cell wall synthesis. Polymyxins and tyrocidins affect the cell membrane. Aminoglycosides, tetracyclines, and others affect protein synthesis. Sulfonamides and quinolones affect cellular metabolism and nucleic acid synthesis.

It's interesting to note that the most common antibiotics are derived from various species of fungi while the greatest variety of antibiotics are derived from species of *Streptomyces*.

92. What are the properties, uses and side effects of antifungal agents, antiviral agents, antiprotozoan agents, and antihelminthic agents.

Antifungal agents such as the imidazoles affect the cell wall of fungal cells. Amphotericin B and nystatin cause cell membrane permeability. Griseofulvin and flucytosine impair synthesis of nucleic acids and affects cellular proteins.

Antiviral agents have been difficult to develop because they must damage viruses inside the cells without affecting the host cell. Antibiotics do not affect viruses. Most antiviral agents are analogs of purines or pyrimidines. These include idoxuridine, ribavirin, and acyclovir. Amantadine seems to prevent influenza A from penetrating cells. Interferon, which is produced and released by infected cells to stimulate other cells to produce antiviral proteins, has been of limited success in treating viral infections.

Antiprotozoan agents such as chloroquine and primaquine interfere with protein synthesis while others such as pyrimethamine affect folic acid synthesis. The mechanism of some agents is not well understood.

Antihelminthic agents such as niclosamide and mebendazole affect carbohydrate metabolism of the worms while others such as piperazine act as powerful neurotoxins.

93. How do resistant hospital infections arise, and how can they be treated and prevented?

Soon after the advent of antibacterial agents, antibiotic-resistant organisms began to appear. Many infections caused by these resistant strains were due to the intensive or inappropriate use of the antibiotics. Since the hospital environment is conducive to the development of these strains and because the hospitalized patients often have a lowered resistance, these infections are more common and are usually severe. Treatment and prevention of these infections are extremely difficult. It is best to use antibiotics only when needed, to apply them aggressively, and to apply them appropriately.

Chapter 14 Host-Microbe Relationships and Disease Processes

INTRODUCTION

Have you ever seen the great American musical Carousel? By the end of this chapter you will know that the lyrics to a song in this show are true: "You'll never walk alone". On your body and in your body there are more prokaryotic cells than you have eukaryotic cells. In a previous chapter, you may recall that we noted that there are more worm infections than there are people on Earth. We do more than passively live together. We form relationships. As in most relationships, all partners benefit. The normal bacterial residents of our skin prevent potentially harmful microbes from taking root and causing disease. The normal flora of our intestine helps to keep us healthy and prevent problems like diarrhea. The author of this book eats yogurt every day because I am more than a teacher and writer of microbiology, I am also a consumer. I hope that the beneficial bacteria in the yogurt will develop into a longtime and mutual relationship with me.

Consider, most microbes are like stray dogs one might encounter in the street. Most stray dogs are likely to be nice and give you a lick and appreciate a pet on the head. However, every once in a while, one may bite off a piece of your head. So there is a dark side to microbes. This chapter will consider the microbial relationships that go rogue, the biology of infectious diseases or pathology.

STUDY OUTLINE

I. Host-Microbe Relationships

A. Symbiosis

Symbiosis means "living together" and refers to an association between two (or more) species. Examples of symbiotic relationships include *mutualism*, *commensalism*, and *parasitism*.

B. Contamination, infection, and disease

Contamination means that microorganisms are present. *Infection* refers to the multiplication of any parasitic organism within or upon the host's body. *Disease* is a disturbance in the state of health wherein the body cannot carry out all its normal functions.

C. Pathogens, pathogenicity, and virulence

A *pathogen* is any organism capable of causing disease in its host. *Pathogenicity* is the capacity to produce disease. *Virulence* is the degree of intensity of the disease produced by a pathogen.

D. Normal (indigenous) microflora

Organisms that live on or in the body but do not cause disease are referred to collectively as *normal microflora* or *normal flora*. Two categories of organisms can be distinguished: *resident microflora* (microbes that are always present on or in the human body) and *transient microflora* (microorganisms that can be present under certain conditions in any of the locations where resident microflora are found). These normal microbes can cause disease when: (1) the host's normal defenses fail; (2) the organisms are introduced into unusual body sites; or (3) the normal microflora is disturbed.

II. Koch's Postulates

Koch's four postulates are: (1) the specific causative agent must be observed in every case of a disease; (2) the agent must be isolated from a diseased host and must be grown in pure culture; (3) when the agent from the pure culture is inoculated into healthy, but susceptible, experimental hosts, the agent must cause the same disease; and (4) the agent must be reisolated from the inoculated, diseased experimental host and identified as identical to the original specific causative agent.

III. Kinds of Disease

A. Infectious versus noninfectious diseases

Infectious diseases are diseases caused by infectious agents such as bacteria, viruses, fungi, protozoa, and helminths. *Noninfectious diseases* are caused by any factor other than infectious organisms.

B. Classification of diseases

Diseases can be classified according to the following scheme: (1) inherited diseases; (2) congenital diseases; (3) degenerative diseases; (4) nutritional deficiency diseases; (5) endocrine diseases; (6) mental diseases; (7) immunological diseases; (8) neoplastic diseases; (9) iatrogenic diseases; and (10) idiopathic diseases.

C. Communicable versus noncommunicable diseases

Communicable infectious diseases can be spread from one host to another. *Noncommunicable infectious diseases* are not spread from one host to another.

IV. The Disease Process

A. How microbes cause disease

Bacterial pathogens often have special structures or physiological characteristics that improve the chances of successful host invasion and infection. Examples include *virulence factors, adhesins, hyaluronidase, coagulase, streptokinase, endotoxins,* and *exotoxins.* Viral pathogens can cause observable *cytopathic effects (CPE)* as they replicate after attaching to and penetrating specific host cells. Viral infections can be *productive, abortive, latent,* or *persistant.* Most fungal diseases result from fungal spores that are inhaled or from fungal cells and/or spores that enter cells through a cut or wound. Pathogenic protozoans and helminths can cause human disease in several ways. Most helminthes are extracellular parasites that inhabit the intestines or other body tissues.

B. Signs, symptoms, and syndromes

A *sign* is a characteristic of a disease that can be observed by examining the patient. A *symptom* is a characteristic of a disease that can be observed or felt only by the patient. A *syndrome* is a combination of signs and symptoms that occur together and are indicative of a particular disease or abnormal condition.

C. Types of infectious diseases

Several important terms are used to describe infectious diseases, including: (1) acute disease; (2) chronic disease; (3) subacute disease; (4) latent disease; (5) local infection; (6) focal infection; (7) systemic infection; (8) septicemia; (9) bacteremia; (10) viremia; (11) toxemia; (12) sapremia; (13) primary infection; (14) secondary infection; (15) superinfection; (16) mixed infection; and (17) inapparent infection.

D. Stages of an infectious disease

Most diseases caused by infectious agents have a fairly standard course, or series of stages. These stages include: (1) the *incubation period;* (2) the *prodromal phase;* (3) the *illness phase* (which includes the *acme*); (4) the *decline phase;* and (5) the *convalescence period.*

SELF-TESTS

Continue with this section only after you have read Chapter 14 of your textbook. Write your answers in the appropriate space provided. Correct answers to all questions can be found at the end of the Self-Test section. A score of 80 percent or better is good. If your score is less than 65, reread the chapter.

True/False

Mark T for True, F for False

______ **1.** The most successful parasites are those that successfully invade a host and quickly kill it.

______ **2.** Koch's postulates can be easily applied to all bacteria and even viruses since they can be cultivated artificially in the laboratory.

______ **3.** Some pathogens such as *Giardia* possess substances called adhesive disks that allow them to attach to specific tissue of the body.

______ **4.** Endotoxins, although poisonous, never cause the death of a host; however, exotoxins can.

______ **5.** Viruses that cause noncytopathic effects generally form inclusion bodies within cells before the cell is killed.

______ **6.** A latent disease may be characterized by periods of inactivity either before or after an attack.

______ **7.** Certain bacteria lyse blood cells because the cells contain magnesium which is essential for microbial replication.

Multiple Choice

Select the best possible answer.

______ **8.** An exotoxin can be characterized by all the following except that it:
a. is found usually with Gram-positive and a few Gram-negative bacteria.
b. is can easily be converted to a toxoid.
c. is relatively unstable.
d. causes a rapid rise in fever.

______ **9.** If a hospitalized patient obtained an infection from a nurse because of poor hand washing it could be:
a. an intoxication.
b. an idiopathic disease.
c. an endogenous disease.
d. an infestation.
e. None of the above are correct.

______ **10.** A substance released by a microbe that can cause it to spread throughout tissues would be:
a. hyaluronidase.
b. coagulase.
c. leukocidins.
d. hemolysins.

______ **11.** All of the following are true except:
a. Any parasite that can cause disease is a pathogen.
b. The degree of intensity of a disease is known as virulence.
c. Attenuation means that the disease-producing ability of the organism has dramatically increased.
d. Increasing animal passage can increase the virulence of a pathogen.

Matching

Select the answer from the right-hand side that corresponds to the term or phrase on the left-hand side of the page. An answer may be used more than once. In some cases, more than one answer may be required.

Topic: Host-Microbe Relationships

______ **12.** Parasitism

______ **13.** Host

______ **14.** Symbiosis

a. an organism harboring another organism

b. a symbiotic relationship in which one member benefits and the other is neither harmed nor benefited

______ **15.** Commensalism

______ **16.** Mutalism

c. both members of a symbiotic association benefit

d. a symbiotic relationship in which one member is harmed and the other benefits

e. the living together of two forms of life

Topic: Contamination, Infection, and Disease

______ **17.** Infection

______ **18.** Infestation

______ **19.** Contamination

______ **20.** Resident microbiota

______ **21.** Disease

______ **22.** Virulence

______ **23.** Transient microbiota

______ **24.** Pathogenicity

______ **25.** Opportunists

______ **26.** Attenuation

a. the presence of an arthropod on the body

b. a disturbance of normal body functions

c. multiplication of a pathogen in a host's tissues

d. a pathogen's capacity to cause disease

e. microbes always present in a host

f. the presence of microbes on inanimate objects

g. the intensity of a disease produced by a pathogen

h. temporary members of the normal microbiota

i. organisms that take advantage of a host's lowered resistance

j. weakening of a pathogen's disease-producing ability

Topic: Disease Categories

______ **27.** Communicable (contagious)

______ **28.** Congenital

______ **29.** Degenerative

______ **30.** Endocrine

______ **31.** Iatrogenic

______ **32.** Idiopathic

______ **33.** Immunological

______ **34.** Inherited

______ **35.** Mental

______ **36.** Neoplastic

______ **37.** Noninfectious

______ **38.** Nosocomial

______ **39.** Nutritional

a. caused by medical procedures and/or treatments

b. result from excesses or deficiencies of hormones

c. refer to diseases that can be spread from one host to another

d. result from structural and functional defects present at birth, caused by drugs, excessive X-ray exposure, or certain infections

e. disorders that develop in one or more body systems as aging occurs

f. caused by malfunction of the host's immune system

g. result from unknown causes

h. caused by a variety of factors, including those of a emotional or psychogenic nature, and consequences of certain infections

i. caused by errors in genetic information

j. result in abnormal growths

k. hospital acquired

l. vitamin deficiency

m. caused by other than infectious disease agents

Topic: Direct Actions of Bacteria

______ **40.** Speading factor

______ **41.** Hyaluronidase

______ **42.** Endotoxins

a. enzyme that digests the gluelike substance holding cells together

b. protein used by bacteria to attach to host cell's surface

______ 43. Adhesins

______ 44. Lipopolysaccharide complexes

______ 45. Streptokinase

______ 46. Coagulase

______ 47. Hemolysins

______ 48. Leukostatin

______ 49. Enterotoxins

c. bacterial enzyme that accelerates blood plasma coagulation

d. bacterial enzyme that dissolves blood clots

e. associated with gram-negative bacterial cell walls

f. blood lysing enzyme

g. act on intestinal tissues

h. interfere with leukocyte activities

Topic: Properties of Endotoxins and Exotoxins

______ 50. Mostly polypeptides

______ 51. Lipopolysaccharide complex

______ 52. Highly specific

______ 53. Mostly produced by gram-negative bacteria

______ 54. Can be converted to toxoid

______ 55. Botulism

______ 56. Tularemia

______ 57. Tetanus

______ 58. Staphylococcal food poisoning

a. associated with endotoxin

b. associated with exotoxin

Topic: Exotoxin Production and Disease States

______ 59. *Streptococcus pyogenes*

______ 60. *Clostridium botulinum*

______ 61. *Bacillus anthracis*

______ 62. *Clostridium perfringens*

______ 63. *Escherichia coli*

______ 64. *Vibrio cholerae*

______ 65. *Corynebacterium diphtheriae*

______ 66. *Shigella dysenteriae*

______ 67. *Staphylococcus aureus*

______ 68. *Clostridium tetani*

a. anthrax cytotoxin

b. cholera enterotoxin

c. erythrogenic toxin

d. traveler's diarrhea

e. botulism

f. gas gangrene

g. bacillary dysentery

h. diphtheria

i. lockjaw

j. scalded skin syndrome

Topic: Viruses and Infections

______ 69. Latent infection

______ 70. Abortive infection

______ 71. Persistent infection

______ 72. Productive infection

a. production of infectious viruses

b. inactive viruses

c. viruses unable to express their genetic properties

d. continued virus production

Topic: Terms Used to Describe Infections

______	73. Mixed infection	a. presence of toxins in the blood
______	74. Septicemia	b. full set of signs and symptoms do not occur
______	75. Inapparent infection	c. an infection following a primary infection
______	76. Systemic infection	d. presence and multiplication of pathogens in the blood
______	77. Chronic disease	e. an infection involving the entire body
______	78. Toxemia	f. a rapidly developing disease
______	79. Acute disease	g. a slowly disappearing disease
______	80. Secondary infection	h. infection caused by two or more pathogens
______	81. Focal infection	i. an infection in confined body areas from which pathogens move to other parts of the body

Topic: Stages of an Infectious Disease

______	82. Prodromal phase	a. time between infection and appearance of signs and symptoms
______	83. Incubation period	b. short period during which nonspecific mild symptoms appear
______	84. Decline phase	c. typical signs and symptoms appear
______	85. Convalescence period	d. time when signs and symptoms reach their greatest intensity
______	86. Illness phase	e. symptoms begin to subside
______	87. Acme	f. body repairs itself

Fill-Ins

Provide the correct term or phrase for the following. Spelling counts.

88. Any parasite capable of causing disease in its host is called a ________________.
89. Any organism that harbors another organism is called a ________________.
90. If two species are living together and both are benefiting from the relationship, the relationship would be an example of ________________.
91. A symbiotic relationship in which one member benefits while the other is not harmed is ________________.
92. If a parasitic organism multiplies in an abnormal location or to abnormal numbers in your body, you have an ________________.
93. A disease organism's ________________ is a measure of its intensity.
94. A pathogen's virulence can be decreased by ________________.
95. The organisms that normally reside on or in a host may be called ________________ ________________ or ________________.
96. Organisms present on or in the body for temporary periods of time are probably ________________.
97. Organisms that cause disease by taking advantage of a reduction in the host's resistance are called ________________.

98. Diseases caused by any factors other than infectious organisms are ________________ diseases.

99. Diseases due to errors in genetic information are called ________________ diseases.

100. Diseases caused by medical treatments are ________________ diseases.

101. Diseases with unknown causes are ________________ diseases.

102. Diseases that are acquired from the environment and not spread from one host to another are ________________ ________________ diseases.

103. Substances that allow bacteria to attach to host cells are called ________________.

104. An enzyme produced by certain bacteria that accelerates the coagulation of blood is a ________________.

105. Soluble, poisonous substances that are secreted by bacteria into host tissues are called ________________.

106. A lipopolysaccharide toxin found mostly with Gram-negative organisms is an ________________.

107. Hemolysins that completely lyse red blood cells are ________________ hemolysins.

108. Enzymes produced by certain bacteria that can kill white blood cells are ________________.

109. Diseases caused by the ingestion of toxins accumulating in a product are ________________ diseases.

110. Toxins that act directly on tissues of the gut are called ________________.

111. Viruses that do not actually kill host cells may be called ________________ viruses.

112. An ________________ infection occurs when a virus invades a cell but is unable to make infectious progeny.

113. A disease characterized by periods of inactivity is a ________________ disease.

114. When a disease causes aftereffects, the condition is known as ________________.

115. A ________________ disease develops slowly and persists for long periods of time.

116. An infection confined to a specific area but from which pathogens can spread is a ________________ infection.

117. Pathogens that are present in the blood and are multiplying characterize a ________________.

118. An infection that fails to produce symptoms can be an ________________ or a ________________ infection.

119. The disease period during which nonspecific symptoms appear such as weakness and fever is the ________________ period.

120. When a disease reaches a period of most intense symptoms it is in the ________________ stage.

121. ________________ are substances produced by bacteria that cause fever.

122. The recovery period of a disease is known as the ________________ period.

Labeling

Questions 123–126: Select the best possible label for each step in the demonstration of Koch's postulates: (a) susceptible organisms are isolated from infected animal and grown in pure culture; (b) one animal sickens and may die. Pure cultures are isolated; (c) organisms isolated are injected into healthy, susceptible animals; (d) isolated pure culture specimens are the same.

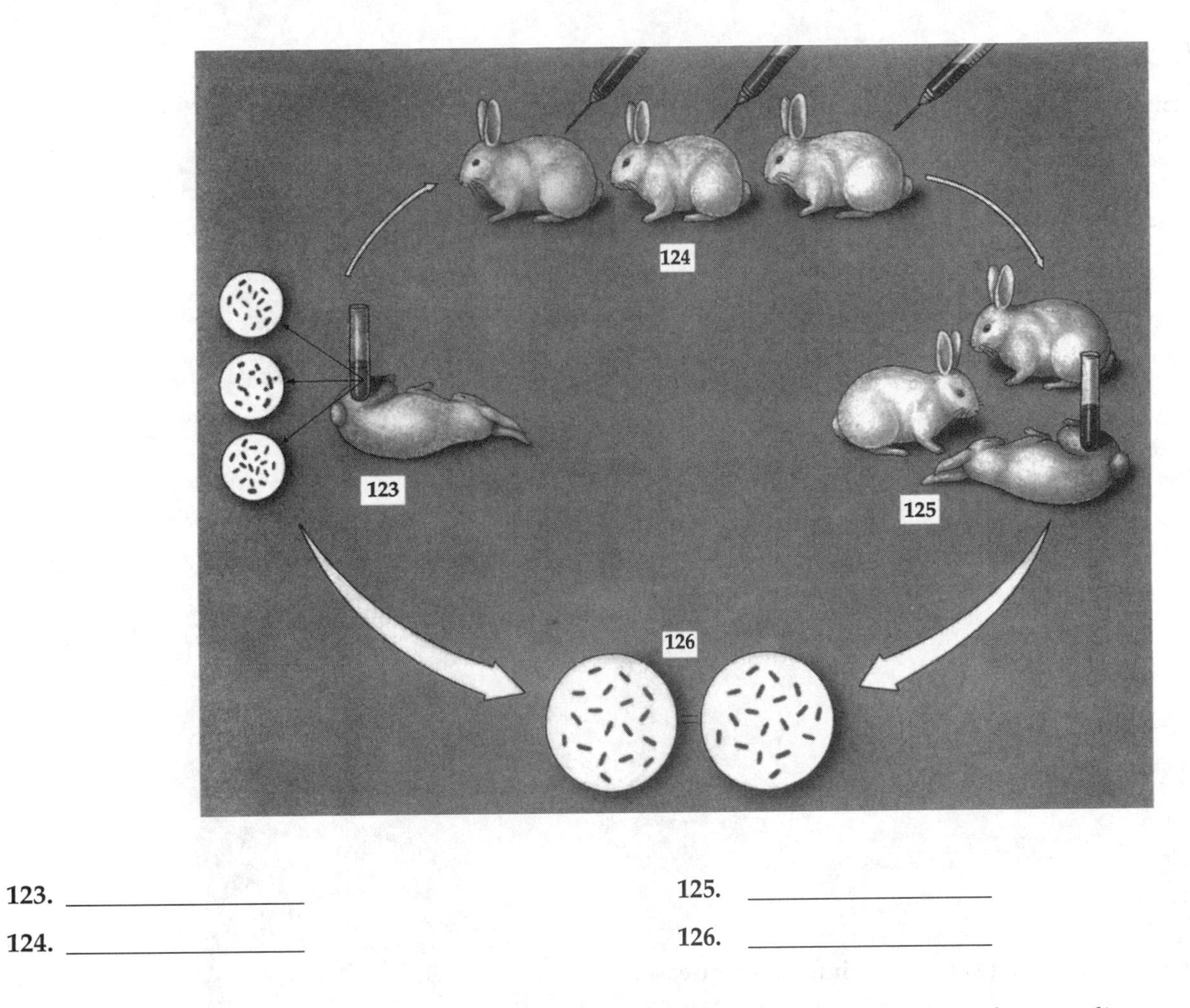

123. ____________________

124. ____________________

125. ____________________

126. ____________________

Questions 127–132: Select the best possible label for each stage in the course of an infectious disease: decline phase; illness phase; incubation period; prodromal phase; acme; convalescence period.

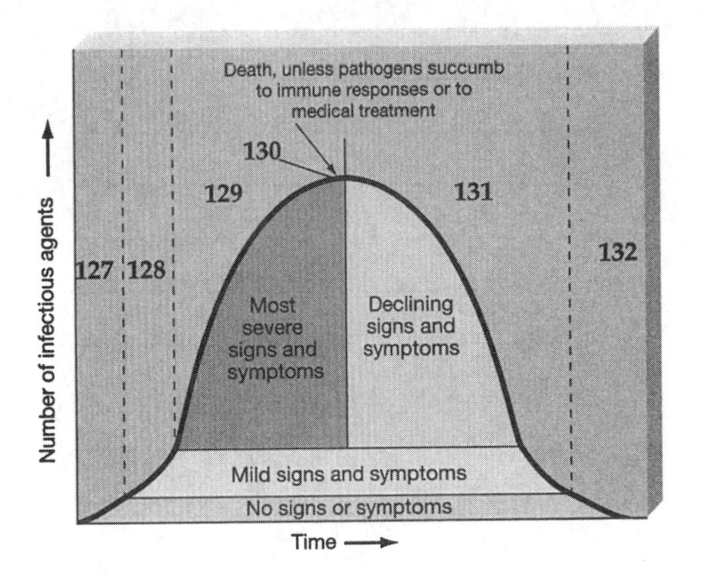

127. ____________________

128. ____________________

129. ____________________

130. ____________________

131. ____________________

132. ____________________

Critical Thinking

133. What terms are used to define host-microbe relationships, and what do they mean?

134. How do Koch's postulates relate to infectious disease?

135. What are the major differences between infectious and noninfectious diseases, between communicable and noncommunicable infectious diseases, and between exogenus and endogenus diseases?

136. How do microbes cause disease?

137. What are the meanings of terms used to describe diseases?

138. What stages occur in the course of an infectious disease?

ANSWERS

True/False

1. **False** A successful parasite is one that can ensure the continued survival of its host without causing too much adverse damage.
2. **False** Viruses cannot be artificially cultured in the laboratory. Furthermore, since they must be grown inside living tissue, it is difficult if not impossible to apply the postulates directly since tissues may harbor other latent viruses that were not detected.
3. **True** Some parasites use pili for attachment (*Neisseria gonorrhoeae*) and others use adhesive disks. The reaction allows them to obtain tissue fluids readily.
4. **False** Endotoxins are weaker than exotoxins; however, some such as found with organisms of typhoid fever and plague can easily kill in relatively large doses.
5. **False** Cytopathic viruses are ones that cause observable changes in cells. They can be cytocidal when the virus kills the cell or noncytocidal if they do not. Noncytopathic means that the virus causes no observable changes.
6. **True** Herpes viruses often cause latent infections because they may not exhibit any symptoms for long periods of time prior to an outbreak.
7. **False** Although bacteria such as streptococci and staphylococci lyse blood cells, they do so in order to obtain iron, which is necessary for their metabolism.)

Multiple Choice

8. d; 9. e; 10. a; 11. c.

Matching

12. d; 13. a; 14. e; 15. b; 16. c; 17. c; 18. a; 19. f; 20. e; 21. b; 22. g; 23. h; 24. d; 25. i; 26. j; 27. c; 28. d; 29. e; 30. b; 31. a; 32. g; 33. f; 34. i; 35. h; 36. j; 37. l; 38. k; 39. l, m; 40. a; 41. a; 42. e; 43. b; 44. e; 45. d; 46. c; 47. f; 48. h; 49. g; 50. b; 51. a; 52. b; 53. a; 54. b; 55. b; 56. a; 57. b; 58. b; 59. c; 60. e; 61. a; 62. f; 63. d; 64. b; 65. h; 66. g; 67. j; 68. i; 69. b; 70. c; 71. d; 72. a; 73. h; 74. d; 75. b; 76. e; 77. g; 78. a; 79. f; 80. c; 81. i; 82. b; 83. a; 84. e; 85. f; 86. c; 87. d.

Fill-Ins

88. pathogen; 89. host; 90. mutualism; 91. commensalism; 92. infection; 93. virulence; 94. attenuation; 95. normal flora, resident; 96. transients; 97. opportunists; 98. noninfectious; 99. inherited; 100. iatrogenic; 101. idiopathic; 102. noncommunicable infectious; 103. adhesions; 104. coagulase; 105. exotoxins; 106. endotoxin; 107. beta; 108. leukocidins; 109. intoxications; 110. enterotoxins; 111. noncytocidal; 112. abortive; 113. latent; 114. sequelae; 115. chronic; 116. focal; 117. septicemia; 118. inapparent, subclinical; 119. prodromal; 120. acme; 121. pyrogens; 122. convalescence.

Labeling

123. a; 124. c; 125. b; 126. d.; 127. incubation period; 128. prodromal phase; 129. illness phase; 130. acme; 131. decline phase; 132. convalescence period.

Critical Thinking

133. **What terms are used to define host-microbe relationships and what do they mean?**
Microorganisms exhibit a variety of complex relationships to other microbes and to the larger organisms that serve as hosts for them. A host is any organism such as a human or an animal that harbors another organism.
Symbiosis refers to any interactions between two organisms; mutualism is where both organisms benefit by the relationship; commensalism is where one organism benefits and the other is not affected; and parasitism is where one organism benefits and the other is harmed.

Various terms are used to illustrate the conditions in which microorganisms have increasingly significant effects on their hosts. Contamination refers to the mere presence of microbes, infection refers to the invasion of the host by the microbes (however, this invasion may go unnoticed, i.e., inapparent infection), and disease refers to the development of symptoms that indicate a disturbance in the health of the host.

A pathogen is a parasite that is capable of causing disease, pathogenicity is the capacity to produce disease, and virulence is the degree of intensity of a disease.

Normal flora represent organisms found normally on or in another organism. Resident flora are always present while transient flora are present under certain conditions. Opportunists are either resident or transient flora that can cause disease when the host has been compromised or weakened. Notice that *Staphylococcus* species are found in every body area. These organisms are very common endogenous infectious agents, and for good reason!

134. How do Koch's postulates relate to infectious disease?
Koch's postulates are used to provide a link between an infectious organism and a disease. When all four of the postulates are met, an organism has been proven to be the causative agent of the disease. Note that the most important part of the postulates is the ability to pure-culture the organisms. Until that is done, there is no way to be absolutely sure that it is the causative agent.

135. What are the major differences between infectious and noninfectious disease, between communicable and noncommunicable infectious diseases, and between exogenous and endogenous diseases?
Infectious diseases are caused by infectious agents such as bacteria and viruses, while noninfectious diseases are caused by other factors such as genetic defects, nutritional deficiency defects, etc. Note that many infectious agents often interact with other factors to cause disease.

Communicable diseases such as diphtheria and whooping cough can be spread from one host to another, while noncommunicable infectious diseases such as tetanus cannot be spread from host to host and usually are acquired from the soil or water.

Exogenous diseases such as tuberculosis and plague are those caused by pathogens or other factors from outside the body, while endogenous diseases such as a urinary tract infection are caused by pathogens found on or in the body.

136. How do microbes cause disease?
Microorganisms act in certain ways that allow them to cause disease. They can cause disease by adherence to a host, by colonization and/or invasion of the host tissues, and by invading the host cells.

Many organisms produce toxins ,which are synthesized inside the cell. Exotoxins are released from the pathogen, are generally very specific, and are highly toxic. Endotoxins are part of the outer portion of the cell wall of bacteria and are released only when the cell dies. Many of the organisms that release exotoxins are Gram-positive whereas those that possess endotoxins are usually Gram-negative.

Bacteria can release other substances such as hemolysins, which lyse red blood cells, leukocidins, which kill white blood cells, hyaluronidases, which allow them to spread throughout tissues, coagulases, which form blood clots, and fibrinolysins, which dissolve fibrin blood clots. It is important to remember that it is not the number of toxins or virulence factors but the type of factors that determines the severity of the microbial attack.

Viruses damage cells and produce a cytopathic effect such as altering the DNA, forming inclusion bodies, or affecting the cell's protein synthesis machinery. The effect is cytocidal if the cell dies and noncytocidal if it does not.

Fungi can progressively digest cells and produce damaging toxins. Protozoa and helminths damage tissues by ingesting cells and tissue fluids, releasing toxic waste products, and by causing allergic reactions.

137. What are the meanings of terms used to describe diseases?
A sign is an observable effect of a disease and a symptom is an effect of a disease that is reported by the patient. A syndrome is a group of signs and symptoms that occur together. Since signs and symptoms are often similar with many diseases, this information is extremely valuable to a physician to determine the exact nature of the patient's illness.

138. What steps occur in the course of an infectious disease?
Most diseases caused by infectious agents have a fairly standard course or stages that can be monitored. These stages include incubation, prodromal, invasive, acme, decline, and convalescent. Even with treatment, the disease still passes through all of these stages.

The incubation period is the time between infection and the appearance of the signs and symptoms of the disease. Note that some such as cholera and diphtheria are short whereas others such as rabies and hepatitis B are fairly long.

The prodromal phase is the period during which the pathogens begin to invade the tissues, which is marked by early nonspecific symptoms such as a cough or sneeze.

The invasive phase is the period during which pathogens invade and significantly damage or affect tissues. This phase is marked by the most classic signs and symptoms of the disease.

The acme, or critical stage, is the period of most intense symptoms.

The decline phase is the period during which the host defenses overcome the pathogens, the signs and symptoms subside, and the body's immune defenses have surmounted the infectious agent.

The convalescent phase is the period when tissue damage is repaired and the patient begins to regain strength. Sometimes during this phase or shortly thereafter, latent effects of the disease called sequelae appear. These could include cranial nerve damage, heart valve damage, hearing loss, and others.

Chapter 15 Epidemiology and Nosocomial Infections

INTRODUCTION

A few years ago, I walked toward my car in the yard and saw an unusual sight. There was a dead bird in front of it. While I am sure that birds die every day, it is not often that I see one. I was in a hurry and began to drive down the highway. I was approaching the underpass to a stone bridge and noted a dark colored bird flying in the same direction approaching the bridge. The bird struck the wall and I noted a puff of feathers seeming to emerge from the wall. These observations were made soon after an epidemic of West Nile Virus began in New England. This disease was first noted in 1999 in North America and is transmitted to humans by the bite of mosquitoes and involves birds as a reservoir for the virus. It is possible that the birds I observed were infected with the West Nile Virus. The disease has spread into all 48 continental states, much of Canada, Mexico, and a number of islands in the Caribbean. Clearly, new diseases are emerging and spreading. The study of the distribution and spread of disease is the province of epidemiology. This is a branch of microbiology concerned with public health.

This chapter will consider three types of diseases: infectious diseases that occurs naturally, infectious diseases that are acquired in medical facilities like hospitals, and infections that can be spread as acts of bioterror. In a world of emerging viruses, antibiotic resistant bacteria, and very nasty people who may have access to lethal organisms, epidemiology is more important than ever to protect public health. There is a reason that during the 2004 Democratic National Convention, various devices were constantly sniffing the air for evidence of bioterror.

Epidemiology also makes us aware of profound historic changes. For example, from 1900 to the year 2000, America changed dramatically with respect to the major causes of disease. We began the 20th century with pneumonia, tuberculosis, and diarrheal diseases as major causes of death in the United States. We ended the 20th century with heart disease, cancer, and stroke as major causes of death. Epidemiology is an essential tool to prevent infectious disease from once again becoming the major cause of death in the United States.

STUDY OUTLINE

I. Epidemiology

A. What is epidemiology?

Epidemiology is the study of factors and mechanisms involved in the frequency and spread of diseases and other health-related problems within populations of humans, other animals, or plants. *Epidemiologists* are concerned with *incidence rates, prevalence rates, morbidity frequencies,* and *mortality frequencies.*

B. Diseases in populations
When studying the frequency of diseases in populations, epidemiologists must consider the geographic areas affected and the degree of harm the diseases cause the population. On the basis of their findings, they classify diseases as *endemic, epidemic, pandemic,* or *sporadic.*

C. Epidemiologic studies
Collecting frequency data and drawing conclusions is the foundation of any *epidemiologic study.* These studies can be *descriptive, analytical,* or *experimental.*

D. Reservoirs of infection
A *reservoir of infection* is a site where microorganisms can persist and maintain their ability to infect. Examples are humans, other animals (including insects), plants, and certain nonliving materials (like water and soil).

E. Portals of entry
A *portal of entry* is a site at which microorganisms can gain access to body tissues. Examples include the skin and the mucous membranes of the digestive, respiratory, and urogenital systems.

F. Portals of exit
A *portal of exit* is a site at which microorganisms can leave the body. Generally, pathogens exit with body fluids or feces.

G. Modes of disease transmission
Diseases can be transmitted in many ways. The three major categories described by the book are: (1) contact transmission; (2) transmission by vehicles (including water, air, and food); and (3) transmission by vectors (including mechanical vectors and biological vectors).

H. Disease cycles
Many diseases occur in cycles. For years or even decades, only a few cases are seen, but then many cases suddenly appear in epidemic or pandemic proportions. Epidemiologists still cannot predict when a cyclic disease will break out and reach epidemic proportions.

I. Herd immunity
Herd immunity is the proportion of individuals in a population who are immune to a particular disease. If herd immunity is high, then the disease can spread only among the small number of susceptible individuals in the population, and the likelihood of disease transmission is small. Thus, a sufficiently high herd immunity protects the entire population.

J. Controlling disease transmission
Several methods are currently available for full or partial control of communicable diseases. They include *isolation, quarantine, immunization,* and *vector control.*

K. Public health organizations
Public health agencies are concerned with controlling infectious diseases and reducing other health hazards in their countries. Examples are the *Centers for Disease Control and Prevention (CDC)* in Atlanta, Georgia, and the *World Health Organization (WHO)* based in Geneva, Switzerland.

L. Notifiable diseases
A *notifiable disease* is a disease that is potentially harmful to the public's health and must be reported by physicians to public health officials.

II. Nosocomial Infections

A. General features
A *nosocomial infection* is an infection acquired in a hospital or other medical facility. Among patients admitted to American hospitals each year, about 2 million (10 percent) acquire a nosocomial infection that increases the risk of death, the duration of the hospital stay, and the cost of treatment.

B. The epidemiology of nosocomial infections
Like the epidemiology of diseases acquired in the community, the epidemiology of nosocomial infections considers sources of infection, modes of transmission, susceptibility to infection, and prevention and control.

C. Preventing and controlling nosocomial infections
Several techniques are available to prevent the introduction and spread of nosocomial infections, including hand washing, the use of gloves, the use of pesticides, and the surveillance of antibiotic use.

III. Bioterrorism

A. Early History
Ancient warriors used infected corpses as weapons. Colonial Americans used blankets containing smallpox to kill native peoples.

B. During World War II the United States, England, and Russia developed anthrax weapons.
C. In 2000–2001 terrorist antrax attack struck the United States.
D. Fear of small pox attacks or attacks to our food supply remain.

SELF-TESTS

Continue with this section only after you have read Chapter 15 of your textbook. Write your answers in the appropriate space provided. Correct answers to all questions can be found at the end of the Self-Test section. A score of 80 percent or better is good. If your score is less than 65, reread the chapter.

True/False

Mark T for True, F for False

_____ **1.** Fortunately, surgical masks, especially the cotton variety, are capable of eliminating the aerosol spread of all microbes during surgery.

_____ **2.** WHO is a health organization that publishes a weekly newsletter called the Morbidity Mortality Weekly Report (MMWR).

_____ **3.** Hand washing and the use of aseptic technique is required by all health personnel when working with patients placed in every type of isolation used in a hospital.

_____ **4.** Endogenous infections are often caused by opportunists from the patient's own normal flora.

_____ **5.** Generally speaking for disease agents, the portal of exit is usually the portal of entrance.

_____ **6.** The most common body site affected by nosocomial infectious agents is the gastrointestinal tract.

Multiple Choice

Select the best possible answer.

_____ **7.** Select the most common pair of nosocomial infectious agents.
a. *Escherichia coli, Proteus.*
b. *Staphylococcus aureus, Klebsiella.*
c. *Escherichia coli, Staphylococcus aureus.*
d. *Proteus, Pseudomonas.*

_____ **8.** With respect to nosocomial infections:
a. Catheterization rarely represents a site of infection because of the use of sterile disposable items.
b. Proper hand washing can ensure sterility and therefore eliminate this procedure as a source of infection.
c. The most common site of nosocomial infections is the skin.
d. The extensive use of antibiotics has greatly increased the chances of nosocomial infections.
e. All of the above are false.

_____ **9.** Transmission of infectious diseases can occur via all the following except:
a. indirect contact via droplets.
b. fomites.
c. mechanical vector transmission
d. airborne via dust particles.
e. All of the above are methods of transmission.

_____ **10.** A disease can be classified as a pandemic when:
a. it reaches worldwide distribution.
b. the mortality rate is very high.
c. it always occurs every year.
d. the disease affects the nervous system?

_____ **11.** Which of the following statements is not true concerning universal precautions.
a. These precautions were the result of concerns about the AIDS virus.

b. The precautions do not apply to feces, sputum, and urine unless they contain visible evidence of blood.
c. The precautions do apply to blood, semen, vaginal fluids, sweat, and tears regardless whether visible blood is present or not.
d. Universal precautions apply to all patients, not just those that are infected with AIDS.

Matching

Select the answer from the right-hand side that corresponds to the term or phrase on the left-hand side of the page. An answer may be used more than once. In some cases, more than one answer may be required.

Topic: Epidemiology and Diseases in Populations

______ **12.** The number of deaths caused by a specific disease as a proportion of the population
______ **13.** The number of new cases of a disease in a specific period
______ **14.** The number of disease cases as a proportion of the population
______ **15.** A disease continually present in low numbers in a geographic area
______ **16.** A worldwide epidemic
______ **17.** A higher -than normal incidence of disease suddenly in a population
______ **18.** Several isolated cases that do not pose a threat to the population
______ **19.** An epidemic arising from contact with contaminated substances
______ **20.** Direct person-to-person contact resulting in an epidemic

a. sporadic
b. epidemic
c. endemic
d. pandemic
e. prevalence rate
f. mortality rate
g. morbidity rate
h. incidence rate
i. common-source outbreak

Topic: Zoonoses and Modes of Transmission

______ **21.** Brucellosis
______ **22.** Leptospirosis
______ **23.** Ringworm
______ **24.** Anthrax
______ **25.** Bubonic plague
______ **26.** Rocky Mountain spotted fever
______ **27.** Relapsing fever
______ **28.** African sleeping sickness
______ **29.** Equine encephalitis
______ **30.** Rabies
______ **31.** Tapeworm

a. inhalation of spores
b. ingestion of milk from infected animals
c. fleas
d. direct contact with urine from an infected animal
e. ticks and/or lice
f. mosquitoes
g. bites of infected animals
h. tsetse flies
i. ingestion of cysts in meat
j. direct contact

Topic: Modes of Disease Transmission

______ **32.** Shaking hands
______ **33.** Fomites
______ **34.** Eating utensils
______ **35.** Bar of soap
______ **36.** Kissing

a. direct contact
b. indirect contact

Fill-Ins

Provide the correct term or phrase for the following. Spelling counts.

37. The study of the factors and mechanisms involved in the spread of diseases within a population is known as ________________.

38. The assignment or study of the causes and origins of a disease is known as ________________.

39. The number of new cases seen in a specific period of time is called the ________________.

40. The number of people infected at any one time is the ________________ rate.

41. The number of cases in relation to the total number of people is the ________________ rate.

42. When a disease occurs at a constant rate in a particular geographic area, it is called an ________________ disease.

43. A worldwide epidemic is known as a ________________.

44. Occurrences of isolated cases that do not appear to pose a threat to the population are known as ________________ cases.

45. An epidemic that arises from contaminated food is a ________________ - ________________ epidemic.

46. A type of epidemiological study that lists the number of cases of a disease, the time period involved, and the population segments affected is a ________________ study.

47. The first case of a disease to be identified is often called the ________________ case.

48. Epidemiological studies designed to test a hypothesis are ________________ studies.

49. A nonmedicated substance given to a patient that has no effect on the patient but is perceived by the patient as a treatment is a ________________.

50. Rodents that can carry and maintain the plague act as ________________ of the disease.

51. A person who has recovered from a disease but continues to be a reservoir of infection for a long time is a ________________ ________________.

52. Diseases that can be transmitted from animals to humans are called ________________ diseases.

53. The sites at which organisms enter the body are the ________________ of ________________.

54. Direct transmission that occurs by shaking hands or kissing is also called ________________ transmission, while diseases that spread from parent to offspring are spread by ________________ transmission.

55. If diseases are transmitted as a result of inanimate objects, this is known as ________________ transmission.

56. Inanimate objects such as pencils, and bar soap act as ________________.

57. Living things, especially arthropods, can act to transmit diseases as ________________.

58. A ________________ ________________ can actively transmit pathogens that use the vector as part of their life cycles.

59. If a large proportion of the population is immune to a disease, the few susceptible individuals can be protected by ________________ or ________________ immunity.

60. A patient with a highly infectious disease such as pneumonic plague or diphtheria must be placed in ________________ isolation.

61. An individual who is separated from others due to a communicable disease is said to be in ________________.

62. Hospital-acquired infections are known as ________________ infections.

63. Infections arising from the patient's own normal flora bacteria are ________________ infections.

64. Special CDC guidelines that have been put into place to guard against the possibility of the AIDS virus being transmitted are known as ________________ ________________.

Labeling

Questions 65–72: Select the best possible label for each picture in the overview of disease transmission: airborne (including dust particles); droplets; direct contact; mechanical (on insect bodies); biological; indirect contact by fomites; waterborne; foodborne.

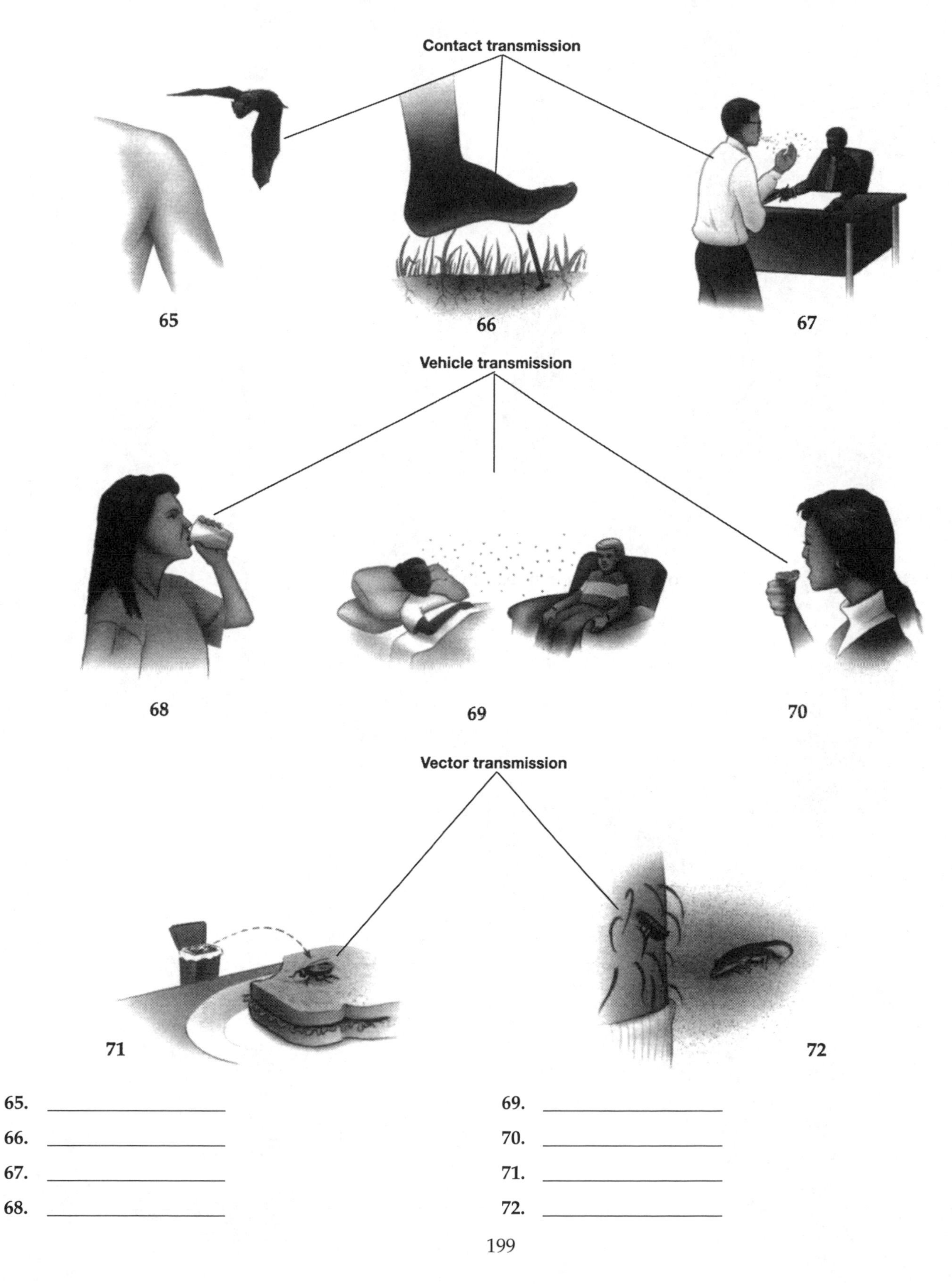

65. ____________________

66. ____________________

67. ____________________

68. ____________________

69. ____________________

70. ____________________

71. ____________________

72. ____________________

Critical Thinking

73. What is epidemiology, and what special terms are used by epidemiologists?

74. How are diseases classified according to their spread in populations?

75. What are the purposes and methods of epidemiologic studies?

76. How do various kinds of reservoirs of infection contribute to human disease?

77. What are the roles of portals of entry and exit and modes of transmission in the spread of human disease?

78. What is a disease cycle, and how is group immunity related to disease cycles?

79. What methods are used to control communicable disease?

80. How do the functions of organizations and the reporting of diseases contribute to public health?

81. What are nosocomial infections, and how are they studied epidemiologically?

82. How can nosocomial infections be prevented and controlled?

ANSWERS

True/False

1. **False** Surgical masks generally can impede the spread of microbes but will not prevent the spread of all droplets following a forceful sneeze. Cotton masks, since they can become moist, can reverberate and actually increase the spread.
2. **False** WHO stands for the World Health Organization. The MMWR is a weekly report published by the Centers for Disease Control in Atlanta, Georgia.
3. **True** Hand transmission is the most common method of transmission of infectious agents in the hospital. Proper hand washing and aseptic technique is mandatory for all types of patients regardless of whether they are under isolation or not.
4. **True** Endogenous organisms arise from one's own normal flora. Since we carry a number of infectious opportunistic agents such as *S. aureus* and *E. coli*, the likelihood of infection by these organisms once our resistance is lowered is very great.
5. **True** Organisms that gain access via the respiratory tract usually exit the same portal and ones affecting the gastrointestinal tract exit via the oral/fecal route.
6. **False** The most common site of infection is actually the urinary tract followed by surgical wounds.

Multiple Choice

7. c; 8. d; 9. e; 10. a; 11. c.

Matching

12. f; 13. h; 14. g; 15. c; 16. d; 17. b; 18. a; 19. i; 20. j; 21. b; 22. d; 23. j; 24. a; 25. c; 26. e; 27. e; 28. h; 29. f; 30. g; 31. i; 32. a; 33. b; 34. b; 35. b; 36. a.

Fill-ins

37. epidemiology; 38. etiology; 39. incidence; 40. prevalence; 41. morbidity; 42. endemic; 43. pandemic; 44. sporadic; 45. common-source; 46. descriptive; 47. index; 48. experimental; 49. placebo; 50. reservoir; 51. chronic carrier; 52. zoonoses; 53. portals, entry; 54. horizontal, vertical; 55. indirect; 56. fomites; 57. vectors; 58. biological vector; 59. herd, group; 60. strict; 61. quarantine; 62. nosocomial; 63. endogenous; 64. Universal Precautions.

Labeling

65. direct contact; 66. indirect contact by fomites; 67. droplets; 68. waterborne; 69. airborne (including dust particles); 70. foodborne; 71. mechanical (on insect bodies); 72. biological.

Critical Thinking

73. **What is epidemiology, and what special terms are used by epidemiologists?**
    Epidemiology is the study of factors and mechanisms involved in the spread of disease within a population. Scientists that study epidemiology are epidemiologists and are concerned with the etiology or cause and transmission of diseases in the population. Incidence rates represent statistics relative to the number of new cases in a specific time period; prevalence rates reflect the number of people infected at any one time; morbidity rates indicate the number of cases as a proportion of the population, and mortality rates indicate the number of deaths as a proportion of the population. These statistics are reported in state health department reports and in the Morbidity and Mortality Weekly Report (MMWR), which is published by the Centers for Disease Control (CDC).
74. **How are diseases categorized according to their spread in populations?**
    If a small number of isolated cases appear in a population, it is called a sporadic disease. If a large number of cases continually appear in a population but the harm to the patients is not too great, it is called an endemic disease. In epidemic diseases, a large number of cases suddenly appear in a population and patients are sufficiently harmed to create a public health problem. Common-source epidemics spread from a single

source such as food or water, and propagated epidemics are spread by person-to-person contact. A pandemic is an epidemic disease that has achieved exceptionally wide geographic distribution.

75. What are the purposes and methods of epidemiological studies?
Epidemiological studies help us to learn more about the spread of diseases in populations and how to control them. Descriptive studies note the number of cases of a disease, which segments of the population were affected, etc. Analytical studies focus on establishing cause and effect relationships in the occurrence of diseases in populations. Experimental studies are designed to test a hypothesis, often about the value of a particular treatment.

76. How do various kinds of reservoirs of infection contribute to human disease?
A reservoir is a site where organisms can persist and maintain their ability to infect. Reservoirs include humans, animals, and nonliving sources. In human reservoirs, carriers often transmit diseases passively, if they are unaware of the disease agent; actively, if they are recovering from the infection; or intermittently, if they release the pathogens periodically. Animal reservoirs can transmit diseases by direct contact or by means of insect vectors. These animal diseases are called zoonoses. Nonliving reservoirs include water, soil, or wastes.

77. What are the roles of portals of entry and exit and modes of transmission in the spread of human disease?
The portals of entry include all those openings into the body such as the eye, ear, nose, and throat. They also include the skin, mucous membrane linings, and even the placenta. Although the latter protects the fetus, some diseases such as syphilis, toxoplasmosis, listeriosis, rubella, and AIDS can cross this barrier.

The portals of exit are generally the same as the portals of entry. Organisms usually exit in body fluids or feces.

Modes of transmission include direct contact, vehicle, or vector. Transmission by carriers, sexual practices, and transmission of zoonoses pose special epidemiologic problems.

78. What is a disease cycle, and how is group immunity related to disease cycles?
Many diseases such as chicken pox and influenza occur in cycles. In many cases, the cycle relates to the resistance status of the population. If many are not resistant and the agent returns, the disease can recur. If a large group within the population is resistant it can provide group or herd immunity and resist the agent even though some in the population are susceptible.

79. What methods are used to control communicable diseases?
Methods of control include isolation, quarantine, active immunization, and vector control. Most hospitals have infection control committees whose responsibility is to set guidelines and policies for isolation procedures of patients. Quarantine is rarely used because of the availability of antibiotics. Active immunization is used to control many diseases including measles, mumps, rubella, and diphtheria. Vector control is effective where vectors can be identified and eradicated.

80. How do the functions of organizations and the reporting of diseases contribute to public health?
Public health organizations exist at city, county, state, federal, and world levels. They help to establish and maintain health standards, cooperate in the control of infectious diseases, collect and disseminate information, and assist with professional and public education. These organizations all contribute to the overall health and productivity of a community.

81. What are nosocomial infections and how are they studied epidemiologically?
Nosocomial infections are acquired in a hospital or other medical facility. They can be exogenous, arising from bacteria outside the patient, or endogenous, arising from bacteria associated with the patient. Most are caused by *Escherichia coli, Staphylococcus aureus, Streptococci,* and *Pseudomonas*. Many strains are antibiotic resistant. Host susceptibility is an important factor in the development of such infections.

82. How can nosocomial infections be prevented and controlled?
Most hospitals have extensive infection control programs to ensure that proper hand washing, use of gloves, scrupulous attention to maintaining sanitary conditions and sterility where possible, and surveillance of antibiotic use and other hospital procedures are followed. By following good common sense, proper handwashing procedures, and the universal precautions guidelines, most infections can be reduced dramatically

Chapter 16 Nonspecific Host Defenses and Host Systems

INTRODUCTION

Have you ever wondered why bacteria and fungi have cell walls? They build walls for the same reason that the ancient people of China built the Great Wall: protection. The world can be a nasty place and every organism including each one of us is subject to attack. Just as a common microbe builds a wall t keep out all possible invaders, each one of has a number of innate host defenses.

This chapter will review the array of defenses common to people and other vertebrates. It will explore all aspects our out immune system from the nonspecific physical defenses such as our skin, to chemical defenses, to cellular defenses. The chapter will also consider inflammation. This process is related to a number of health issues in the United States and the process itself can be responsible for medical difficulties. Consider, when people die of severe burns, it is usually not the secondary infections that kill them. It is an immune system that goes out of control in response. At present, we have few tools to modulate the immune system to save the burn victims. Once, I studied the disease trichinosis. Commonly, Americans acquire this disease by eating infected pork that is not properly cooked. If we place 14,000 infectious worm larvae into a normal mouse, it dies from hemorrhage in its intestine. On the other hand, if we place the same number of larvae into a nude mouse, that lacks part of its immune response, it does not die. In fact, one could increase the dose of larvae by a factor of ten and it can survive. Thus, as in burn victims, it is not infection that kills the host, but the inability of a normal immune response to deal with the insult. Clearly, a better understanding of the vertebrate immune response in the future will result in the saving of many lives.

STUDY OUTLINE

I. Nonspecific and Specific Host Defenses

Specific defenses are host defenses that operate in response to particular invading pathogens like viruses and bacteria. *Nonspecific defenses* are host defenses against pathogens that operate regardless of the invading agent. Examples include *physical barriers, chemical barriers, cellular defenses, inflammation, fever,* and *molecular defenses.*

II. Physical Barriers

The skin and mucous membranes protect the body and internal organs from injury and infectious agents, as do physical reflexes like coughing, sneezing, vomiting.

III. Cellular Defenses

A. Defensive cells

Cellular defense cells include the *granulocytes* and the *agranulocytes.*

B. Phagocytes

Phagocytes are cells that literally eat or engulf other materials. Examples of phagocytes include neutrophils and macrophages.

C. The process of phagocytosis

If an infection occurs, neutrophils and macrophages use a four-step process to destroy the invading microorganisms: (1) find; (2) adhere to; (3) ingest; and (4) digest the microorganisms.

D. Extracellular killing

Phagocytosis is an example of *intracellular killing*. Other microbes, such as viruses and parasitic worms, are destroyed *extracellularly*. Eosinophils and natural killer cells are examples of cells involved in extracellular killing.

E. The lymphatic system

The lymphatic system has three major functions: (1) it collects excess fluid from the spaces between body cells (*lymphatic circulation*); (2) it transports digested fats to the cardiovascular system; and (3) it provides many of the specific and nonspecific defense mechanisms against infection and disease. The process of *lymphatic circulation*, the specific *lymphatic organs*, and *other lymphoid tissues* are important to understand and know.

IV. Inflammation

A. Characteristics of inflammation

The *cardinal signs* or symptoms of inflammation are: (1) an increase in temperature; (2) redness; (3) swelling; and (4) pain at the infected or injured site.

B. The acute inflammatory process

In an infection, *acute inflammation* functions to: (1) kill invading microbes; (2) clear away tissue debris; and (3) repair injured tissue. This process relies on the actions of *histamine, bradykinin, prostaglandins* in causing *edema, leukocytosis,* and *diapedesis.*

C. Repair and regeneration

During the entire inflammatory reaction, the healing process is also underway. Healing depends on the actions of *fibroblasts*, which are part of the *granulation tissue* seen at the cut site. Several factors affect the healing process, including age, blood circulation, and vitamins.

D. Chronic inflammation

A *chronic inflammation* is an inflammation in which neither the host nor the agent of inflammation is the decisive winner of the battle. Such inflammations can last for years. This can lead to the formation of *granulomas* and *tubercles*.

V. Fever

A. Normal body temperature

Normal body temperature is about 37 °C (98.6 °F). Fever is defined clinically as an oral temperature above 37.8°C (100.5°F) or a rectal temperature above 38.4 °C (101.5 °F).

B. Pyrogens

Pyrogens are substances that act on the hypothalamus to set the body's "thermostat" at a higher-than-normal temperature. Such substances can be *exogenous* or *endogenous*.

C. Effects of fever

Fevers can: (1) raise the body temperature above the optimum temperature for the growth of many pathogens; (2) raise the body temperature to a temperature at which microbial enzymes or toxins may be inactivated; (3) heighten the level of immune responses by increasing the rate of chemical reactions in the body; and (4) cause a patient to feel ill, causing the patient to rest to prevent further damage to the body and to allow energy to be used in fighting the infection.

D. Clinical approaches to fever

Today, many physicians recommend allowing fevers to run their course. However, antipyretics are still used when fevers reach extreme temperatures and when patients have disorders that might be worsened by fevers (such as severe heart disease or fluid and electrolyte imbalances).

VI. Molecular Defenses

A. Interferon

Interferon is a small protein often released from virus-infected cells that binds to adjacent uninfected cells, causing them to produce antiviral proteins that interfere with viral replication. They can be used therapeutically (in the form of *recombinant interferon*) to block virus replication and stimulate specific immune defenses against viral infections and tumors.

B. Complement

Complement, or the *complement system*, is a set of more than 20 large regulatory proteins that circulate in the plasma and form a nonspecific defense mechanism against many different microorganisms when activated. The two complement pathways that have been identified are the *classical pathway* and the *alternative pathway*. These pathways both cause certain proteins to participate in three kinds of molecular defenses: (1) *opsonization*; (2) *inflammation*; and (3) *membrane attack complexes*.

C. Acute phase response

The *acute phase response* is a response to an acute illness that produces specific blood proteins called acute phase proteins. Examples of these proteins include *C-reactive protein (CRP)* and *mannose-binding protein (MBP)*.

SELF-TESTS

Continue with this section only after you have read Chapter 16 of your textbook. Write your answers in the appropriate space provided. Correct answers to all questions can be found at the end of the Self-Test section. A score of 80 percent or better is good. If your score is less than 65, reread the chapter.

True/False

Mark T for True, F for False

_____ **1.** The sebum from oil glands creates a salty, alkaline environment, which is inhibitory to microbes.

_____ **2.** Some pathogens can actually multiply within white blood cells.

_____ **3.** Interferon seems to interfere directly with viral penetration of uninfected cells.

_____ **4.** Fever is now found to be beneficial in many infections and should not always be inhibited unless it goes above 40°C.

_____ **5.** Part of the "complement cascade" is designed to actually kill foreign cells by forming holes in cell membranes.

_____ **6.** The eyes are protected by enzyme amylase, which is produced by the lacrimal gland and which causes lysis of cell membranes.

_____ **7.** Acute inflammations are often characterized by the formation of granulomatous tissue followed by the release of histamine.

Multiple Choice

Select the best possible answer.

_____ **8.** In the development of the immune system:

a. All animals and a few plants have been found to possess a specific immune system.
b. Specific antibodies have been found in all vertebrates and in all types of fish.
c. Although all vertebrates have been found to be capable of rejecting grafts of foreign tissue, all invertebrates seem to lack this capacity.
d. Virtually all invertebrates appear to lack specific and nonspecific defenses but compensate for this by reproducing at extremely high rates.

_____ **9.** Prostaglandins:

a. cause capillaries to dilate.
b. are released by virus-infected cells.
c. may intensify pain in an injured area.
d. are produced only by T-cells.

_____ **10.** Which of the following is not a part of the effects of the complement proteins?

a. opsonization.
b. inflammation.
c. membrane lysis.
d. interferon production.
e. all of the above are major parts.

_____ 11. In terms of inflammation:
  a. The process is always detrimental to the host.
  b. Aspirin should always be given, especially for children, to reduce the fever.
  c. Inflammation should be suppressed to allow healing to occur.
  d. Bradykinin injections should be given to alleviate the symptoms.
  e. None of the above is true.

_____ 12. Which of the following is not a useful nonspecific defense mechanism?
  a. Presence of a mucous lining of the respiratory system.
  b. Lysozyme present in tear fluids.
  c. Release of large quantities of histamine.
  d. Pepsin release in the stomach.

Matching

Select the answer from the right-hand side that corresponds to the term or phrase on the left-hand side of the page. An answer may be used more than once. In some cases, more than one answer may be required.

Topic: Nonspecific Defenses

_____ 13. Interferon
_____ 14. Saliva
_____ 15. Skin
_____ 16. Gastric juices
_____ 17. Mucous membranes
_____ 18. Cells that engulf foreign cells
_____ 19. Body temperature elevation to kill pathogens
_____ 20. Complement
_____ 21. Reddening, swelling, and increased temperature at infection site

a. physical barriers
b. chemical barriers
c. cellular defenses
d. inflammation
e. fever
f. molecular defenses

Topic: Defensive Cells

_____ 22. Lymphocyte
_____ 23. Eosiniphile
_____ 24. Platelet
_____ 25. Monocyte
_____ 26. Erythrocyte
_____ 27. Neutrophil
_____ 28. Mast cell

a. important component of blood-clotting mechanism
b. granulocyte
c. agranulocyte
d. none of the above

Topic: Inflammation

_____ 29. Bradykinin
_____ 30. Leukocytosis
_____ 31. Histamine
_____ 32. Redness
_____ 33. Abscess
_____ 34. Gumma
_____ 35. Increase in temperature

a. cardinal sign or symptom
b. vasodilation
c. pain
d. increase in white blood cells
e. local collection of pus
f. granuloma

Fill-Ins

Provide the correct term or phrase for the following. Spelling counts.

36. A host defense that operates in response to a particular invading pathogen is a ________________ ________________ response.

37. Viruses and pathogenic bacteria, agents that stimulate a specific defense response, are examples of ________________.

38. Host defenses against pathogens that operate regardless of the invading agent are ________________ ________________ responses.

39. The skin and mucous membranes that protect your body and internal organs from injury and infectious agents are examples of ________________ ________________.

40. White blood cells are also called ________________.

41. ________________ are granulocytes that release histamine and are responsible for allergic symptoms.

42. ________________ are granulocytes that release defensive chemicals to damage parasites (worms) and are present in large numbers during allergic reactions.

43. Cells that eat other materials are called ________________.

44. ________________ are "big eaters" that can be fixed or wandering.

45. Certain cells can digest and generally destroy invading microbes and foreign particles by a process called ________________.

46. Phagocytes can move towards a chemical stimulus, a process known as ________________.

47. A ________________ is a vacuole that forms around a microbe within the phagocyte that engulfed it.

48. A ________________ is a structure resulting from the fusion of lysosomes with a phagosome.

49. Microbes such as viruses and parasitic worms that are destroyed without being ingested by a defensive cell are destroyed ________________.

50. The leukocytes responsible for killing intracellular viruses are NK cells, or ________________.

51. The thymus gland and the spleen are examples of ________________ organs.

52. The cardinal signs of ________________ are calor, rubor, tumor, and dolor.

53. Histamine can cause the walls of vessels to dilate, a process known as ________________.

54. When tissues are injured, a small peptide known as ________________ is produced at the injured site, causing pain.

55. Neutrophils can pass out of the blood by squeezing between endothelial cells lining the vessel walls in a process known as ________________.

56. ________________ are new connective tissue cells that replace fibrin as blood clots dissolve, forming granulation tissue.

57. A ________________ is a pocket of tissue that surrounds and walls off inflammatory agents.

58. A systemic increase in body temperature that often accompanies inflammation is known as ________________.

59. The small, soluble protein that is responsible for viral interference is called ________________.

60. ________________ refers to a set of more than 20 large regulatory proteins that play a key role in host defense.

Labeling

Questions 61–69: Select the best possible label for each step in the phagocytosis process: formation of phagolysosome; lysosomes; adherence; excretion; phagosome; undigested material; ingestion; residual body; digestion.

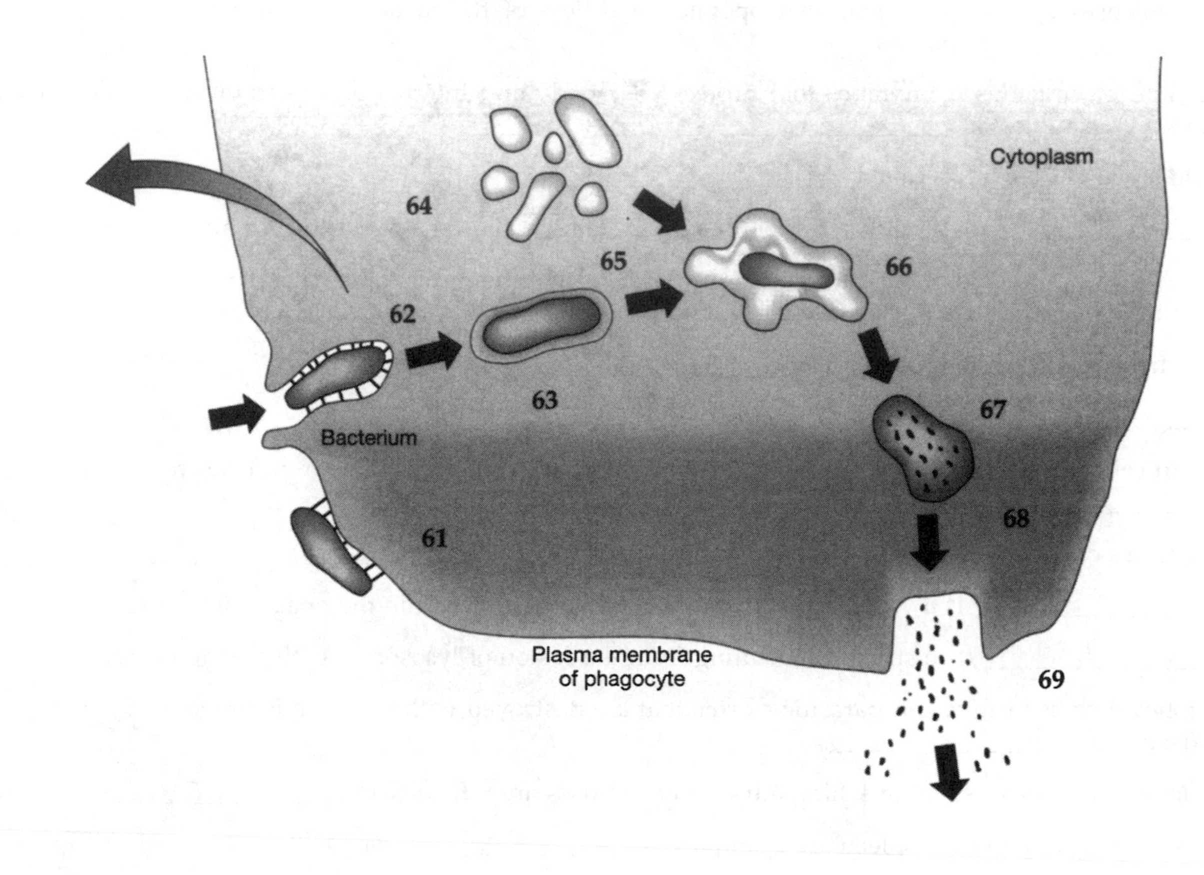

61. ________________
62. ________________
63. ________________
64. ________________
65. ________________
66. ________________
67. ________________
68. ________________
69. ________________

Questions 70–76: Select the best possible label corresponding to the descriptions and examples of the body's non-specific defenses given below: inflammatory response; interferons; physical barriers; extracellular killing; complement system; phagocytes; fever.

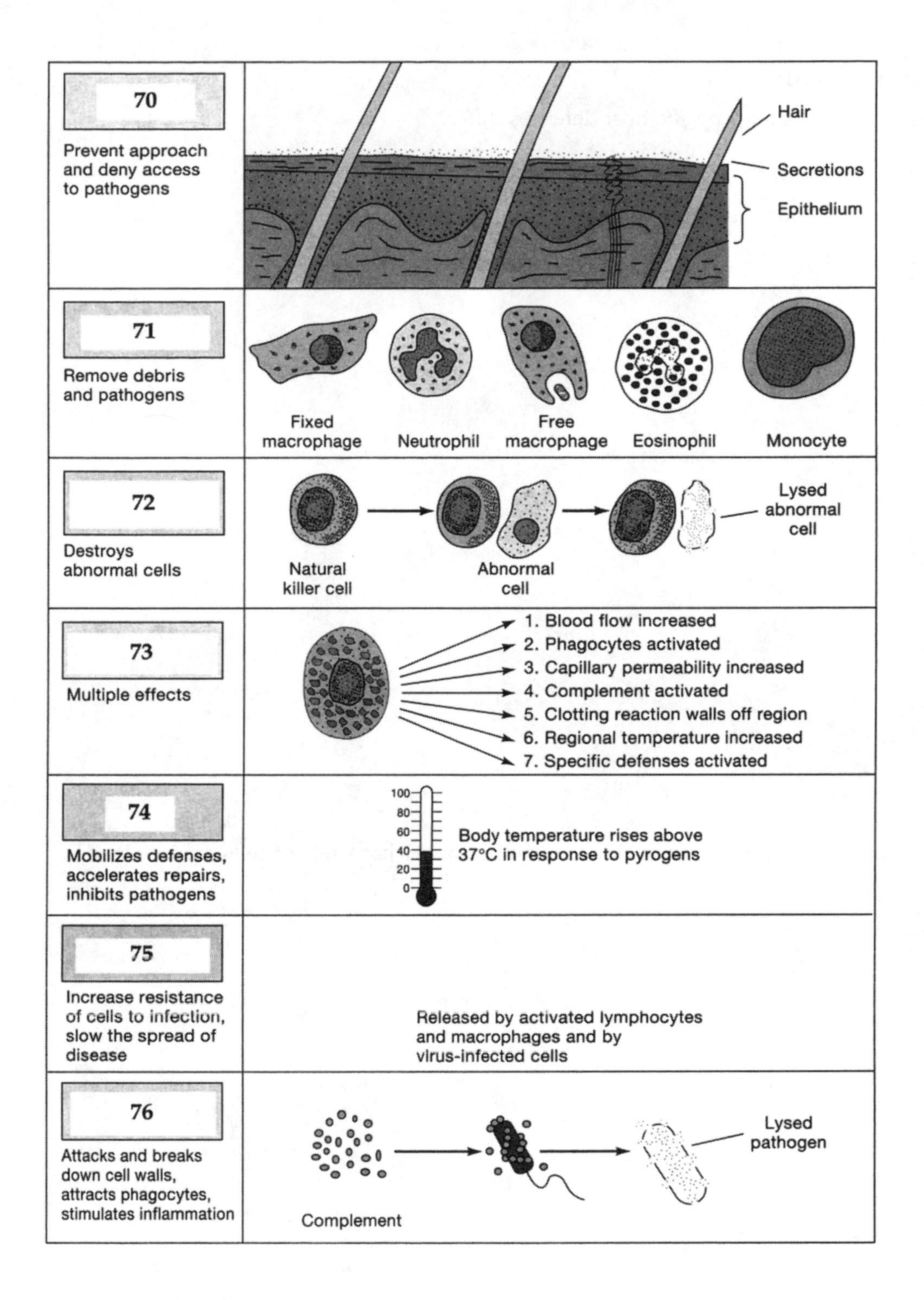

70. ____________________

71. ____________________

72. ____________________

73. ____________________

74. ____________________

75. ____________________

76. ____________________

Critical Thinking

77. How do nonspecific and specific host defenses differ?

78. What are the stages in the process of phagocytosis, and what kinds of cells are involved?

79. What is inflammation?

80. What are the steps in the acute inflammatory process and their functions?

81. How do repair and regeneration occur following acute inflammation?

82. How do the causes and effects of chronic and acute inflammation differ?

83. How does fever function as a nonspecific defense?

84. How do interferon and complement function in nonspecific defenses?

ANSWERS

True/False

1. **False** Although sebum is inhibitory, the secretions are actually acidic (pH 3-5). The salt is produced by the sweat glands.
2. **True** White blood cells can phagocytize foreign particles and bacteria. Unfortunately some bacteria such as Rocky Mountain Spotted Fever rickettsia and the bacteria that cause tuberculosis can multiply within the white blood cells.
3. **False** Interferon acts indirectly and is produced by virus-infected cells. This protein is then absorbed by adjacent uninfected cells which stimulates them to produce antiviral proteins.
4. **True** Fever seems to increase the level of immune response and seems to inhibit the growth of microorganisms.
5. **True** The complement cascade functions to assist phagocytosis, inflammation, and to initiate cell lysis.
6. **False** The enzyme lysozyme is produced by the lacrimal gland which affects bacterial cell walls. amylase is produced by salivary glands and breaks down starch.
7. **False** Acute inflammations are characterized by cellular damage followed by the release of histamine, however, the formation of granulomatous tissue or granulomas is more related to chronic inflammations.

Multiple Choice

8. b; 9. c; 10. d; 11. e; 12. c.

Matching

13. f; 14. b; 15. a; 16. b; 17. a; 18. c; 19. e; 20. f; 21. d; 22. c; 23. b; 24. a; 25. b; 26. d; 27. b; 28. b; 29. c; 30. d; 31. b; 32. a; 33. e; 34. f; 35. a.

Fill-Ins

36. specific; 37. antigens; 38. non-specific; 39. physical barriers; 40. leukocytes; 41. basophils; 42. eosinophils; 43. phagocytes; 44. macrophages; 45. phagocytosis; 46. chemotaxis; 47. phagosome; 48. phagolysosome; 49. extracellularly; 50. natural killer cells; 51. lymphoid; 52. inflammation; 53. vasodilation; 54. bradykinin; 55. diapedesis; 56. fibroblasts; 57. granuloma; 58. fever; 59. interferon; 60. complement.

Labeling

61. adherence; 62. ingestion; 63. phagosome; 64. lysosomes; 65. digestion; 66. formation of phagolysosome; 67. residual body; 68. excretion; 69. undigested material.; 70. physical barriers; 71. phagocytes; 72. extracellular killing; 73. inflammatory response; 74. fever; 75. interferons; 76. complement system.

Critical Thinking

77. **How do nonspecific and specific host defenses differ?**
The human body is protected by two basic defense systems. Nonspecific defenses operate regardless of the invading agent and constitute a first line of defense. These include the skin as an anatomical barrier, body fluids with antimicrobial substances, phagocytosis where a cell can engulf a foreign particle, and inflammation which accelerates the protective and healing processes.
The second line of defense is a specific one, which involves the immune defense system. It will be described in the next chapter.

78. **What are the stages in the process of phagocytosis, and what kinds of cells are involved?**
Phagocytes are cells that ingest and digest foreign particles. Phagocytosis is the process by which this is accomplished. Phagocytic cells (macrophages and monocytes) are found fixed in the organs of the body, wandering in the tissues, and circulating in the blood stream.
The process of phagocytosis occurs when the invading microorganisms cause the release of chemical at-

tractants called lymphokines, which aid the macrophage in locating them (chemotaxis). The phagocyte then engulfs the particle and ingestion begins. A phagosome is formed. Lysosomes migrate to the phagosome and empty their digestive enzymes to begin the digestive process. Following digestion, the wastes are released by exocytosis. Microbes can resist phagocytosis by producing resistant capsules, preventing release of lysosomal enzymes, and by producing toxins to kill the phagocyte.

79. What is inflammation?

Inflammation is a very complex process and is the body's response to tissue damage, characterized by redness, swelling, heat, and pain. Inflammation can be either acute (short-term) or chronic (long-term).

80. What are the steps in the acute inflammatory process and their functions?

Inflammation is initiated by some kind of cellular injury or death. When this occurs, the damaged cells release histamine and bradykinin. Capillaries then dilate (vasodilation), bringing more blood to the injured tissue. The skin becomes reddened and warmer. Capillaries also become more permeable at this time, allowing fluids to accumulate and cause swelling (edema).

After all of these events occur, blood clotting occurs and a scab forms. The bacteria, however, continue to multiply in the cut, so phagocytes enter the infected tissue by moving through the walls of blood vessels (diapedesis). Phagocytic cells are attracted to bacteria and tissue debris (chemotaxis) and engulf them. Larger blood vessels then dilate, further increasing the blood supply to the infected tissue and adding to the injury's heat and redness.

As dead cells and debris are removed by phagocytes, epithelial cells proliferate and begin to grow under the scab. Scar tissue (connective tissue) replaces cells that cannot replace themselves, and the injury is healed.

81. How do repair and regeneration occur following acute inflammation?

The end result of inflammation is the repair and/or restoration of the damaged tissue. This occurs as capillaries grow into the site of injury and fibroblasts replace the dissolving blood clot. The resultant granulation tissue is strengthened by connective tissue fibers and the overgrowth of epithelial cells.

82. How do the causes and effects of chronic and acute inflammation differ?

Acute inflammations such as bee stings, tend to occur very rapidly, reach a height very quickly, and subside quickly with very little tissue damage. On the other hand, chronic inflammations such as poison ivy tend to develop very slowly, reach a height very slowly, and take a long time to subside. It also exhibits a significant degree of tissue damage. Chronic inflammations occur when the host defenses fail to completely overcome the agent and may persist for years.

83. How does fever function as a nonspecific defense?

Fever is an increase in body temperature caused by pyrogens increasing the setpoint of the temperature-regulating center in the hypothalamus. Exogenous pyrogens such as bacteria come from outside the body and endogenous pyrogens come from inside the body. Fever augments the immune response system and inhibits the growth of many microbes. It also increases the rate of chemical reactions and raises the temperature above the optimum growth rate for some pathogens. Antipyretics such as aspirin are recommended only for very high fevers and for patients with disorders that would adversely be affected by fever.

84. How do interferon and complement function in nonspecific defenses?

Nonspecific defenses also include the molecules interferon and complement. Interferons are proteins produced by virus-infected cells that act nonspecifically to cause adjacent cells to produce antiviral proteins. In other words, interferon itself does not inhibit viruses but its action on other cells causes the development of the resistance. Interferon is now made by recombinant DNA technology and may be very useful for certain therapeutic applications.

Complement refers to a series of serum proteins that when activated by an antigen/antibody reaction cause a sequence of events that includes increasing phagocytosis, inflammation, and cellular lysis. It is interesting to note that antibodies do not actually kill but only bind to antigen surfaces. Complement proteins are responsible for the cellular lysis. The complement system includes two pathways, the classic complement pathway and the properdin pathway. The action of the system is rapid and nonspecific. Deficiencies in any of the complement proteins can greatly reduce resistance to infection.

A third recently discovered nonspecific defense mechanism is acute phase response, which involves several proteins including C-reactive protein (CRP). CRP seems to initiate an inflammatory response or accelerate an ongoing one. It also seems to activate the complement system, stimulate migration of phagocytes, and initiate platelet aggregation. CRP may also prevent death in certain otherwise fatal bacterial infections.

Chapter 17 Immunology I: Basic Principles of Specific Immunity and Immunization

INTRODUCTION

While all muticellular animals have nonspecific defense mechanisms, it thought that only vertebrates have a specific immune response. That is, in addition to having nonspecific defenses, humans can respond to many specific parasites. A specific immune response has two basic characteristics. First, when a specific parasite is seen for the second time, the response is accelerated relative to the primary response or initial experience. This is the basis for vaccinations. Second, this secondary response involves memory. It is worth noting that as we age the primary or first immune response ages much faster than the secondary response. Thus, vaccinations or previous experiences with specific diseases experienced in youth can protect us in old age, but when an elderly person sees an infection for the first time, they have a poor ability to respond compared to when they were young. Hence, influenza is more likely to be more serious in older people compared to younger people.

While a true immune response is usually described as being very specific, there can be overlapping protection, which can be quite curious at times. You may recall from Chapter 1 of our textbook that Edward Jenner used cowpox to protect people from smallpox. This worked because the two viruses are very similar. More recently, we have learned of a relationship between protection from bubonic plague and HIV AIDS. This is very odd because bubonic plague is caused by a bacterium and a virus causes HIV AIDS. What these two diseases have in common is that both must enter white blood cells in order to carry out their life cycles. A study of people in England found a population of people who would never get bubonic plague because of a mutated gene (a CCR5 gene). The same mutation prevents both plague bacteria and HIV virus from entering white blood cells. Such people are protected from both diseases. Perhaps a better understanding of this mutation will lead to prevention or treatment for both diseases? Clearly, the history of medicine shows that vaccination is the most effect way to prevent disease and protect the public health.

STUDY OUTLINE

I. Immunology and Immunity

Immunity refers to the ability of an organism to recognize and defend itself against infectious agents. Important words to understand are *susceptibility, nonspecific immunity, specific immunity, immunology,* and *the immune system.*

II. Types of Immunity

A. Innate immunity

Innate immunity is immunity to infection that exists in an organism because of genetically determined characteristics. One kind of innate immunity is *species immunity,* which is common to all members of a species.

B. Acquired immunity

Acquired immunity is immunity obtained in some manner other than by heredity. *Naturally acquired immunity* and *artificially acquired immunity* are two types of acquired immunity.

C. Active and passive immunity

Active immunity is immunity created when an organism's own immune system produces antibodies or other defenses against an agent recognized as foreign; it can be *naturally acquired* or *artificially acquired*. *Passive immunity* is immunity created when ready-made antibodies are introduced into, rather than created by, an organism; it can be *naturally acquired* or *artificially acquired*.

II. Characteristics of the Immune System

A. Antigens and antibodies

Antigens, or *immunogens*, are substances the body identifies as foreign and towards which it mounts an immune response. Antigens have *antigenic determinants*, or *epitopes*, to which antibodies can bind. An *antibody* is a protein produced in response to an antigen that is capable of binding specifically to the antigen.

B. Cells and tissues of the immune system

Specific immune responses are carried out by lymphocytes, such as *B lymphocytes* (or *B cells*), *T lymphocytes* (or *T cells*) and *null cells*. Tissues of the immune system include the bone marrow, the gut-associated lymphoid tissues (GALT), and the thymus.

C. Dual nature of the immune system

Lymphocytes give rise to two major types of immune responses, *humoral immunity* and *cell-mediated immunity*.

D. General properties of immune responses

Both humoral and cell-mediated responses have certain common attributes that enable them to confer immunity: (1) recognition of self versus nonself; (2) specificity; (3) heterogeneity; and (4) memory.

III. Humoral Immunity

A. General characteristics

Humoral immunity depends, first, on the ability of B lymphocytes to recognize specific antigens, and, second on their ability to initiate responses that protect the body against foreign agents. Recognition depends on the specific antibody on the membrane of each kind of B cell that can bind immediately to a specific antigen. The initiation of a response depends on the binding of an antigen (which *sensitizes* a B cell) and the activity of certain T cells.

B. Properties of antibodies (immunoglobulins)

Antibodies are Y-shaped protein molecules composed of four polypeptide chains (two identical *light chains* and two identical *heavy chains*). Antibodies possess *constant regions* and *variable regions*, *Fab* (antibody binding fragment) pieces and an *Fc* (crystallizable fragment) region.

C. Classes of immunoglobulins

The five classes of immunoglobulins in humans and other higher vertebrates are IgG, IgA, IgM, IgE, and IgD.

D. Primary and secondary responses

In humoral immunity, the *primary response* to an antigen occurs when host B cells first recognize the antigen. The *secondary response* occurs when memory cells recognize an antigen; it is more rapid and stranger than a primary response.

E. Kinds of antigen-antibody reactions

Humoral immunity is most effective against bacterial infections, toxins, and some viruses. These microorganisms can be neutralized by *neutralization*, *opsonization* (along with phagocytosis), or *complement* activation.

F. Monoclonal antibodies

Monoclonal antibodies are antibodies produced in the laboratory by a clone of cultured cells that make one specific antibody. Specific monoclonal antibodies can be used in diagnostic procedures and are being used for their possible uses in therapy.

IV. Cell-Mediated Immunity

A. General characteristics

Cell-mediated immunity involves the direct action of T cells, which produce chemical mediators called *lymphokines*, against other cells that display foreign antigens.

B. The cell-mediated immune reaction

The cell-mediated immune response is the immune response to an antigen carried out at the cellular level by T cells. T cells involved include *helper T cells*, *suppressor T cells*, *delayed hypersensitivity cells*, *cytotoxic (killer) T cells*, and *natural killer cells*.

C. How killer cells kill
Recent research shows that cytotoxic T cells and natural killer cells kill other cells by making a lethal protein called *perforin* and firing it at target cells.

D. Role of activated macrophages
The lymphokine macrophage activating factor, released by delayed hypersensitivity cells, cause macrophages to increase production of toxic hydrogen peroxide and enzymes that attack phagocytized organisms and accelerate the inflammatory response.

V. Factors that Modify the Immune Response
A variety of disorders, injuries, medical treatments, environmental factors, diet, exercise, and even age can affect resistance to infectious diseases.

VI. Immunization

A. Active immunization
Active immunization is the use of *vaccines* to control disease by increasing herd immunity through stimulation of the immune response. There are hazards in using vaccines, but the overwhelming benefits outweigh the risks, leading to the regular administration of immunizations.

B. Passive immunization
Passive immunization is the process of inducing immunity by introducing ready-made antibodies (such as *immune serum globulin, hyperimmune sera,* or *antitoxins*) into, rather than being created by, an organism.

C. Future of immunization
Immunologists are looking for vaccines that: (1) are protective against the disease for which they were designed; (2) are safe and do not have adverse side effects; (3) provide sustained, long-term protection against infection; (4) generate neutralizing antibodies or protective T cells to vaccine antigens; and (5) are practical in terms of stability and use. *Whole-cell killed vaccines, subunit vaccines, attenuated vaccines,* and *recombinant DNA vaccines* are vaccines that are currently being studied.

VII. Immunity to Various Kinds of Pathogens

A. Bacteria
Antibodies can work with complement to opsonize bacteria for later phagocytosis or lysis by other cells of the immune system. Or they can neutralize bacterial toxins or inactivate bacterial enzymes.

B. Viruses
The components of the immune system that act against viruses include interferons, secretory IgA, IgG and IgM antibodies, cytotoxic T cells, and natural killer cells. Fevers also provide a nonspecific response to viruses.

C. Fungi
Fungal skin infections are probably combated by IgA antibodies and T helper cells, which release certain cytokines that activate macrophages. The macrophages, in turn, engulf and digest fungi.

D. Protozoa and helminths
The host's immune system combats parasitic protozoa and helminths by forming antibodies against these parasites and by using cell-mediated processes.

SELF-TESTS

Continue with this section only after you have read Chapter 17 of your textbook. Write your answers in the appropriate space provided. Correct answers to all questions can be found at the end of the Self-Test section. A score of 80 percent or better is good. If your score is less than 65, reread the chapter.

True/False

Mark T for True, F for False

_____ **1.** Acquired immunities obtained by natural passive means generally last a lifetime.

_____ **2.** The upper ends of the Y of an antibody contain the combining sites.

_____ **3.** Although B and T cells are greatly involved in the immune response, macrophages and other phagocytes are essential to act upon and destroy bacteria and virus-infected cells.

_____ **4.** The MMR series contains live attenuated viruses.

_____ **5.** In most cases, the antibody molecules will actually kill the invading microbe immediately upon contact.

_____ 6. The BCG vaccine for tuberculosis is now being used in the United States and is given to school children in inner city schools due to the increase in tuberculosis.

Multiple Choice

Select the best possible answer.

_____ 7. Which of the following is not located on the Fc portion of the IgG molecule?
- **a.** Opsonization site.
- **b.** Complement binding site.
- **c.** Secretory transport unit.
- **d.** Allergic reaction site.
- **e.** All of the above are found on the Fc portion.

_____ 8. An important vaccine that is given to young children to prevent meningitis is:
- **a.** Hib.
- **b.** MMR.
- **c.** DPT.
- **d.** BCG.

_____ 9. In human lymphoid tissue, B cells are most abundant in the ______ tissue while T cells are most abundant in the ______ tissue.
- **a.** Thymus/Peyer's patchs
- **b.** Spleen/blood
- **c.** Lymph nodes/spleen
- **d.** Peyer's patchs/thymus

_____ 10. Which of the following is not a good characteristic of an antigen?
- **a.** protein composition.
- **b.** molecular weight of 1500.
- **c.** multiple epitopes.
- **d.** having charged groups such as amine and carboxyl groups.
- **e.** Actually, all of the above are good characteristics.

_____ 11. IgG is characterized by:
- **a.** having four combining sites.
- **b.** having a G-shape to the protein chains.
- **c.** having 2H and 2L chains.
- **d.** acting as the secretory antibody.
- **e.** None of the above is true.

_____ 12. TD cells can perform all the following except:
- **a.** stimulate phagocytes by producing MAF.
- **b.** act in a delayed-type hypersensitive response.
- **c.** produce specific antibodies against foreign cells.
- **d.** produce lymphokines to help macrophages find microbes.

_____ 13. Vaccines made of live viruses:
- **a.** are always the best choice to develop immunity.
- **b.** always provide permanent immunity.
- **c.** do not cause any adverse symptoms to recipients.
- **d.** can be dangerous to pregnant women.

_____ 14. In the cell-mediated immune reaction:
- **a.** The foreign antigens are processed and presented on the cell membrane surface of a macrophage.
- **b.** The foreign antigen is first engulfed by a null cell prior to processing by a macrophage.
- **c.** When T cells bind to macrophages, they quickly divide and transform into plasma cells.
- **d.** Macrophages lack histocompatibility proteins, which allow them to bind with any foreign substance.

Matching

Select the answer from the right-hand side that corresponds to the term or phrase on the left-hand side of the page. An answer may be used more than once. In some cases, more than one answer may be required.

Topic: Types of Immunity

_____ **15.** Colostrum
_____ **16.** Nonspecific defense mechanisms
_____ **17.** Placental passage of antibodies
_____ **18.** Toxoid injection
_____ **19.** Phagocytic cell activities
_____ **20.** A case of influenza
_____ **21.** Snake antivenin injection
_____ **22.** An infection of attenuated organisms
_____ **23.** Recovery from measles
_____ **24.** An injection of killed organisms

a. innate immunity
b. naturally acquired active immunity
c. naturally acquired passive immunity
d. artificially acquired active immunity
e. artificially acquired passive immunity

Topic: Classes of Immunoglobulins

_____ **25.** Crosses placenta
_____ **26.** Main class of Ig found in blood
_____ **27.** Contains secretory piece
_____ **28.** Reagin
_____ **29.** Largest Ig
_____ **30.** Special affinity for basophil receptors
_____ **31.** Associated with allergies and helminths
_____ **32.** Main Ig in tears
_____ **33.** FC recognized by phagocytes
_____ **34.** First Ig formed by a fetus
_____ **35.** Antigen receptor in early immune response stages

a. IgG
b. IgM
c. IgA
d. IgE
e. IgD

Topic: Properties of Immune Responses

_____ **36.** The ability of T cells to recognize cells exposed to previously
_____ **37.** The ability to produce a large variety of antibodies
_____ **38.** Clonal selection theory
_____ **39.** An amnestic response
_____ **40.** An ability to distinguish among different epitopes
_____ **41.** Distinguishing between host and foreign tissues

a. recognition of self versus nonself
b. specificity
c. heterogeneity
d. immunological memory

Topic: Properties of B Cells, T Cells; NK Cells and Macrophages

_____ **42.** Bone marrow site of production
_____ **43.** IgD antigen receptor
_____ **44.** Cell-mediated

a. B cell
b. T_H1 cell
c. T_H2 cell

_____ 45. Lack membrane surface markers
_____ 46. Humoral immunity
_____ 47. Perforin and granzyme
_____ 48. APC
_____ 49. Thymus site of production or thymic hormonal control

d. macrophage
e. NK cell
f. T_C cell

Topic: Types of Materials Used for Active Immunization

_____ 50. Diphtheria
_____ 51. Hepatitis B
_____ 52. Lobar pneumonia
_____ 53. Tuberculosis
_____ 54. Tetanus
_____ 55. Yellow fever
_____ 56. Typhoid fever
_____ 57. Pertussis
_____ 58. Meningococcal meningitis
_____ 59. Measles
_____ 60. Rabies

a. polysaccharide
b. live (attenuated) virus
c. viral antigen
d. toxoid
e. killed bacteria
f. killed viruses
g. live (attenuated) bacteria

Fill-Ins

Provide the correct term or phrase for the following. Spelling counts.

61. The ability of a host to recognize and respond to an infectious agent is called _______________.
62. The study of the immune response is known as _______________.
63. An immunity common to all members of a species is _______________ or _______________.
64. An immunity obtained by causing the host to produce antibodies is called _______________.
65. An immunity obtained by receiving preformed antibodies is called _______________.
66. A substance that the body identifies as foreign is an _______________.
67. Antibody binding sites on the surface of antigens are called _______________ _______________ or _______________.
68. A molecule that is generally too small to qualify as an antigen may be called a _______________.
69. Cells that are processed and matured in the bursal equivalent tissue are called _______________ - _______________.
70. Cells that undergo differentiation in the thymus gland are called _______________ - _______________.
71. Undifferentiated lymphocytes that are neither B nor T cells may be called _______________ cells.
72. The two major types or "arms" of our immune system are the _______________ and the _______________ - _______________ immunities.
73. Normal host tissues and substances can be identified by our immune response system as _______________.
74. The most plausible theory to explain how the body can recognize self from nonself is the _______________ _______________ theory.
75. The ability to remove those cells programmed to destroy host tissues is known as _______________.
76. The ability of the immune system to recognize and respond quickly to a foreign substance due to a previous encounter is called _______________.

77. B-cells that actually produce antibodies are ________________ cells.

78. The H and L chains of antibody molecules are held together by ________________ bonds.

79. The component of IgA molecules that facilitates transport across cell borders is the ________________ component.

80. The quantity of a substance needed to produce a reaction involving antibodies and antigens is measured as a ________________.

81. An antibody-antigen reaction involving large antigens such as bacteria or blood cells results in a visible response called ________________.

82. An antibody-antigen reaction involving bacterial toxins can usually be detected by a test called a ________________.

83. Specific antibodies produced by a single clone of cultured cells are ________________.

84. A lymphokine that appears to activate helper T-cells is called ________________ - ________________.

85. B-cells generally require a ________________ T-cell to aid in the production of antibodies.

86. T-cells that are capable of a direct response against foreign cells by killing them are called ________________ T-cells.

87. Killer cells contain granules of a lethal protein called ________________.

88. A substance that contains an antigen to which the immune system responds is a ________________.

89. A vaccine prepared by neutralizing the toxin is a ________________.

90. In the DPT vaccine, the P stands for ________________, which is the disease of ________________ ________________.

91. In the MMR vaccine, the R stands for ________________, which is the disease of ________________ ________________.

92. A vaccine used to protect against tuberculosis is the ________________ vaccine.

93. An ________________ is a serum that contains antibodies toward a particular antigen.

94. Antibodies made against specific toxins are called ________________.

95. Two recently-developed types of vaccine preparations that should reduce unwanted side effects are ________________ and ________________ vaccines.

Labeling

Questions 96–103: Select the best possible label for each of the following descriptions: acquired; artificial; natural; passive; natural; innate; active; artificial. *Use the blanks provided for your answers on page 227.*

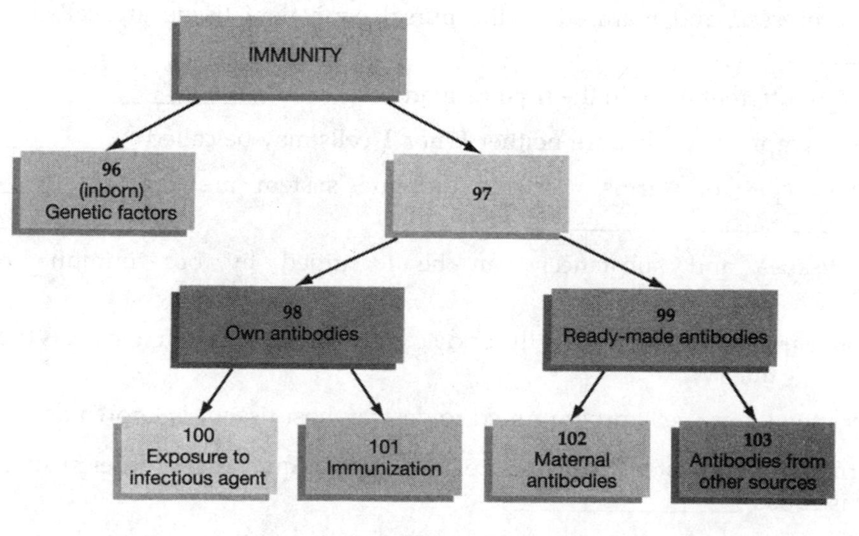

96. ____________________

97. ____________________

98. ____________________

99. ____________________

100 ____________________

101. ____________________

102. ____________________

103. ____________________

Questions 104–106: Select the best possible label for each of the following descriptions: neutralization; immune complexes; opsonization.

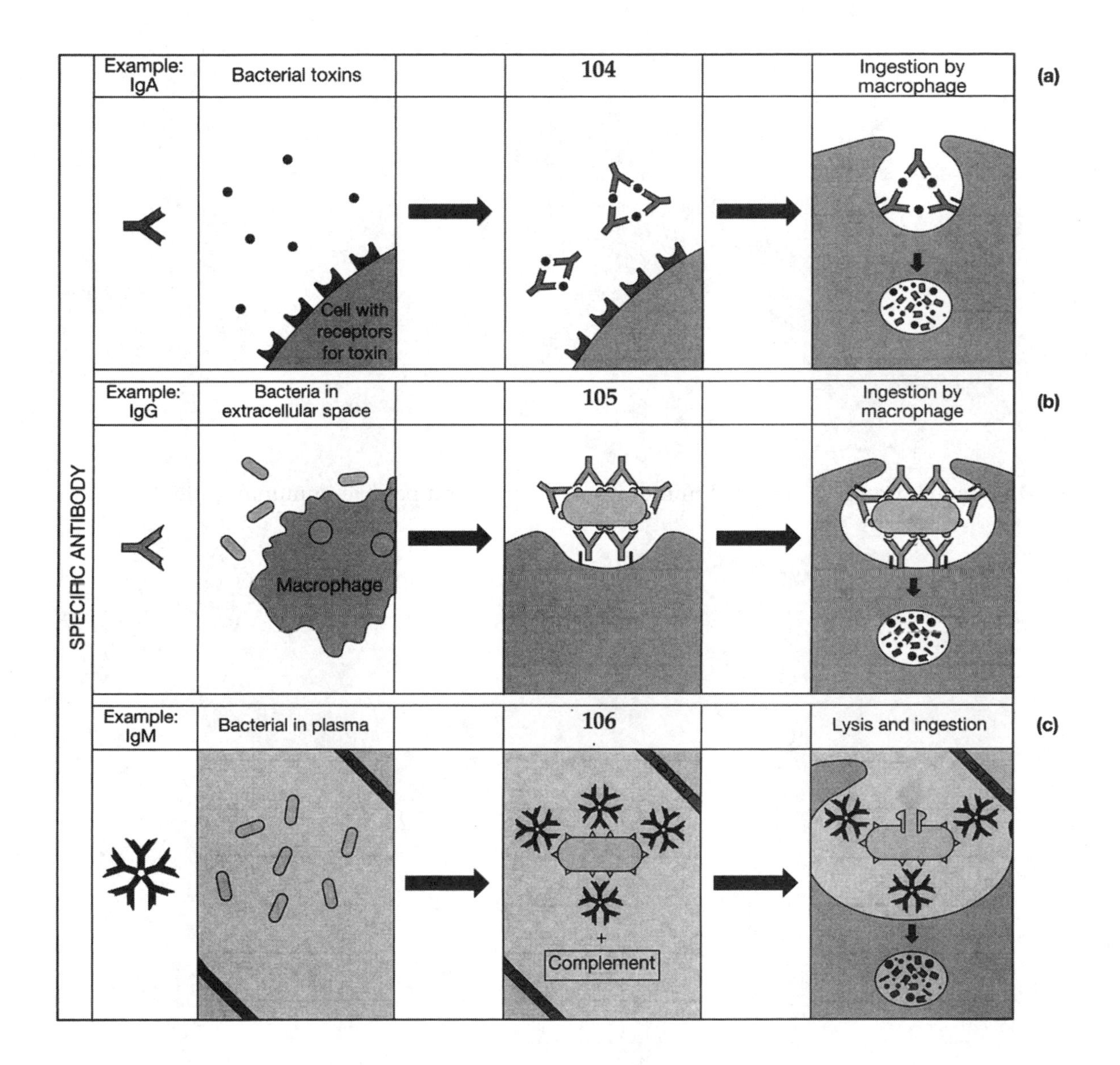

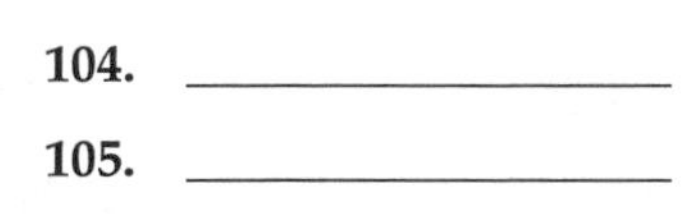

104. ____________________

105. ____________________

106. ____________________

Critical Thinking

107. What do immune, immunity, susceptibility, nonspecific immunity, specific immunity, immunology, and immune system mean?

108. How do innate immunity, acquired immunity, and active and passive immunity differ?

109. What are the properties of antigens and antibodies, and how do cells and tissues function in the dual branches of the immune system?

110. How do recognition of self, specificity, heterogeneity, and memory function in the immune system?

111. How do B cells and antibodies function in humoral immunity?

112. What are monoclonal antibodies, and how are they made and used?

113. How does cell-mediated immunity differ from humoral immunity, and how do reactions of cell-mediated immunity occur?

114. What are the special roles of killer cells and activated macrophages?

115. What factors modify the immune responses?

116. What are the mechanisms of immunization, recommended immunizations, and the benefits and hazards of immunization?

ANSWERS

True/False

1. **False** Immunities of this type last only a few days or weeks. The antibodies are pre-made, that is, made by someone else. Therefore the body is not stimulated to develop a response to the foreign agent.
2. **True** The upper ends of all antibody molecules consist of variable regions of the heavy and light chains and differ from antibody to antibody. Therefore they can act as the combining sites.
3. **True** Macrophages are generally the first cells to respond to a foreign antigen and act to process it. The cells then seek out appropriate T or B cells for cellular or antibody response. Once a response is made, macrophages also become involved to specifically engulf and destroy the invading agent.
4. **True** The MMR vaccine is for measles, mumps, and rubella and is given routinely to children after 15 months of age.
5. **False** Antibodies bind to antigen but do not actually "kill" the foreign substance. Macrophages, killer T-cells, or the complement cascade are responsible for the "killing" effect.
6. **False** The BCG vaccine is not licensed for use in the United States. If it were used, all recipients would skin test positive for tuberculosis.

Multiple Choice

7. c; 8. a; 9. d; 10. b; 11. c; 12. c; 13. d; 14. a.

Matching

15. c; 16. a; 17. c; 18. d; 19. a; 20. b; 21. e; 22. d; 23. b; 24. d; 25. a; 26. a; 27. c; 28. d; 29. b; 30. d; 31. d; 32. c; 33. a; 34. b; 35. e; 36. d; 37. c; 38. a; 39. d; 40. b; 41. a; 42. a; 43. a; 44. b; 45. e; 46. c; 47. e, f; 48. d; 49. b, c, f; 50. d; 51. c; 52. a; 53. g; 54. d; 55. b; 56. g; 57. e; 58. a; 59. b; 60. f.

Fill-Ins

61. immunity; 62. immunology; 63. innate, species; 64. active; 65. passive; 66. antigen; 67. antigenic determinants, epitopes; 68. hapten; 69. B-cells; 70. T-cells; 71. null; 72. humoral, cell-mediated; 73. self; 74. clonal selection; 75. tolerance; 76. memory, 77. plasma; 78. disulfide; 79. secretory; 80. titer; 81. agglutination; 82. neutralization; 83. monoclonal; 84. interleukin-1; 85. helper; 86. cytotoxic; 87. perforin; 88. vaccine; 89. toxoid; 90. pertussis, whooping cough; 91. rubella, German measles; 92. BCG; 93. antiserum; 94. antitoxins; 95. subunit, recombinant.

Labeling

96. innate; 97. acquired; 98. active; 99. passive; 100. natural; 101. artificial; 102. natural; 103. artificial; 104. neutralization; 105. opsonization; 106. immune complexes.

Critical Thinking

108. **What do immune, immunity, susceptibility, nonspecific immunity, specific immunity, immunology, and immune system mean?**
The word immune means free from burden. Immunity refers to the ability of an organism to recognize and defend itself against infectious agents. In other words, to be free of "burden" of these infectious agents. Susceptibility means to be vulnerable to disease agents. Nonspecific immunity is a defense against any infectious agent without having any experience with it, while specific immunity is a defense against specific agents generally because of some type of experience with the agent. Immunology is the study of specific immunity, and the immune system is the body system that provides the host with specific immunity to a particular infectious agent.

109. **How do innate immunity, acquired immunity, and active and passive immunity differ?**
Innate or genetic immunity exists because of genetically determined characteristics. In other words, we are simply born with it and play no role in obtaining it. We do not get diseases that dogs or horses get simply because we are human. Acquired immunity is immunity obtained in some manner other than by heredity.

We acquire this after we are conceived or after we are born. Active immunity means our body must make the antibodies while passive means that we receive the preformed antibodies that were made by someone else.

Natural immunities are derived naturally such as actually getting the disease while artificial immunity is usually obtained by getting an injection of the material.

110. What are the properties of antigens and antibodies, and how do cells and tissues function in the dual roles of the immune system?

Antigens are substances that our immune system can recognize and, if foreign, cause a specific immune response. To be recognized as an antigen, it must be greater than 10,000 molecular weight, it must have an electrical charge, and it must have multiple antigenic determinants or epitopes. The best antigens are protein in nature but those that are made of glycoproteins, nucleoproteins, lipoproteins, or complex polysaccharide are quite good as well. Epitopes are found on large molecules such as bacterial toxins, viruses, bacterial cells, animal cells, and, of course, human cells.

Antibodies are proteins produced in response to the presence of a foreign antigen. Antibodies are capable of binding specifically to the epitopes found on the antigen surface.

Lymphocytes differentiate into B-cells in the bursal-equivalent (gut-associated lymphoid) tissues, or into T-cells in the thymus. Null cells remain undifferentiated.

The immune system consists of a dual or two-arm system. The humoral or antibody producing arm is carried out by B-cells, and the cell-mediated arm is carried out mainly by T-cells.

111. How do recognition of self, specificity, heterogeneity, and memory function in the immune system?

Both the humoral and cell-mediated response systems based on the clonal selection theory have the ability to recognize self antigens from nonself antigens and to respond to them. However, the ability to respond to self antigens was probably destroyed during embryonic development by deleting those lymphocytes that were capable of responding to self antigens. Specificity refers to the ability of lymphocytes to respond to each antigen in a different and particular way. Heterogeneity refers to the ability of the immune system to produce many different substances such as antibodies in accordance with the many different antigens they encounter. The immune system also has the property of memory, that is, it can recognize substances it has previously encountered often at a faster rate.

112. How do B cells and antibodies function in humoral immunity?

Humoral or antibody immunity represents one arm of the immune defense system. Upon sensitization of appropriate B-cells, specific antibodies are produced to the antigen that sensitized the B-cell. The appropriate B-cell is selected on the basis of the antibody present on the B-cell's membrane. This membrane antibody is somewhat like a sign board of a restaurant or bakery that indicates what it can do (in this case what type of antibody it can produce). When a B-cell is presented with an antigen it can react with, it binds with it, and divides many times to produce a clone of antibody-producing plasma cells and some memory cells. Helper T-cells are often required to assist in the binding of the antigen, while suppressor cells probably limit the duration of antibody production. The memory cells remain in the lymph tissue to respond to a later exposure to the same antigen.

Antibodies are Y-shaped protein molecules composed of four amino acid chains: two identical short or light chains and two identical long or heavy chains. The chains are held together by disulfide bonds. Note that the IgG molecule represents the basic format. It's also the body's primary antibody molecule.

The primary response is the response to the first exposure of the antigen. The secondary response is the increased response due to the presence of memory cells present as a result of the first exposure.

Antibody-mediated immunity is most effective against toxins produced by bacteria and acute bacterial and viral infections. Agglutination, lysis by complement, or neutralization reactions are most effective against these agents.

113. What are monoclonal antibodies, and how are they made and used?

Monoclonal antibodies are antibodies produced in the laboratory by hybrid cells called hybridomas. These cells contain the genetic information from both a myeloma (cancer) cell and a sensitized antibody-producing B-cell. The cancer cell provides the hybrid with the ability to produce large quantities of specific antibodies for virtually an infinite amount of time. These antibodies are used in diagnostic tests and for experimental methods in the treatment of certain diseases and cancers.

114. How does cell-mediated immunity differ from humoral immunity, and how do reactions of cell-mediated immunity occur?

Cell-mediated immunity relates to the direct actions of sensitized T-cells that defend the body against chronic bacterial and virus infections, that reject tumors and transplanted tissues, and that cause delayed hypersensitivities. This immune response involves the differentiation and activation of several kinds of T-cells such as cytotoxic (T_c), delayed hypersensitivity (T_D), helper (T_H), and suppressor (T_s), and the secretion of lymphokines. T-cells do not make antibodies but do have membrane receptors for antigens and for histocompatibility proteins.

The reaction begins with the processing of an antigen by a macrophage, with the subsequent insertion of antigenic molecules into its own cell membrane. This antigen binds with T-cell receptors and with proteins on the membrane of macrophages. These macrophages then secrete interleukin-1 (IL-1) that activates helper T-cells. IL-1 and IL-2 (from helper t-cells) then activate cytotoxic T-cells, suppressor T-cells, and delayed hypersensitivity T-cells. IL-1, IL-2, and gamma interferon activate natural killer (NK) cells. The process is very complex, but the main idea is to specifically activate the proper type of T-cell so that the response to the foreign substance can be rapid and controlled.

T_D cells secrete several lymphokines that stimulate various activities of macrophages. T_c cells can kill virus-infected cells and NK cells kill tumor cells.

115. What are the special roles of killer cells and activated macrophages?

Cytotoxic T (T_c) and natural killer (NK) cells destroy target cells by releasing the lethal protein perform. Activated macrophages can specifically seek out a foreign substance such as a bacterium and phagocytize it. However, some pathogens can grow inside the macrophage, but an activating factor may help to stimulate antimicrobial processes so that the macrophage can kill the pathogens. Failure to do so can result in the macrophage being walled off in the formulation of granulomas.

116. What factors modify the immune responses?

The nonspecific host defenses in healthy adults are usually sufficient to prevent nearly all infectious diseases. Individuals with reduced resistance are called compromised hosts. Factors that can reduce this are very young or old age, poor nutrition, lack of exercise, injury, pollution, and radiation. Immune system deficiencies such as a lack of certain complement proteins, infections such as AIDS, and genetic defects can also modify the response.

117. What are the mechanisms of immunization, recommended immunizations, and the benefits and hazards of immunization?

Active immunization occurs by the same mechanism as having a disease since it challenges the immune system to develop specific defenses and memory cells. However, it usually does not result in a permanent immunity with the first exposure and therefore requires subsequent "booster" shots. Vaccines made from live attenuated or dead organisms can be used as well as toxoids, which are made from inactivated toxins. The benefits of active immunization outweigh any of the hazards in their use. Vaccines made from cell wall material (whooping cough vaccine) are very difficult to purify and generally result in the greatest number of problems.

Passive immunization occurs by the same mechanism as natural passive transfer of antibodies in that the antibodies are pre-made. Passive immunity is provided by using gamma globulin shots, convalescent sera, or antitoxins made in other animals. Since the antibodies are pre-made, the protection is short-lived but is immediate. The main hazard comes from hypersensitivity.

Nonspecific and antibody mediated defenses are most important in controlling bacterial infections. Viral infections are also controlled by nonspecific and antibody mediated defenses and by interferon. Fungi, protozoa, and helminths are mostly controlled by cell-mediated defenses.

Efforts are being made to produce and purify vaccines that will not produce any unwanted side effects. These new preparations include subunit and recombinant vaccines, which should be much safer than the currently used attenuated vaccines.

Chapter 18 Immunology II: Immunological Disorders and Tests

INTRODUCTION

While we have been focusing on how our immune system can protect us from infectious microbes, there are times when our immune system functions defectively and the autoimmune disease adversely affects our health. For some, this may express itself as seasonal allergies or the skin lesions associated with poison ivy. At times, the answer in clinical medicine is to use drugs like antihistamines or steroids to suppress the immune system. While the first rule of clinical medicine is to do no harm, mistakes can occur. I once listened to a physician who presented a case of clinical schistosomiasis. This flatworm infection results in an enlargement of the liver in tens of millions of people in poor countries. The physician felt they he could treat the symptoms of the disease by giving the patient steroids to reduce the inflammation of the liver. The patient died soon after treatment. Autopsy revealed that the liver was a liquefied necrotic mass. He concluded that sometime inflammation is a good thing.

In the last chapter we learned how vaccinations prevent infectious diseases. On the other hand, our zeal to prevent disease by vaccination may have gone too far in some cases. The battles we fight during childhood may make us stronger and help our immune system to develop. Consider, the current incidence of asthma in the United States is at an all time high. At the time of the writing of this essay, the State of Massachusetts is number one in baseball and football. It is also number one in asthma. Interestingly enough, the author of this story grew up many years ago in Brooklyn N.Y. where asthma was almost unknown. The air of my childhood was more polluted than today. Cars used gasoline mixed with lead and the houses in my neighborhood burned coal for heat. How can we explain the increase in asthma? While the answer to this increase is not completely known, many scientists believe that the explanation is linked to vaccinations that have reduced the infectious diseases of childhood. Diseases like mumps and chickenpox were once rites of childhood passage. Now they are almost vanished from our population. It is possible that the lack of infections causes our immune system to develop improperly, and the autoimmune disease of asthma is but one result.

Lastly, this chapter reveals how we can exploit antibody-antigen reactions to diagnosis diseases, sometimes by joining antibodies to fluorescent probes that light up by an organism of interest.

STUDY OUTLINE

I. Overview of Immunologic Disorders

A. Hypersensitivities

In a *hypersensitivity*, or *allergy*, the immune system reacts in an exaggerated or inappropriate way to a foreign substance. There are four types of hypersensitivity:

(1) immediate hypersensitivity (Type I);
(2) cytotoxic hypersensitivity (Type II);
(3) immune complex hypersensitivity
(Type III); (4) and cell-mediated, or delayed, hypersensitivity (Type IV).

B. Immunodeficiency

In *immunodeficiency*, the immune system responds inadequately to an antigen, either because of inborn or acquired defects in B cells or T cells.

II. Types of Hypersensitivities

A. Immediate (Type I) hypersensitivity

Immediate hypersensitivity, or *anaphylactic hypersensitivity*, typically produces an immediate response upon exposure to an *allergen*. The mechanism involves *sensitization* and *degranulation* mediated by *prostaglandins* and *leukotrienes*. *Anaphylaxis* is an immediate, exaggerated allergic reaction to antigens; the reaction can occur as *localized anaphylaxis* and *generalized anaphylaxis*. In many cases, genetic factors are thought to contribute to the development of allergies. Allergy treatments include avoiding contact with allergens, allergy shots, and antihistamines.

B. Cytotoxic (Type II) hypersensitivity

In cytotoxic hypersensitivity, specific antibodies react with cell surface antigens recognized as foreign by the immune system, leading to phagocytosis, killer cell activity, or complement-mediated lysis. Examples of cytotoxic reactions include *transfusion reactions* and *hemolytic disease of the newborn*.

C. Immune complex (Type III) hypersensitivity

Immune complex hypersensitivity results from the formation of antigen-antibody complexes that persist or are continuously formed. Examples of immune complex disorders include *serum sickness* and the *Arthus reaction*.

D. Cell-mediated (Type IV) hypersensitivity

Cell-mediated hypersensitivity, or *delayed hypersensitivity*, takes more than 12 hours to develop. T cells, not antibodies, mediate these reactions. Examples of cell-mediated disorders include *contact dermatitis, tuberculin hypersensitivity*, and *granulomatous hypersensitivity*.

III. Autoimmune Disorders

A. General features

Autoimmune disorders occur when individuals become hypersensitive to specific antigens on cells or tissues of their own bodies and produce *autoantibodies*.

B. Autoimmunization

Autoimmunization is the process by which hypersensitivity to "self" develops. Several different factors probably play a role in autoimmunity: (1) genetic factors; (2) antigenic or molecular mimicry; (3) clonal deletion; (4) mutations; (5) viral components; and (6) the sympathetic nervous system.

C. Examples of disorders

Myasthenia gravis is an autoimmune disorder specific to skeletal muscle, especially muscles of the limbs and those involved in eye movements, speech, and swallowing. *Rheumatoid arthritis* is an autoimmune disorder that affects mainly the joints but can extend to other tissues. *Systemic lupus erythematosus (SLE)* is a widely disseminated, systemic autoimmune disease resulting from production of antibodies against DNA and other body components.

IV. Transplantation

A. General features

Transplantation is the transfer of tissue, called *graft tissue*, from one site to another. Important concepts to understand include *isograft, allograft, xenograft, transplant rejection*, and *graft-versus-host (GVH) disease*.

B. Histocompatibility antigens

All human cells, and those of many other mammals, have a set of self-antigens called *histocompatibility antigens*. These antigens are unique in all individuals except identical twins. If donor and recipient histocompatibility antigens are different, recipient T cells recognize these cells as foreign and destroy the donor tissue.

C. Transplant rejection

Transplant rejection displays specificity and memory. T cells are responsible for rejection of grafts of solid tissue, such as a kidney, a heart, skin, or other organs. The time required for rejection to occur determines whether the rejection is a *hyperacute rejection*, an *accelerated rejection*, an *acute rejection*, or a *chronic rejection*.

D. Immunosuppression

Immunosuppression is the minimizing of immune reactions using *radiation* or *cytotoxic drugs*.

V. **Drug Reactions**
   A. General features
      If small drug molecules combine with proteins, the protein-drug complexes can sometimes induce any of the four types of hypersensitivity.
   B. Types of reactions
      Type I hypersensitivity can be caused by various drugs. Type II hypersensitivity can occur when the drug binds to a plasma membrane directly. Type III hypersensitivity appears as serum sickness and can be caused by any drug that participates in the formation of immune complexes. Type IV hypersensitivity usually occurs as contact dermatitis after topical application of drugs.

VI. **Immunodeficiency Diseases**
   A. General features
      Immunodeficiency diseases are diseases of impaired immunity caused by a lack of lymphocytes, by defective lymphocytes, or by the destruction of lymphocytes.
   B. Primary immunodeficiency diseases
      Primary immunodeficiency diseases are caused by genetic defects in embryological development. Examples include *agammaglobulinemia, DiGeorge syndrome,* and *severe combined immunodeficiency (SCID).*
   C. Secondary (or acquired) immunodeficiency diseases
      Secondary immunodeficiency diseases are acquired as a result of infections, malignancies, autoimmune diseases, or other conditions. The most well-known example is the *acquired immune deficiency syndrome (AIDS).*

VII. **Immunologic Tests**
   A. General features
      Laboratory tests used to detect and quantify antigens and antibodies make up the branch of immunology called *serology,* so named because many of the tests are performed on serum samples.
   B. The precipitin test
      One of the first serologic tests to be developed was the *precipitin test,* which can be used to detect antibodies or antigens based on *precipitation reactions. Immunodiffusion tests, immunoelectrophoresis,* and *radial immunodiffusion* are tests that are based on the same principle as the precipitin test.
   C. Agglutination reactions
      When antibodies react with antigens on cells, they can cause *agglutination* (clumping together of cells). One application of *agglutination reactions* is to determine whether the *antinbody titer* against a particular pathogen in the patient's blood is increasing. The *tube agglutination test,* the *hemagglutination inhibition test,* the *Coomb's antiglobulin test,* the *complement fixation test,* the *Schick test,* and the *viral neutralization test* are important applications of the agglutination reaction.
   D. Tagged antibody tests
      The most sensitive immunological tests used to detect antibodies or antigens use antibodies that have a "molecular tag" that is easy to detect even at very low concentrations. Examples of such tagged antibody tests include *immunofluorescence, radioimmunoassay (RIA), enzyme-linked immunosorbent assay (ELISA),* and *Western blotting.*

SELF-TESTS

Continue with this section only after you have read Chapter 18 of your textbook. Write your answers in the appropriate space provided. Correct answers to all questions can be found at the end of the Self-Test section. A score of 80 percent or better is good. If your score is less than 65, reread the chapter.

True/False

Mark T for True, F for False

______ **1.** Atopic allergies are often genetically based.

______ **2.** Immunodeficiency diseases are always inherited.

______ **3.** A major difference between hay fever and a common cold is that hay fever exhibits greater numbers of eosinophils in nasal secretions.

______ **4.** A useful treatment for individuals with severe allergies to insect stings are syringes filled with epinephrine and SRS-A.

______ 5. HLA antigens are determined by a set of genes designated A, B, C, D, and DR.

______ 6. A granulomatous response is usually associated with Type I hypersensitivities.

______ 7. An elevation of the CD4 cell count can be used to predict the onset of disease symptoms of AIDS.

Multiple Choice

Select the best possible answer.

______ 8. Women who receive Rhogam shots are attempting to control:
- a. severe asthma attacks.
- b. hemolytic disease of newborn.
- c. serum sickness
- d. Arthus reactions.

______ 9. All the following relate to Type IV hypersensitivities except:
- a. responses elicited in 2-3 days.
- b. granulomatous responses.
- c. mediators including lymphokines and macrophages.
- d. passive transfer of circulating reagins.
- e. All of the above relate to Type IV.

______ 10. Autoimmunization:
- a. occurs when we vaccinate ourselves.
- b. involves GVH responses.
- c. is a process of hypersensitivity to pollen.
- d. is the basis for rheumatoid arthritis and systemic lupus erythematosus.

______ 11. All the following are true concerning immunosuppression during organ transplants except:
- a. may require radiation of the recipient to destroy the immune system.
- b. may leave the patient susceptible to infections.
- c. may result in the transplant tissue rejecting the recipient.
- d. may result in an autoimmune disease called myasthenia gravis.

Matching

Select the answer from the right-hand side that corresponds to the term or phrase on the left-hand side of the page. An answer may be used more than once. In some cases, more than one answer may be required.

Topic: Immunological Disorders

______ 12. Blood transfusion reaction

______ 13. Poison ivy reaction

______ 14. Anaphylaxis

______ 15. Transplant rejection

______ 16. Formation of antigen–antibodycomplexes

______ 17. Harmful reaction due to IgE response to allergens

______ 18. Atopy

a. immediate hypersensitivity

b. cytotoxic hypersensitivity

c. immune complex hypersensitivity

d. cell-mediated hypersensitivity

Topic: Examples of Hypersensitivity Reactions

______ 19. Localized Arthus reaction

______ 20. Contact dermatitis

______ 21. Hay fever

______ 22. Mismatched blood transfusions

______ 23. Rheumatoid arthritis

a. atopy

b. cytotoxic hypersensitivity

c. immune complex hypersensitivity

d. cell-mediated hypersensitivity

e. autoimmune disorder

______ 24. Hemolytic disease of the newborn

______ 25. Seasonal allergic rhinitis

______ 26. Systemic lupus erythematosus

______ 27. Tuberculin hypersensitivity

______ 28. Granulomatous hypersensitivity

______ 29. Myasthenia gravis

Topic: Transplantation

______ 30. A skin graft from a patient's chest to repair burn damage on a leg

______ 31. A graft between two individuals not genetically identical

______ 32. A transplant between two highly inbred animals

______ 33. A transplant between identical twins

______ 34. A transplant between two different animal species

a. isograft

b. allograft

c. autograft

d. xenograft

Topic: Immunodeficiency Diseases

______ 35. Severe combined immunodeficiency disease

______ 36. DiGeorge syndrome

______ 37. AIDS

______ 38. Agammaglobulinemia

______ 39. Tuberculosis

______ 40. Multiple myeloma

______ 41. Adenosine deaminase (ADA) deficiency

a. primary immunodeficiency disease

b. secondary immunodeficiency disease

Topic: Immunological Tests

______ 42. Coomb's antiglobulin test

______ 43. Precipitin test

______ 44. Radioimmunoassay

______ 45. Neutralization

______ 46. ELISA

______ 47. Hemagglutination

______ 48. Immunodiffusion

a. use of radioactivate antibodies to detect antigens

b. viral clumping of red blood cells

c. detects bacterial toxins and antibodies to viruses

d. antigen–antibody complexes produce a precipitate reaction

e. detects anti-Rh antibodies

f. use of antibodies tagged with enzymes

g. precipitin test performed in solidified agar

Fill-Ins

Provide the correct term or phrase for the following. Spelling counts.

49. An exaggerated response to an antigen that the immune system usually ignores is a ________________.

50. When a person has defect in the B or T cell population it is called an ________________.

51. Another term for immediate hypersensitivity is ________________.

52. An ordinary foreign substance that elicits an adverse immune response in a sensitized person is a ________________ .

53. The initial recognition of an allergen that promotes IgE production is ________________.

54. When mast cells release histamine, the process is known as ________________.

55. Localized, "out of place" allergies are referred to as ________________.

56. The current method of treatment for allergies involves ________________.

57. The treatment for allergies involves the formation of antibodies called ________________ antibodies.

58. A reaction that occurs when matching antigens and antibodies are present in the blood at the same time is called ________________ reaction.

59. Type III hypersensitivity results in the formation of large antibody-antigen complexes called ________________ complexes.

60. A hyperimmune reaction to horse protein used in treatment of diseases such as diptheria and tetanus is called ________________ ________________.

61. Cell-mediated immunity is also known as ________________ because it may require 2-3 days to elicit a response.

62. Antibodies directed against self antigens are called ________________.

63. An autoimmune disease that affects the skeletal muscle is ________________ ________________.

64. An autoimmune disease that affects the joints is ________________ ________________.

65. Grafts made between genetically identical individuals are called ________________.

66. The use of pigskin to cover a burned area on a patient would involve a graft called a ________________.

67. When cells from a tissue graft react negatively against the host it is called ________________.

68. The genetic complex that encodes the genes for histocompatibility is known as the ________________ ________________ ________________.

69. Human histocompatibility antigens identified specifically on leukocytes are called ________________ ________________ antigens.

70. In order to minimize transplant rejections in recipients, a process of ________________ is carried out.

71. A drug that suppresses the immune system, specifically T-cell population is ________________ ________________.

72. Abnormalities in the embryonic development of the host, such as failure of the thymus gland to develop, are ________________ deficiencies.

73. The absence of B-cells and therefore antibodies in a patient is called ________________.

74. A type of cancer common to AIDS patients is ________________ ________________.

75. A branch of immunology that involves laboratory testing of serum samples is ________________.

76. An antibody-antigen reaction that involves the formation of tiny lattice-like networks of molecules is a ________________ reaction.

77. A process of separating antigen molecules in a gel using an electric current and then reacting them with antibody is called ________________.

78. An antibody-antigen reaction involving the agglutination of red blood cells is ________________.

79. An antibody-antigen reaction that indirectly detects antibodies by determining whether complement added to the test is used up in the reaction is the ________________ test.

80. A neutralization test involving diphtheria toxin injected under the skin is the ________________ test.

81. The use of fluorescent-tagged antibodies or antigens is the basis of ________________.

82. The use of an enzyme as an indicator after an antibody/antigen reaction has taken place is the basis of the ________________ test.

Critical Thinking

83. What are the different types of immunologic disorders?

84. What are the causes and effects of immediate hypersensitivity?

85. What are the causes and effects of cytotoxic reactions?

86. What are the causes and effects of immune complex disorders?

87. What are the causes and effects of cell-mediated reactions?

88. How does autoimmunization occur, and how is hypersensitivity involved with it?

89. Why are organ transplants sometimes rejected, and how can rejection be prevented?

90. How is hypersensitivity involved in some drug reactions?

91. What are the causes, mechanisms, and effects of immunodeficiency diseases?

92. How can antigens and antibodies be detected and measured?

93. What is AIDS and why is it causing an epidemic?

94. How can AIDS be diagnosed, treated, and prevented?

ANSWERS

True/False

1. **True** Atopy means "out of place" and is often genetically related. At least 60% of those with atopy have a family history of the disorder.
2. **False** Immunodeficiency diseases may be acquired as a result of infection, malignancies, autoimmune diseases, or other.
3. **True** An elevated eosinophil count can also suggest allergy or infection with eukaryotic parasites such as trichinella worms.
4. **False** Severe allergic attacks can be treated with antihistamine or epinephrine but not SRS-A since this substance is one of the mediators in the allergic response.
5. **True** HLA refers to human leukocyte antigens, which are determined by a set of genes designed A, B, C, and D. Some of the genes are highly variable and may specify a variety of antigens.
6. **False** Granulomatous hypersensitivities are the most serious of the cell-mediated Type IV hypersensitivities
7. **False** AIDS viruses specifically damage T4 (helper) cells because they bear a CD4 receptor. A drop in this cell count indicates the onset of disease symptoms.

Multiple Choice

8. b; 9. d; 10. d; 11. d.

Matching

12. b; 13. d; 14. a; 15. d; 16. c; 17. a; 18. a; 19. c; 20. d; 21. a; 22. b; 23. c, e; 24. b; 25. a; 26. c, e; 27. d; 28. d; 29. e; 30. c; 31. b; 32. a; 33. a; 34. d; 35. a; 36. a; 37. b; 38. a; 39. b; 40. b; 41. a; 42. e; 43. d; 44. a; 45. c; 46. f; 47. b; 48. g.

Fill-Ins

49. hypersensitivity; 50. immunodeficiency; 51. anaphylaxis; 52. allergen; 53. sensitization; 54. degranulation; 55. atopy; 56. desensitization; 57. blocking; 58. transfusion; 59. immune; 60. serum sickness; 61. delayed; 62. autoantibodies; 63. myasthenia gravis; 64. rheumatoid arthritis; 65. isografts; 66. xenografts; 67. graft versus host; 68. major histocompatibility complex; 69. human leukocyte; 70. immunosuppression; 71. cyclosporine A; 72. primary; 73. agammaglobulinemia; 74. Kaposi's sarcoma; 75. serology; 76. precipitation; 77. immunoelectrophoresis; 78. hemagglutinatin; 79. complement fixation; 80. Schick; 81. immunofluorescence; 82. ELISA.

Critical Thinking

83. **What are the different types of immunologic disorders?**
An immunologic disorder results from an inappropriate or an inadequate immune response. Most disorders involve either hypersensitivities which are inappropriate reactions to an antigen, or allergen, and immunodeficiencies, which are inadequate immune responses.

Immunodeficiencies can be primary in which the patient lacks T-cells or B-cells or has defective ones due to a genetic or developmental defect, or secondary in which the patient has defective T- or B-cells after they have developed normally.

84. **What are the causes and effects of immediate hypersensitivity?**
Immediate (Type I) hypersensitivity (anaphylaxis in its worst form) is the result of an inappropriate immune response to a harmless substance called an allergen. The patient must have an initial exposure or sensitization which causes the release of IgE antibodies. The second exposure provokes the release of histamine and other chemicals, which mediates the response.

Atopy is a localized reaction to an allergen such as pollen in which histamine and other mediators elicit sneezing and congestion of sinuses and other signs typical of this allergy. A skin reaction would exhibit a wheel and flare pattern.

Generalized anaphylaxis is a life-threatening systemic reaction in which blood pressure is greatly decreased or the airway is occluded.Treatment requires desensitization. Symptoms can be relieved with antihistamines or epinephrine.

85. **What are the causes and effects of cytotoxic reactions?**
Cytotoxic or Type II hypersensitivity involves specific antibodies reacting with cell-surface antigens detected as foreign by the immune system. Transfusion reactions and hemolytic diseases of the newborn are examples of this type of hypersensitivity.

86. **What are the causes and effects of immune complex disorders?**
Immune complex (Type III) hypersensitivity results from the formation of antigen-antibody complexes. Serum sickness and the Arthus reaction are examples of this type of hypersensitivity.

87. **What are the causes and effects of cell-mediated reactions?**
Cell-mediated (Type IV) is also known as delayed hypersensitivity because the reactions take more than 12 hours to develop. This hypersensitivity is mediated by T_D cells rather than IgE antibodies. A sensitized T_D cell releases lymphokines which in turn activate macrophages to mediate the response.

Contact dermatitis such as seen with poison ivy, metals, rubber material, etc., is common. Other examples include hypersensitivity to tuberculin material, leprosy, fungal agents, and granulomatous hypersensitivity.

88. **How does autoimmunization occur, and how is hypersensitivity involved with it?**
Autoimmunization is the development of hypersensitivity to self; it occurs when the immune system responds to a body component as if it were foreign. Normally all those cells capable of responding to self antigens are destroyed prenatally.

Tissue damage from these autoimmune disorders can by caused by cytotoxic, immune complex, and cell-mediated hypersensitivity reactions. Examples include myasthenia gravis, rheumatoid arthritis, and systemic lupus erythematosus. The mechanism for the development of autoimmunities is not clear and may be different in different types of conditions.

89. **Why are organ transplants sometimes rejected, and how can rejection be prevented?**
The ability to recognize self from nonself is the fundamental basis of the immune system. Genetically determined histocompatibility antigens (HLA's) produced by genes of the major histocompatibility complex (MHC) are found on the surface membranes of all cells and can be recognized by immune surveillance cells. These antigens in graft tissue are the main cause of transplant rejection, but immunocompetent cells in bone marrow grafts sometimes can destroy host tissue, resulting in a dangerous condition called graft vs. host.

Transplant rejection can be controlled by immunosuppression, which is a lowering of the responsiveness of the immune system to materials it recognizes as foreign. This is accomplished by radiation and by cytotoxic drugs such as cyclosporin. The use of these techniques can also reduce the host's immune response to infectious agents.

90. **How is hypersensitivity involved in drug reactions?**
Most drug molecules are too small to act as allergens. However, if a drug combines with cellular proteins, it can be large enough to be recognized and a response can occur. All four types of hypersensitivity reactions have been observed in immunologic drug reactions.

91. **What are the causes, mechanisms, and effects of immunodeficiency diseases?**
Immunodeficiency diseases can arise from defects in any part of the immune system and include deficiencies in T-cells, B-cells, complement proteins, macrophages, and others. T-cell deficiencies lead to a lack of cell-mediated immunity; B-cell deficiencies lead to a lack of antibody-mediated immunity; complement deficiencies lead to problems in opsonization, inflammation, and cellular destruction of pathogens.

Immunodeficiencies can be hereditary or acquired as a result of infections, malignancies, and autoimmune disorders.

92. **How can antigens and antibodies be detected and measured?**
Serology is the use of laboratory tests to detect antigens and/or antibodies. Detection is possible because of the high degree of specificity of the antigen/antibody reaction and because the combination of these molecules forms a lattice network that can be observed. Diagnosis can be made on the basis of seroconversion.

Precipitation reactions involve tiny soluble antigens that precipitate from solutions or in agar gels. They must reach an appropriate equilibrium to be visible. Examples include immunodiffusion, immunoelectrophoresis and radial immunodiffusion.

Agglutination reactions involve large insoluble antigens such as blood cells or bacteria that can form large clumping, visible particles.

Complement fixation tests involve the use of complement proteins and blood cells to indirectly determine the presence of certain antibodies.

Immunofluorescence allows detection of products of immune reactions within tissues or with cells by means of fluorescent dyes.

Other special tests include radioimmunoassay, which uses radioactivity and enzyme-linked immunosorbant assay tests, which use the high degree of specificity of enzymes in the detection of antigen/antibody reactions.

93. What is AIDS and why is it causing an epidemic?

Most AIDS cases are caused by human immunodeficiency viruses (HIV-1 and HIV-2), which have evolved to attack the immune system. It is an infectious disease that gradually destroys the victim's T_4 or helper cell population. This leaves the patient susceptible to infections caused by opportunistic agents. The agent is transmitted by contact with infected blood and body fluids and can also be acquired by a fetus carried by an HIV positive woman. Because it can be easily spread via blood transfusions, sexual contact, infected needles, etc., it has reached epidemic proportions in many countries.

94. How can AIDS by diagnosed, treated, and prevented?

AIDS infections are detected by immunologic tests such as the ELISA test for the presence of antibodies. The disease is treated currently with AZT, ddC, and interferon but cannot be cured. Prevention is to avoid exposure to infected body fluids. A vaccine has not yet been developed.

Chapter 19 Diseases of the Skin and Eyes, Wounds, and Bites

INTRODUCTION

We now begin a series of six chapters that will explore the microbial diseases associated with different organ systems. There is a useful way of thinking about this subject. Microbes are to a person the way that plants and animals are to the Earth. In this analogy, a person is like a world, and every organ is like a habitat. The biosphere of the Earth contains forests, oceans, and mountains. Each habitat hosts a series of distinct organisms that are specifically adapted for each specific habitat. In this manner we think about diseases as an ecologists think of an ecosystem.

Our first chapter in this series considers skin. If skin were an ecosystem, which would it be? Some years ago there was an article in Scientific American that had a scanning electron micrograph of the surface of human skin and an aerial photograph of an ecosystem on Earth. The photograph was of a desert and it looked very similar to skin. Both are dry habitats with little shade. Skin consists mostly of the protein keratin, and this is a protein that few organisms can eat. Hence, it is a good protective surface. Skin is made of the same type of intermediate filament protein that makes up our hair, and just about nothing eats hair. Consider, did one of your parents every cut a lock from your head when you were a baby? Did you every see it many years later? It would appear the same. Bacteria and molds would not have eaten it. If you are every in Salem, Massachusetts, you might visit the Peabody-Essex Museum. There you can see a lock of Napoleon's hair. It looks just fine. This makes skin a challenge and few parasites are up to the challenge. This chapter will review those organisms that can infect our skin. You may even learn that when people think that there is something crawling under their skin, they just might be correct.

STUDY OUTLINE

I. **The Skin, Mucous Membranes, and Eyes**
   A. The skin
      The skin provides a nonspecific physical barrier against most microbes. It possesses an *epidermis*, a *dermis, keratin, sebaceous glands* that secrete *sebum*, and *sweat glands.*
   B. Mucous membranes
      Mucous membranes line tissues and organs that open to the exterior of the body. The cells of mucous membranes block the penetration of microbes and secrete *mucus*.

C. The eyes

The eyes have several external protective structures that lessen the chances of infection. These include the *conjunctiva,* the *cornea,* and *lysozyme* from the lacrimal gland.

D. Normal microflora of the skin

The acidic pH of the skin produced by the normal microflora colonized there presents a first-line, nonspecific defense against infection. Other first-line, nonspecific defenses of the skin include the secretion of salt-rich sweat and the continuous shedding of dead skin cells.

II. Diseases of the Skin

A. Bacterial skin diseases

Bacterial and other skin infections usually arise from the failure of the nonspecific defense mechanisms of the skin.

1. Staphylococcal infections

Staphylococcal infections can cause *folliculitis,* other skin lesions, and *scalded skin syndrome.*

2. Streptococcal infections

Streptococcal infections can cause *scarlet fever* and *erysipelas.*

3. Pyoderma and impetigo

Staphylococci, streptococci, and corynebacteria, singly or in combination, cause *pyoderma.* Staphylococci and streptococci, singly or in combination, cause *impetigo.*

4. Acne

Microorganisms feeding on sebum cause *acne.* Bacteria, especially *Propionibacterium acnes,* may be involved.

5. Burn infections

Pseudomonas aeruginosa is the prime cause of life-threatening burn infections, but *Serratia marcescens* and species of *Providencia* also often infect burns. The thick crust or scab that forms over a severe burn is called *eschar;* some of the eschar can be removed by a surgical scraping technique called *debridement.*

B. Viral skin diseases

1. Rubella

Rubella, or *German measles,* is the mildest of several human viral diseases that cause *exanthema,* or skin rash. *Congenital rubella syndrome* results from the infection of a developing embryo across the placenta. Rubella and congenital rubella syndrome have almost been eliminated due to a rubella vaccine. Laboratory tests diagnose rubella and the MMR vaccine is the only means of preventing rubella.

2. Measles

Measles, or *rubeola,* is a febrile disease with a rash caused by the rubeola virus and characterized by *Koplik's spots.* Complications include *measles encephalitis* and *subacute sclerosing panencephalitis (SSPE).* The disease is serious but rarely fatal, can be diagnosed by its symptoms, and is prevented by administration of the MMR vaccine.

3. Chicken pox and shingles

The *varicella-zoster virus,* a herpesvirus, causes both *chickenpox* and *shingles.* Chickenpox is usually first contracted between the ages of 5 and 9, while shingles usually appears in people over 45 years old. Doctors usually diagnose the disease by asking patients about their history of exposure and by studying the lesions. A vaccine is available for the virus, although lifelong immunity usually follows a childhood case of chickenpox.

4. Other pox diseases

Other pox diseases include *smallpox, cowpox,* and *molluscum contagiosum.*

5. Warts

Human *warts* are caused by *human papillomaviruses.* A few are malignant. Papillomaviruses are transmitted by direct contact or by fomites and can be treated using cryotherapy or caustic chemical agents, although recurrences are still common.

C. Fungal skin diseases

The fungi that invade keratinized tissue are called *dermatophytes* and fungal skin diseases are called *dermatomycoses.*

1. Ringworm

Ringworm occurs in several forms, including *tinea pedis, tinea corporis, tinea cruris, tinea unguium, tinea capitis,* and *tinea barbae.* Observation of the skin is usually sufficient for diagnosis.

2. Subcutaneous fungal infections

   Subcutaneous fungal infections include *sporotrichosis,* which is caused by *Sporothrix schenckii,* and *blastomycosis,* which is caused by *Blastomyces dermatitidis.*

3. Opportunistic fungal infections

   Examples of opportunistic fungal infections include *candiasis,* caused by *Candida albicans; aspergillosis,* caused by various species of *Asperigillus;* and *zygomycoses,* caused by certain zygomycetes of the genera *Mucor* and *Rhizopus.*

D. Other skin diseases

Madura foot (or *maduromycosis*), *swimmer's itch,* and *dracunculiasis* are other skin diseases caused by a variety of microorganisms.

III. Disease of the Eyes

A. Bacterial eye diseases

1. Ophthalmia neonatorum

   Opthalmia neonatorum, or conjunctivitis of the newborn, is a pyogenic infection of the eyes caused by organisms such as *Neisseria gonorrhoeae* and *Chlamydia trachomatis.* It can cause *keratitis,* an inflammation of the cornea that can progress to blindness. Penicillin, tetracycline, and erythromycin are used to combat this disease.

2. Bacterial conjunctivitis

   Bacterial conjunctivitis, or *pinkeye,* is an inflammation of the conjunctiva caused by organisms such as *Staphylococcus aureus, Streptococcus pneumoniae, Neisseria gonorrhoeae, Pseudomonas* species, and *Haemophilus influenzae* biogroup *aegyptius.* Bacterial conjunctivitis is extremely contagious. Topically applied sulfonamide ointment is an effective treatment.

3. Trachoma

   Trachoma is caused by specific strains of *Chlamydia trachomatis.* The disease can lead to blindness, although most cases of blindness are due to secondary bacterial infections developing in trachoma-infected tissues. Flies are important mechanical vectors and close mother-child contact facilitates human transfer.

B. Viral eye diseases

1. Epidemic keratoconjunctivitis

   Epidemic keratoconjunctivitis, caused by an adenovirus, can cause inflamed conjunctiva, eyelid edema, pain, tearing, sensitivity to light, and corneal clouding.

2. Acute hemorrhagic conjunctivitis

   Acute heorrhagic conjunctivitis is caused by an enterovirus. It causes severe eye pain, abnormal sensitivity to light, blurred vision, hemorrhage under conjunctival membranes, and sometimes transient inflammation of the cornea.

C. Parasitic eye diseases

1. Onchoceriasis (river blindness)

   The filarial larvae of the roundworm *Onchocerca volvulus* cause *onchocerciasis,* or *river blindness.* Total blindness can result. Diagnosis involves finding microfilariae in thin skin samples or adult worms visible through the skin. The drug ivermectin is effective in killing both adults and larvae, but toxins from dying microfilariae can cause anaphylactic shock.

2. Loaiasis

   A filarial worm called *Loa loa* causes *loaiasis.* It is diagnosed by finding microfilariae in blood or by finding worms in the skin or eyes and is treated by excising adult worms or by using drugs to eradicate microfilariae.

IV. Wounds and Bites

A. Wound infections

1. Gas gangrene

   Gas gangrene is often a mixed infection caused by two or more species of *Clostridium.* It causes a "snap, crackle, and pop" sound in *crepitant tissue* and can be prevented by adequately cleaning wounds, delaying wound closure, and providing drainage when wounds are closed.

2. Other anaerobic infections

   Other anaerobic infections include *cat scratch fever,* caused by *Afipia felis* and *Bartonella henselae;* and *rat bite fever,* caused by *Streptobacillus moniliformis.*

B. Arthropod bites and infections

1. Tick paralysis
   Ticks can attach themselves to the skin of a host and produce toxins that cause *tick paralysis*. Removal of ticks when symptoms first appear prevents permanent damage.
2. Chigger dermatitis
   Chiggers, the larvae of certain species of *Trombicula* mites, can cause a violent allergic reaction called *chigger dermatitis* in sensitive adults.
3. Scabies and house dust allergy
   Scabies, or *sarcoptic mange*, is caused by the itch mite *Sarcoptes scabiei*. Scabies is spread by close human contact and can be transmitted by sexual activities.
4. Flea bites
   The sand flea, *Tunga penetrans*, can cause extreme itching, inflammation, and pain. Wearing shoes and avoiding sandy beaches help protect against sand fleas.
5. Pediculosis
   Pediculosis, or lice infestation, results in reddened areas at bites, dermatitis, and itching. Insecticides can be used to eradicate lice, but sanitary conditions and good personal hygiene must be maintained to prevent their return.
6. Other insect bites
   Other diseases caused by insects include *blackfly fever* and *myiasis*.

SELF-TESTS

Continue with this section only after you have read Chapter 19 of your textbook. Write your answers in the appropriate space provided. Correct answers to all questions can be found at the end of the Self-Test section. A score of 80 percent or better is good. If your score is less than 65, reread the chapter.

True/False

Mark T for True, F for False

_____ **1.** Once a person develops scarlet fever due to streptococcal infection, he will never get a streptococcal infection again.

_____ **2.** Following severe burns, it is essential to remove the eschar to avoid serious bacterial infections.

_____ **3.** Chicken pox and shingles are actually caused by the same virus.

_____ **4.** Ringworm is actually caused by a fungus.

_____ **5.** The genus name for the most common fungal agents that affect the skin is *Tinea*.

_____ **6.** Ophthalmia neonatorum is an eye infection caused by species of *Chlamydia* similar to the one that causes trachoma.

Multiple Choice

Select the best possible answer.

_____ **7.** Molluscum contagiosum:
- **a.** causes warts.
- **b.** affects soft-bodied animals with hard shells.
- **c.** is a member of the pox virus family.
- **d.** can be treated with antibiotics.

_____ **8.** Pygmies of Uganda attained their short status by:
- **a.** being vegetarians.
- **b.** being infected with a tapeworm.
- **c.** having a growth hormone deficiency due to a nematode infection.
- **d.** having an addiction to hallucinogenic drugs.

_____ 9. Dermatophycoses:
   a. are all caused by the same fungus.
   b. often have a high mortality rate.
   c. treatment usually requires amphotericin B.
   d. can cause infection in laboratory workers who are culturing the infectious agents.
   e. None of the above is true.

_____ 10. The virus that causes warts:
   a. can only be transmitted sexually.
   b. can sometimes spontaneously disappear.
   c. can be treated with acyclovir.
   d. can only invade the respiratory tract.

_____ 11. An appropriate treatment for chickenpox in young children could be:
   a. tetracyclines.
   b. aspirin.
   c. amantadine.
   d. ivermectin.
   e. None of the above.

_____ 12. Which of the following would represent the least likely way of contracting an eye infection by pseudomonad or other opportunistic pathogen?
   a. frequent use of a hot tub.
   b. wearing mascara.
   c. wearing contact lenses.
   d. using testers at cosmetic counters.
   e. All of the above are good ways to contract eye infections.

Matching

Select the answer from the right-hand side that corresponds to the term or phrase on the left-hand side of the page. An answer may be used more than once. In some cases, more than one answer may be required.

Topic: Skin Lesions

_____ 1. Pustules
_____ 2. Ulcers
_____ 3. Folliculitis
_____ 4. Crusts
_____ 5. Abscess

a. inflammation of hair follicles
b. small elevated skin lesions with pus
c. rounded or irregular-shaped depressions
d. a collection of pus in any part of the body
e. dried accumulation of pus, blood, or serumcombined with cellular or bacterial remains

Topic: Bacterial Skin Diseases

_____ 6. *Propionibacterium acnes*
_____ 7. Exfoliatins
_____ 8. *Staphylococcus aureus*
_____ 9. *Pseudomonas aeruginosa*
_____ 10. *Streptococcus pyogenes*
_____ 11. *Clostridium* species

a. scalded skin syndrome
b. scarlet fever
c. impetigo
d. acne
e. pyoderma
f. common cause of burn infections
g. gas gangrene

Topic: Viral Skin Diseases

______	**12.** Dermal warts	**a.** human papillomaviruses
______	**13.** Koplik's spots	**b.** rubella virus
______	**14.** Shingles	**c.** varicella–zoster virus
______	**15.** German measles	**d.** molluscum contagiosum virus
______	**16.** Subacute sclerosing panencephalitis	**e.** rubeola virus
______	**17.** Genital warts	
______	**18.** Chickenpox	
______	**19.** Pearly white to pink painless tumors	
______	**20.** *Condylomata acuminata*	

Topic: The Dermatomycoses

______	**21.** *Tinea unguium*	**a.** body ringworm
______	**22.** *Tinea corporis*	**b.** groin ringworm
______	**23.** *Tinea barbae*	**c.** barber's itch
______	**24.** *Tinea cruris*	**d.** ringworm of the nails
______	**25.** *Tinea capitis*	**e.** scalp ringworm
______	**26.** *Tinea pedis*	**f.** ringworm of the foot

Topic: Microbial Eye Diseases

______	**27.** *Ophthalmia neonatorum*	**a.** *Haemophilus influenzae* biogroup *aegyptius*
______	**28.** Brazilian purpuric fever	**b.** *Neisseria gonorrhoeae*
______	**29.** Trachoma	**c.** enterovirus
______	**30.** Epidemic keratoconjunctivitis	**d.** *Chlamydia trachomatis*
______	**31.** Acute hemorrhagic conjunctivitis	**e.** adenovirus

Topic: Wounds and Bites

______	**32.** Cat scratch fever	**a.** *Clostridium perfringens*
______	**33.** Spirillar fever	**b.** *Bartonella henselae*
______	**34.** Myiasis	**c.** *Spirillum minor*
______	**35.** Gas gangrene	**d.** *Afipia felis*
______	**36.** Rat bite fever	**e.** *Streptobacillus moniliformis*
______	**37.** Chigger dermatitis	**f.** *Sarcoptes scabies*
______	**38.** Scabies	**g.** *Trombicula*
______	**39.** Pediculosis	**h.** *Tunga penetrans*
______	**40.** Flea bites	**i.** *Pediculus humanus*
		j. *Gastrophilus intestinalis*

Fill-Ins

Provide the correct term or phrase for the following. Spelling counts.

53. The outermost, thin layer of the skin is the ________________.

54. The underlying, thicker layer of the skin is the ________________.

55. The dead epithelial cells of the epidermis contain a waterproofing protein called ________________.

56. The acidic nature of the oily secretion of the sebaceous glands, called ________________, acts to maintain an acidic skin pH that discourages pathogen growth.

57. ________________ is a thick but watery secretion produced by goblet cells and certain glands that prevents drying and cracking of the mucous membrane.

58. An infection at the base of an eyelash is commonly called a ________________.

59. The toxin responsible for scarlet fever is called the ________________ toxin.

60. A highly infectious pyoderma caused by a staphylococcus or streptococcus is ________________.

61. Blackheads that become infected by the bacterium *Propionibacterium* can result in ________________.

62. The removal of dead tissue following a severe burn is called ________________.

63. Rubella is also known as ________________.

64. Two to three days before the onset of symptoms of the red measles, red spots with bluish central specks, called ________________ spots, appear on the oral mucosa.

65. Red measles viruses can cause two severe complications, measles ________________ and ________________ ________________ ________________.

66. An adult case of chicken pox often results in painful lesions called ________________.

67. Human papillomaviruses cause skin eruptions called ________________.

68. Fungi that invade keratinized tissue are called ________________.

69. *Tinea pedis* is also known as ________________ ________________.

70. An oral infection caused by *Candida albicans* is ________________.

71. A foot infection caused by fungi of the genus *Madurella* is ________________ foot.

72. Schistosome larvae that normally infect birds can cause a skin infection in humans called ________________.

73. An inflammation of the cornea of the eye is called ________________.

74. Bacterial conjunctivitis, which causes the eye to become red and inflamed, is commonly known as ________________.

75. A serious eye infection typified by a "pebbled" or rough appearance is ________________.

76. Epidemic keratoconjunctivitis, sometimes called "shipyard eye," is most likely caused by a ________________.

77. Onchocerciasis caused by nematode larvae is known as ________________ ________________.

78. Deep wound infections caused by species of *Clostridia* that result in the production of gas are ________________ ________________.

79. Tissues filled with gas bubbles that snap, crackle, and pop are called ________________ tissues.

80. Spirillar fever is also known as ________________ in Japan.

81. Scabies is caused by a ________________.

82. A louse infestation is called ________________.

83. "Nit-picking" means to remove ________________.

84. An infection caused by fly larvae is ________________.

Critical Thinking

85. What kinds of pathogens cause skin diseases?

86. What are the important epidemiologic and clinical aspects of skin diseases?

87. What kinds of pathogens cause eye diseases?

88. What are the important epidemiologic and clinical aspects of eye diseases?

89. What kinds of pathogens infect wounds and bites?

90. What are the important epidemiologic and clinical aspects of wound and bite infections?

ANSWERS

True/False

1. **False** A person will not get another case of scarlet fever due to antibodies produced against the toxin. However, because there is a great variety of strains of streptococci, anyone acquiring a streptococcal infection will be just as likely to get another one.
2. **True** The scab or eschar that forms over a severe burn can represent a site for infection beneath the tissue. Pathogens could then gain access to the blood.
3. **True** The varicella-zoster virus causes both diseases. Chicken pox usually occurs in children and zoster in adults.
4. **True** Ringworm is not caused by a worm but by a fungus such as *Epidermophyton, Microsporum,* and *Trichophyton.*
5. **False** Tinea is actually a synonym for cutaneous infections. It is used along with another term to identify a cutaneous fungal infection on some portion of the body.
6. **False** Ophthalmia neonatorum is caused by a pyogenic diplococcus, *Neisseria gonorrhoeae.*

Multiple Choice

7. c; 8. c; 9. d; 10. b; 11. e; 12. e.

Matching

13. b; 14. c; 15. a; 16. e; 17. d; 18. d; 19. a; 20. a, c, e; 21. f; 22. b, c, e; 23. g; 24. a; 25. e; 26. c; 27. b; 28. e; 29. a; 30. c; 31. d; 32. a; 33. d; 34. a; 35. c; 36. b; 37. e; 38. f; 39. b, d; 40. a; 41. d; 42. e; 43. a; 44. b, d; 45. c; 46. j; 47. a; 48. e; 49. g; 50. f; 51. i; 52. h.

Fill-Ins

53. epidermis; 54. dermis; 55. keratin; 56. sebum; 57. mucus; 58. sty; 59. erythrogenic; 60. impetigo; 61. acne; 62. debridement; 63. German; 64. Koplik's; 65. encephalitis, subacute sclerosing panencephalitis; 66. shingles; 67. warts; 68. dermatophytes; 69. athlete's foot; 70. thrush; 71. madura; 72. swimmer's itch; 73. keratitis; 74. pinkeye; 75. trachoma; 76. adenovirus; 77. river blindness; 78. gas gangrene; 79. crepitant; 80. sodoku; 81. mite; 82. pediculosis; 83. lice; 84. myiasis.

Critical Thinking

85. **What kinds of pathogens cause skin diseases?**
Various bacterial, viral, fungal, and parasitic organisms cause skin diseases.

86. **What are the important epidemiologic and clinical aspects of skin diseases?**
Bacterial skin diseases are usually transmitted by direct contact, droplets, or fomites. Many of the agents are part of the normal flora of the skin that have gained access via cuts or scratches. Most of these diseases can be treated with penicillin or other antibiotics. Some bacteria such as *Staphylococcus* have become very resistant to many antibiotics because of their inappropriate use. Burn infections often are nosocomial and are commonly caused by antibiotic-resistant organisms such as *Pseudomonas* and *Staphylococcus.*

Viral skin diseases such as rubella, measles, and chicken pox are usually transmitted via nasal secretions. Some, such as small pox and warts are commonly transmitted by direct contact. Treatment of these skin diseases is difficult and usually confined to relieving the symptoms and not curing the disease. Warts can be excised but commonly recur.

Fungal skin diseases are commonly acquired by direct contact and by fomites. Some fungal agents such as *Candida* are opportunists. Healthy, intact skin and mucous membranes are usually adequate to resist most of these agents. Topical antifungal agents are usually effective in controlling the dermatophytes such as athlete's foot and ringworm. Subcutaneous fungal infections often persist in spite of treatment with topical fungicides. Systemic infections such as blastomycosis and systemic candidiasis are very difficult to treat and often require Amphotericin B.

87. **What kinds of pathogens cause eye diseases?**
Diseases of the eyes are caused by bacteria, viruses, and parasites.

88. **What are the important epidemiologic and clinical aspects of eye diseases?**
Bacterial eye diseases can be transmitted by direct contact, fomites, insect vectors, and to infants during delivery. Most are treated with antibiotics and prevented by common sense and good sanitation.

Viral eye disases are transmitted by dust particles or direct contact. Since viruses do not respond to antibiotics, there is no real effective treatment available, but good common sense and proper sanitation can help to prevent these diseases.

Parasitic eye diseases are usually transmitted by insect vectors. By avoiding the insect bites and by controlling them, the incidence of these diseases can be reduced.

89. **What kinds of pathogens infect wounds and bites?**
Wound infections can be caused by various types of bacteria such as *Clostridium*. Bite infections can be caused by ticks, chiggers, mites, fleas, and others. Often the bite wounds can be susceptible to further bacterial infection.

90. **What are the important epidemiologic and clinical aspects of wound and bite infections?**
Wound infections such as gas gangrene and other anaerobic infections can be caused by various species of soil bacteria entering uncleaned lacerated wounds. Proper cleaning and draining of these wounds followed by treatment with antibiotics such as penicillin or antitoxins can reduce or eliminate the possibility of severe complications. Cat scratch and rat bite fevers can be prevented by avoiding the animals that can cause the injuries and by proper treatment of the affected areas. Laboratory workers who must work with these animals should be aware of the dangers of these infections and should take all necessary precautions including the use of protective clothing.

Arthropod bites and infections can be prevented by good sanitation and proper hygiene along with protection of the skin to avoid the bites.

Chapter 20 Urogenital and Sexually Transmitted Diseases

INTRODUCTION

The journey of a parasite into the urogenital system may seem unlikely. Entering the urethra is definitely to go against the flow. To face the full brunt of a stream of acidic urine cannot be easy. As you may know too well, many organisms and viruses succeed. Some like *E. coli* are able to swim against the current and hold on to the epithelial walls when flagella would fail. Obviously, the path to the urinary bladder is much shorter in males than females, but success happens all too frequently. For your generation, the health of your urogenital system is much more questionable than when I was in college.

I am one of a relatively few lucky people who spent their youth with an unusual gift. While I was in college I knew that as I journeyed through my exploratory social life there was one thing that I could really depend upon. For those of us lucky enough to have been born soon after World War II, that was penicillin. From just after the end of World War II to 1980, young people who were sexually active in the United States had the knowledge that it was very unlikely that they wo9uld die from a sexually transmitted disease. Your generation is not so fortunate. Risks abound.

More has changed than HIV AIDS, and this chapter documents the array of difficulties facing the sexually active. Even bacterial infections have changed. When I was in college, gonorrhea was easily treated with penicillin. Today, many strains are resistant. If the disease becomes entrenched in the prostate gland, perhaps remission is the only hope. Clearly, people must seek help quickly is there is a problem with their plumbing.

STUDY OUTLINE

I. Components of the Urogenital System

A. The urinary system

The *urinary system* consists of paired *kidneys* (both of which have *nephrons*) and *ureters*, the *urinary bladder*, and the *urethra*.

B. The female reproductive system

The *female reproductive system* consists of the *ovaries, uterine tubes, uterus, vagina,* and *external genitalia.*

C. The male reproductive system

The *male reproductive system* consists of the *testes, ducts*, specific glands, and the *penis.*

D. Normal microflora of the urogenital system

In healthy individuals, all parts of the urinary tract (except for the portion of the urethra closest to the urethral opening) are sterile. This is because the acidic pH and high salt and urea concentrations retard the growth of bacteria in urine. *Mycobacterium smegmatis* and *Mycobacterium tuberculosis*, however, are normal microflora of different parts of the urogenital system

II. Urogenital Diseases Usually Not Transmitted Sexually

A. Bacterial urogenital diseases

Examples of bacterial urogenital diseases include *urinary tract infections, prostatis, pyelonephritis, glomerulonephritis, leptospirosis, bacterial vaginitis*, and *toxic shock syndrome.*

B. Parasitic urogenital diseases

The example of parasitic urogenital diseases given in the book is *trichomoniasis.*

III. Sexually Transmitted Diseases

A. Bacterial sexually transmitted diseases

The only means of preventing sexually transmitted diseases (STDs) is to avoid exposure to them.

1. Gonorrhea

Gonorrhea is caused by endotoxins of *Neisseria gonorrhoeae.* As many as half of infected females develop *pelvic inflammatory disease* after contracting gonorrhea. Diagnosis is made by identifying *N. gonorrhoeae* in laboratory cultures. Increasing resistance to drugs makes it very difficult to combat gonorrhea, and no vaccine is available.

2. Syphilis

Syphilis is caused by the spirochete *Treponema pallidum.* The disease typically undergoes certain stages: (1) the incubation stage; (2) the primary stage; (3) the primary latent stage; (4) the secondary stage; (5) the secondary latent stage; and (6) the tertiary stage. Diagnostic tests include fluorescent antibody and treponemal immobilization tests. Syphilis is usually treated with penicillin, but tetracycline and erythromycin are also effective. No vaccine is available and recovery does not confer immunity.

3. Chancroid

Chancroid is caused by *Haemophilus ducreyi.* Chancroid begins with the appearance of chancres and can form *buboes* if it spreads to the groin. The disease is diagnosed by identifying the organism in scrapings from a lesion or in fluids from a bubo. It is treated with antibiotics.

4. Nongonococcal urethritis

Nongonococcal urethritis is a gonorrhea-like STD caused by organisms other than gonococci. *Chlamydia trachomatis, Mycoplasma hominis,* or *Ureaplasma urealyticum* can cause this disease.

5. Lymphogranuloma venereum

Lymphogranuloma venereum is an STD caused by *Chlamydia trachomatis.* It is diagnosed by finding chlamydias as inclusions stained with iodine in pus from lymph nodes and can be treated with tetracycline, erythromycin, and sulfamethoxasole.

6. Granuloma inguinale

Granuloma inguinale is caused by *Calymmatobacterium granulomatis,* which is related to *Klebsiella.* Diagnosis from scrapings of lesions is confirmed by finding large mononuclear cells called *Donovan bodies.* Antibiotics such as ampicillin, tetracycline, erythromycin, and gentamicin offer effective treatment.

B. Viral sexually transmitted diseases

1. Herpesvirus infections

Herpes simplex virus 1 typically causes fever blisters (cold sores) and *herpes simplex virus* 2 typically causes *genital herpes.* Other herpes infections include *neonatal herpes, gingivostomatitis, herpes labialis, keratoconjunctfvitis, herpes meningoencephalitis, herpes pneumonia, eczema herpeticum, traumatic herpes,* and *herpes gladiatorium.* Herpesvirus cells from the base of a lesion can be grown for diagnostic purposes in a laboratory. The best available means of preventing HSV infections is to avoid contact with individuals with HSV-1 or HSV-2 lesions.

2. Genital warts

Condylomas, or *genital warts,* are caused by the human papillomavirus. A special hazard of genital warts is their epidemiologic association with cervical carcinoma.

3. Laryngeal papillomas

Laryngeal papillomas, usually caused by HPV-6 and HPV-11, are benign growths that can be dangerous if they block the airway.

4. Cytomegalovirus infections

The *cytomegaloviruses* can cause fetal and infant CMV infections like *cytomegalic inclusion disease* or disseminated CMV infections. Definitive diagnosis of CMV infections requires identification of the virus from clinical specimens, although monoclonal antibodies can be used to detect viral antigens much more quickly. No effective treatment for CMV infections in infants is available, although interferon and hyperimmune gamma globulin given before and after organ transplantation reduce the incidence and severity of CMV infections in transplant patients. No effective vaccine exists.

SELF-TESTS

Continue with this section only after you have read Chapter 20 of your textbook. Write your answers in the appropriate space provided. Correct answers to all questions can be found at the end of the Self-Test section. A score of 80 percent or better is good. If your score is less than 65, reread the chapter.

True/False

Mark T for True, F for False

______ **1.** Toxic shock syndrome seems to occur only with menstruating women.

______ **2.** Strains of gonococci that lack pili usually are nonvirulent.

______ **3.** Once infected, a patient retains herpes simplex viruses for life.

______ **4.** CMV infections seem to be most severe in teenagers, causing fever, malaise, and abnormal liver function.

______ **5.** NGU infections appear to be caused most often by papilloma viruses because of their ease of transmission.

______ **6.** Chancroid is caused by a *Hemophilus* species, which forms soft, painful lesions on the genitalia.

______ **7.** Women are 40 to 50 times more likely to acquire a UTI.

______ **8.** Although *Trichomonas hominis* is considered a commensal, *T. tenax* causes serious genital infections in cattle and is the leading cause of spontaneous abortion in these animals.

Multiple Choice

Select the best possible answer.

______ **9.** Which of the following venereal diseases is easiest to treat?
- **a.** Genital warts.
- **b.** HSV-2.
- **c.** Tertiary syphilis.
- **d.** Gonorrhea.

______ **10.** Syphilis is characterized by all the following except:
- **a.** The primary stage is characterized by a painless, hard, crusty chancre.
- **b.** The secondary stage is characterized by a noninfectious rash similar to measles.
- **c.** Neurosyphilis can occur in the tertiary stage.
- **d.** Congenital syphilis may cause all types of deformities including saber shin, saddle nose, and Hutchinson's incisors.

______ **11.** Select the most correct statement concerning UTI's.
- **a.** UTI's generally are very rare in the United States but rather common in the tropics.
- **b.** UTI's can rapidly spread throughout the urinary tract by "ascending" or "descending".
- **c.** Most UTI's cannot be treated with antibiotics.
- **d.** Once a patient contracts a UTI, they are usually immune to further infections.

______ **12.** In terms of venereal diseases:
- **a.** *Lymphogranuloma venereum* is fairly common in the United States especially in the Northeast.
- **b.** Donovanosis is a rare disease and does not respond to antibiotic therapy.
- **c.** Genital warts seems to have a close link with cervical carcinoma.
- **d.** CMV infections are only transmitted sexually.

_____ 13. In cases of leptospirosis, all the following are true except:
a. The disease is zoonosis.
b. The spirochaetes specifically invade the bladder and eventually infect the kidneys of humans resulting in a severe pyelonephritis.
d. The spirochaetes readily respond to most antibiotics if given early.
e. A vaccine is available for dogs and cats.

Matching

Select the answer from the right-hand side that corresponds to the term or phrase on the left-hand side of the page. An answer may be used more than once. In some cases more than one answer is required.

Topic: Pathogens of the Urogenital System

_____ 14. Syphilis	a. *Trichomonas vaginalis*
_____ 15. Nongonococcal urethritis	b. *Treponema pallidum*
_____ 16. Chancroid	c. *Neisseria gonorrhoeae*
_____ 17. Inclusion conjunctivitis	d. *Gardnerella vaginalis*
_____ 18. Gonorrhea	e. *Chlamydia trachomatis*
_____ 19. Trichomoniasis	f. *Haemophilus ducreyi*
_____ 20. Inclusion blenorrhea	g. *Mycoplasma hominis*
_____ 21. Lymphogranuloma	h. *Ureaplasma urealyticum*
_____ 22. Granuloma inguinale	i. *Calymmatobacterium granulomatis*
_____ 23. Weil's disease	j. *Leptospira interrogans*
_____ 24. Vaginitis	k. *Candida albicans*

Topic: Characteristics of Disease Agents

_____ 25. *Escherichia coli*	a. spirochete
_____ 26. *Haemophilus ducreyi*	b. gram-negative rod
_____ 27. *Mycoplasma hominis*	c. lack of cell wall
_____ 28. *Treponema pallidum*	d. gram-negative diplococcus
_____ 29. *Chlamydia trachomatis*	e. gram-positive coccus
_____ 30. *Ureaplasma urealyticum*	
_____ 31. *Neisseria gonorrhoeae*	
_____ 32. *Calymmatobacterium granulomatis*	
_____ 33. *Staphylococcus aureus*	
_____ 34. *Leptospira interrogans*	

Topic: Disease Signs and Symptoms

_____ 35. Severe liver damage	a. chancroid
_____ 36. Clue cells	b. gonorrhea
_____ 37. Ophthalmia neonatorum	c. syphilis
_____ 38. Hard chancre	d. *chlamydia trachomatis* infection
_____ 39. Gummas	e. granuloma inguinale
_____ 40. Soft chancre	f. cystitis

_____ 41. Donovan bodies

_____ 42. Urinary bladder inflammation

_____ 43. Kidney inflammation

g. pyelonephritis

h. *Gardnerella* vaginitis

i. Leptospirosis (Weil's disease)

Topic: Viral STDs

_____ 44. Gingivostomatitis

_____ 45. Herpes labialis

_____ 46. Genital herpes

_____ 47. Neonatal herpes

_____ 48. Laryngeal papillomas

_____ 49. Cervical carcinoma

_____ 50. Herpes gladiatorium

_____ 51. Cytomegalic inclusion disease

a. HPV-8

b. HSV-2

c. HPV-6

d. HSV-1

e. CMV

Fill-Ins

Provide the correct term or phrase for the following. Spelling counts.

52. The _______________ _______________ includes organs of both the urinary system and the reproductive system.

53. _______________ is a laboratory analysis of urine that can reveal imbalances in pH or water concentration, the presence of substances such as glucose or proteins, and other conditions associated with infections, metabolic disorders, and other diseases.

54. _______________ _______________ is an acid-fast staining bacillus that comprises normal microflora on the external genitalia of both males and females.

55. UTI stands for _______________ _______________ infection.

56. An inflammation of the urethra is called _______________.

57. An infection that spreads from the urethra to the bladder would best be termed _______________.

58. An inflammation of the prostate gland is _______________.

59. Women with painful, burning urination, indicative of urethral infections, have a condition called _______________.

60. An inflammation of the kidneys is known as _______________.

61. Liver damage and jaundice caused by leptospiras is indicative of _______________ disease.

62. Symptomatic epithelial cells covered with bacteria that suggest infections caused by *Gardnerella* are called _______________ cells.

63. Toxigenic strains of staphylococci produce an _______________ _______________ that enhances the effect of its endotoxin.

64. STD refers to _______________ _______________ diseases.

65. PMNL's refer to _______________ _______________.

66. Gonorrhea infections that spread throughout the pelvic cavity indicate a _______________ _______________ disease.

67. A hard, crusty, painless lesion of syphilis is a _______________.

68. Noninfectious granulomatous inflammations seen in tertiary syphilis are known as _______________.

69. The complement fixation or VDRL test for syphilis is named for the _______________ _______________ _______________ _______________.

70. Notched incisors seen in congenital syphilis are ________________ incisors.

71. Inflamed lymph nodes are called ________________.

72. NGU stands for ________________ urethritis.

73. Self-inoculation with chlamydial infections from the genitals to the eyes can result in ________________ ________________.

74. LGV stands for ________________ ________________.

75. The finding of large mononuclear cells called ________________ bodies is diagnostic for *Granuloma inguinale*.

76. Lesions of the mucous membranes of the mouth, especially in young children, are called ________________.

77. Herpes infections of the eye are called herpes ________________.

78. A herpetic lesion on a finger can be called a herpes ________________.

79. The condition characterized by genital warts is ________________.

80. A generalized infection in babies caused by the cytomegalovirus is ________________ ________________ disease.

Labeling

Select the best possible label for each of the body areas that can show lesions from a herpes simplex virus infection: herpes of the eye; sacral ganglia; whitlow; brachial plexus ganglia; trigeminal ganglion; fever blister (labial herpes); genital herpes. *Place your answers in the spaces provided on page 269.*

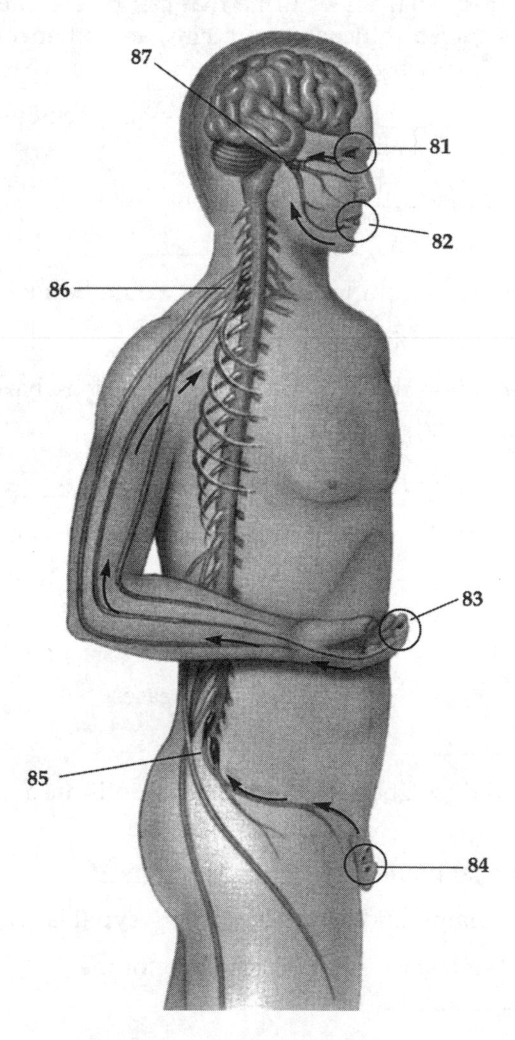

81. ____________________

82. ____________________

83. ____________________

84. ____________________

85. ____________________

86. ____________________

87. ____________________

Critical Thinking

88. What bacteria cause urogenital diseases that are not usually sexually transmitted, and what are the important epidemiologic and clinical aspects of these diseases?

89. What parasites cause urogenital diseases that are not usually sexually transmitted, and what are the important epidemiologic and clinical aspects of these diseases?

90. What bacteria cause sexually transmitted urogenital diseases, and what are the important epidemiologic and clinical aspects of these diseases?

91. What viruses cause sexually transmitted urogenital diseases, and what are the important epidemiologic and clinical aspects of these diseases?

ANSWERS

True/False

1. **False** Infections with toxigenic strains of *Staphylococcus aureus* can affect both males and females and may occur following surgery, boils, or furuncles.
2. **True** The gonococci possess sex pili, which are essential for conjugation and transmission of antibiotic resistance and common pili, which are used for attaching to epithelial cells lining the urinary tract to avoid being swept out with the elimination of urine.
3. **True** The HSV viruses appear to become latent in the cells they infect and periodically recur. This reactivation can be spontaneous or be the result of UV light, stress, or other factors.
4. **False** CMV infections generally go unnoticed in older children and adults but are most severe and even life-threatening in fetuses and infants.
5. **False** NGU appears to be caused by *Chlamydia trachomatis.* Papilloma viruses cause genital warts.
6. **True** *Hemophilus ducreyi* is the causative agent of chancroid or soft chancre to distinguish it from the lesions of syphilis.
7. **True** The female urethra is shorter (4 cm) than the male urethra (20 cm) thus greatly increasing the chances of infection.
8. **False** *Trichomonas tenax* and *T. hominis* are both considered commensals. *T. vaginalis* causes serious vaginal infections in humans and *T. foetus* causes serious vaginal infections in cattle.

Multiple Choice

9. d; 10. b; 11. b; 12. c; 13. c.

Matching

14. b; 15. e, g, h; 16. f; 17. e; 18. c; 19. a; 20. e; 21. e; 22. i; 23. j; 24. a, d, k; 25. b; 26. b; 27. c; 28. a; 29. b; 30. c; 31. d; 32. b; 33. e; 34. a; 35. i; 36. h; 37. b, d; 38. c; 39. c; 40. a; 41. e; 42. f; 43. g; 44. d; 45. d; 46. b; 47. b; 48. c; 49. a; 50. d; 51. e.

Fill-Ins

52. urogenital system; 53. urinalysis; 54. *Mycobacterium smegmatis*; 55. urinary tract; 56. urethritis; 57. urethrocystitis; 58. prostatitis; 59. dysuria; 60. pyelonephritis; 61. Weil's; 62. clue; 63. exotoxin C; 64. sexually transmitted; 65. polymorphonuclear leukocytes; 66. pelvic inflammatory; 67. chancre; 68. gummas; 69. Venereal Disease Research Laboratory; 70. Hutchinson's; 71. buboes; 72. nongonococcal; 73. inclusion conjunctivitis; 74. *Lymphogranuloma venereum*; 75. Donovan; 76. gingivostomatitis; 77. keratoconjunctivitis; 78. whitlow; 79. condylomas; 80. cytomegalic inclusion.

Labeling

81. herpes of the eye; 82. fever blister (labial herpes); 83. whitlow; 84. genital herpes; 85. sacral ganglia; 86. brachial plexus ganglia; 87. trigeminal ganglion.

Critical Thinking

88. **What bacteria cause urogenital diseases not usually sexually transmitted, and what are the important epidemiologic and clinical aspects of these diseases?**
    Urinary tract infections or UTI's are among the most common of all infections seen in clinical practice. They can ascend from the lower urethra to affect the bladder and kidneys or they can descend from systemic kidney infections to affect the bladder. They are diagnosed from urine cultures, treated with antibiotics, and prevented by good personal hygiene and complete emptying of the bladder. Drinking plenty of fresh water can greatly help to reduce these infections. Various organisms can affect this system.

    Urethritis, or inflammation of the urethra, and cystitis, or inflammation of the bladder, are commonly caused by *Escherichia coli* and others. Prostate infections may result from a urinary tract infection (UTI) and usually can be treated with antibiotics. Dysuria or painful urination is often indicative of an infection and is very common in the elderly. Kidney infections also may result from a UTI and are usually more difficult to treat because of renal failure. Infections of the glomeruli of the kidneys may follow streptococcal or viral infections and may result from immune complexes that deposit in the tissue. Treatment of the initial agent can reduce the risk of these infections.

Leptospirosis is considered a zoonosis since it is associated with dogs, rodents, and other wild animals. It is transmitted through the animal's urine and can be treated with antibiotics. Pets can be vaccinated to prevent its transmission to humans. Animal control workers and zookeepers are at risk for this disease.

Vaginal infections or vaginitis is usually caused by opportunistic bacteria and other microbes that multiply when the normal vaginal flora is disturbed. *Gardnerella* vaginitis is quite common and can be diagnosed by finding "clue cells." It can be treated with metronidazole to eradicate the anaerobes and allow restoration of normal lactobacilli.

Toxic shock syndrome most often arises from the use of high-absorbency tampons, although it can follow surgeries that were contaminated with the toxigenic strain of *Staphylococcus*. Whenever a high fever, rash, and drop in blood pressure occur, treatment should be made promptly.

89. What parasites cause urogenital diseases not usually sexually transmitted, and what are the important epidemiologic and clinical aspects of these diseases?

The most important parasite is *Trichomonas,* which can be transmitted sexually and by indirect contact from linens and contaminated toilet seats. It is a very plump flagellate and is easily diagnosed from smears of the secretions. It can be treated with metronidazole. If the disease is transmitted sexually, both partners must be treated or reinfection will most likely occur.

90. What bacteria cause sexually transmitted urogenital diseases, and what are the important epidemiologic and clinical aspects of these diseases?

Sexually transmitted diseases or STD's have become an ever increasingly serious health problem in recent years, partly because of changing sexual behavior and because of the development of antibiotic-resistant strains. No vaccines have ever been developed to control any of the STD's and the only methods of prevention are good common sense and avoidance of exposure. Several types of bacteria are responsible for many of the STD's.

Gonorrhea is diagnosed by finding Gram-negative intracellular diplococci in pus samples and in culture. It is normally treated with penicillin but the emergence of resistant strains has required the use of other, more potent drugs. Because of the difficulty in recognizing the disease in pregnant women, antibiotic ointments are placed in the eyes of all newborns to prevent possible eye infections. As many as one-half of all infected females will develop pelvic inflammatory disease (PID).

Syphilis has seen a slow, but steady increase, possibly because of a lack of concern for the disease. The incidence of congenital cases has also increased dramatically. It is diagnosed by immunological tests and is treated with penicillin or other broad-spectrum antibiotics.

Chancroid occurs primarily in underdeveloped countries; however, cases have occurred as a result of prostitutes "visiting" the homes of illegal immigrants. It is diagnosed by finding the Gram-negative bacteria in lesions or buboes and is treated with tetracyclines.

Nongonococcal urethritis or NGU has increased dramatically and appears to be caused primarily by species of *Chlamydia*. It can be diagnosed by culturing samples from discharges and treated with broad-spectrum antibiotics. Many victims have inapparent infections for years and only later discover that serious damage has occurred to their reproductive organs.

Lymphogranuloma venereum and *Granuloma inguinale* are venereal diseases most common to tropical and subtropical areas and to Africa and Asia. They are diagnosed by finding the agents in pus or other discharges from lesions or buboes. They are treated with broad-spectrum antibiotics.

91. What viruses cause sexually transmitted urogenital diseases, and what are the important epidemiologic and clinical aspects of these diseases?

Several viruses can be transmitted sexually and because they do not respond to antibiotics, they are often difficult to treat.

Herpes simplex virus Type 1 (HSV-1) causes most of the oral herpes infections or cold sores and herpes simplex virus type 2 (HSV-2) causes most of the genital infections. Genital herpes is the most common and most severe of the herpes simplex virus infections. It is diagnosed by finding the viruses in vesicular fluids, by immunologic tests, and by clinical symptoms. Acyclovir can help to reduce the symptoms but cannot cure the disease. Once patients are infected the viruses invade the sensory nerves and establish a permanent latency. Reactivation of these viruses may occur spontaneously or by some external influence.

Genital warts have seen an alarming increase in recent years and must be distinguished from cervical dysplasia and carcinoma. Cryosurgery is usually necessary to remove them; however, they often recur.

Cytomegalovirus infections are most severe in fetuses, newborns, and in the immunocompromised. Monoclonal antibody tests can be used for diagnosis but no effective treatment is available.

Chapter 21 Diseases of the Respiratory System

INTRODUCTION

Microbes often enter our respiratory system in our mouths. As a habitat it is quite favorable indeed. It has plenty of food and water. A few risks to the microbes exist. For example, microbes can be swallowed and face the acids and enzyme of the stomach. Some organism can make their way into the larynx and lower regions, but we can usually prevent them from moving into our lungs.

In the year that 3,000 Americans died on 9/11, 36,000 died of influenza. Upper respiratory infections are a major cause of death in the United States. In is worth noting, that in 2004, America was not able at first to provide influenza vaccine to those at risk. Clearly, we do not place appropriate priority on diseases of the respiratory system. Perhaps we take breathing for granted, and we do so at our peril.

Under normal circumstances, microbes are not present in our lungs. Cilia and mucus in our upper respiratory tract prevent most them from entering. Usually infections begin in our upper respiratory system and involve our nasal epithelium and throat. The upper respiratory infections can overwhelm our defense and bacteria or viruses can invade our lower respiratory track. It is bacterial pneumonia that consisting ranks in the top ten causes of death. Bacteria are becoming resistant to antibiotics like penicillin and clearly we lose many patients, especially the elderly, despite all efforts.

New risks to the public health are emerging. In the East, avian flu is a common disease of poultry. A small number of cases spread from birds to people but it did not spread from person to person. Now that is not the case and a pandemic of avian flu is possible. Great Britain in 2005 has committed to stock enough antiviral drugs for 25% of its population. The United States will stock enough for 10% of its population. The potential for a flu pandemic to kill was demonstrated in 1918, when millions died all over the world, including young and healthy people in America. Clearly, more resources to prevent and fight reparatory infections are needed.

STUDY OUTLINE

I. **Components of the Respiratory System**

   A. The upper respiratory tract

   Air from the environment first passes through the upper respiratory tract: air first passes through the *nasal cavity*, then through the *pharynx*, then through the *trachea*, and lastly through the primary *bronchi*.

B. The lower respiratory tract

After passing through the upper respiratory tract, air passes through the lower respiratory tract. First come the secondary bronchi, then the smaller *bronchioles*, then the *respiratory bronchioles*, which end in a series of saclike *alveoli*.

C. The ears

Air entering the ear first passes through the *outer ear*, composed of the flaplike *pinna* and the *auditory canal*. The *tympanic membrane* separates the outer and middle ears. The *middle ear* is a small, air-filled cavity containing small bones called *ossicles*. If the tympanic membrane is intact, most middle ear infections arise from microorganisms that move up the *auditory (eustachian) tube* from the nasopharynx. The *inner ear* converts sound waves into nerve impulses carried by the vestibulocochlear cranial nerve.

D. Normal microflora of the respiratory system

Although the lungs are usually sterile, *Mycobacterium tuberculosis* and *coccidioides immitis* can remain airborne until they reach the alveoli. The pharynx has a normal microflora similar to that of the mouth. Potential pathogens such as *Klebsiella pneumoniae* and species of *Haemophlius* can survive in and colonize the pharynx. The upper respiratory tract has a normal microflora similar to that of the skin, with *Staphylococcus epidermidis* and corynebacteria being most numerous. *Staphylococcus aureus* is also often found here.

II. Diseases of the Upper Respiratory Tract

A. Bacterial upper respiratory diseases

Bacterial infections of the *upper respiratory tract* are easily acquired and exceedingly common.

1. Pharyngitis and related infections

Pharyngitis, or sore throat, is an infection of the pharynx sometimes caused by *Streptococcus pyogenes*. Other related infections include: (1) *laryngitis*, caused by *Haemophilus influenzae* or *Streptococcus pneumoniae* or viruses; (2) *epiglottitis*, caused by *H. influenzae;* (3) *sinusitis*, caused by *Streptococcus pneumoniae, Haemophilus influenzae, Staphylococcus aureus*, or *Streptococcus pyogenes;* (4) *bronchitis*, caused by *Streptococcus pneumoniae, Mycoplasma pneumoniae*, and various species of *Haemophilus, Streptococcus*, and *Staphylococcus*; and (5) *tonsilitis*.

2. Diphtheria

Diphtheria is caused by strains of *Corynebacterium diphtheriae* infected with a prophage that carriers an exotoxin-producing gene. *C. diphtheriae* is usually spread by respiratory droplets; infection usually begins in the pharynx 2 to 4 days after exposure. Diphtheria is treated by administering both antitoxin to counteract toxin and antibiotics (such as penicillin or erythromycin) to kill the organism. The DTP vaccine can prevent the disease.

3. Ear infections

Ear infections commonly occur as *otitis media* in the middle ear and as *otitis externa* in the external auditory canal. *Streptococcus pneumoniae, S. pyogenes*, and *Haemophilus influenzae* account for about half of acute cases. Treatment for otitis media infections involve antibiotics that must be taken until all organisms are eradicated to prevent complications.

B. Viral upper respiratory diseases

1. The common cold

The common cold, or *coryza*, is not life-threatening. Different cold viruses predominate during different seasons. *Rhinoviruses* are the most common cause of colds, while *conornaviruses* are the second most common cause. Cold viruses are spread by fomites and by close contact with infected persons, and treatment involves remedies that alleviate some of the symptoms.

2. Parainfluenza

Rhinitis, pharyngitis, bronchitis, and sometimes pneumonia characterize *parainfluenza*, a disease caused by parainfluenza viruses. Two of the four parainfluenza viruses can cause *croup*. By age 10, most children have antibodies to all four parainfluenza viruses, so incidence of clinically apparent disease is much lower than incidence of infection.

III. Diseases of the Lower Respiratory Tract

A. Bacterial lower respiratory diseases

1. Whooping cough

Whooping cough, also called *pertussis*, is a highly contagious disease known only in humans. *Bordetella pertussis, Bordetella parapertussis*, or *B. bronchiseptica* can cause it. After an incubation period of 7 to 10 days, the disease progresses through three stages: *catarrhal, paroxysmal*, and *convalescent*. Whooping cough is diagnosed by obtaining organisms from the posterior nasal passages and culturing them.

Pertussis vaccine has saved many lives and recovery from the disease confers immunity, though not for a lifetime.

2. Classic pneumonia

   In the U.S., more deaths result from *pneumonia* (an inflammation of lung tissue) than from any other infectious disease. Bacteria, viruses, fungi, certain helminths, chemicals, radiation, and some allergies can cause the disease. Pneumonias are classified by site of infection as *lobar* or *bronchial*. Both are transmitted by respiratory droplets and, in winter, by carriers. After a few days of mild upper respiratory symptoms, the onset of pneumonia is sudden. Diagnosis is based on clinical observations, X-rays, or sputum culture. Penicillin is the drug of choice for treatment, while artificial immunity can be induced with the polyvalent vaccine Pneumovax.

3. Mycoplasma pneumonia

   Mycoplasma pneumoniae cause *mycoplasma pneumonia*, a disease that is also known as *primary atypical pneumonia* and *walking pneumonia*. Transmission is by respiratory secretions in droplet form, and onset of symptoms follows an incubation period of 12 to 14 days. Diagnosis can be made by isolating *M. pneumoniae* from sputum or from a nasopharyngeal swab. Erythromycin and tetracycline are the drugs of choice in treatment and no vaccine is currently available.

4. Legionnaires' disease

   Legionella pneumophila causes *Legionnaires' disease*. After an incubation period of 2 to 10 days, Legionnaires' disease appears with fever, chills, headache, diarrhea, vomiting, fluid in the lungs, pain in the chest and abdomen, and, less frequently, profuse sweating and mental disorders. *Pontiac fever* is a mild legionellosis. Direct fluorescent antibody tests, ELISA, and commercially available genetic probes are used to diagnose *Legionella* infections, and erythromycin is used to treat the disease.

5. Tuberculosis

   Mycobacterium tuberculosis and other members of the genus *Mycobacterium* cause tuberculosis (formerly called *consumption*). The organism is acquired by the inhalation of droplet nuclei of respiratory secretions or by contact with particles of dry sputum containing tubercle bacilli. Tuberculosis causes *caseous tubercles* to form, and *miliary tuberculosis* or *disseminated tuberculosis* can follow. Tuberculosis can be diagnosed by sputum cultures, genetic probes, or skin tests. Treatment usually involves isoniazid and rifampin, while prevention involves vaccination with attenuated organisms in the vaccine BCG.

6. Psittacosis and ornithosis

   Chlamydia psittaci causes *psittacosis*, or *parrot fever*, and *ornithosis*. Humans usually acquire the disease from birds. Definitive diagnosis is made by inoculation into tissue culture. Tetracycline can be used to treat ornithosis pneumonia, and no vaccine is available.

7. Q fever

   Coxiella burnetti causes *Q fever*. Symptoms are very similar to those of primary atypical pneumonia. Diagnosis is made by serologic testing or by direct immunofluorescent antibody staining, treatment is with antibiotics such as tetracycline or fluoroquinolone, and a vaccine is available for workers with occupational exposure.

8. Nocardiosis

   Nocardia asteroides causes *nocardiosis*. Disease is usually initiated by inhalation of *N. asteroides*. Diagnosis is based on finding the organism in sputum or other specimes and on culture results.

B. Viral lower respiratory diseases

1. Influenza

   Orthomyxoviruses cause *influenza*. Antigenic variation in influenza viruses occurs by two processes, *antigenic drift* and *antigenic shift*. The disease is relatively superficial. The best specimens for isolation of viruses are throat swabs taken as early in the illness as possible; amantadine, rimantidine, and ribavirin can be used to treat the disease. Annual immunization is recommended for high-risk persons and those with chronic conditions.

2. Respiratory syncytial virus infection

   The *respiratory syncytial virus* causes the most common and the most costly lower respiratory tract infections in children under 1 year. The virus can be identified by enzyme immunoassay performed on nasal secretions or by a direct immunofluorescent test on cells from such secretions. Ribavirin shortens the duration of the disease and there is no vaccine.

3. Hantavirus pulmonary syndrome

   PCR techniques and RNA sequencing identified *Hantavirus pulmonary syndrome*. Ribavirin has been somewhat useful in treatment, but further studies are needed.

4. Acute respiratory disease
   Acute respiratory disease is a lower respiratory tract viral disease that ranges from mild to severe. Adenoviruses cause about 5 percent of cases; adenovirus penumonia accounts for about 10 percent of all childhood pneumonia and is occasionally fatal.

C. Fungal respiratory diseases
   1. Coccidioidomycosis
      The soil fungus *Coccidioides immitis* causes *coccidioidomycosis,* or San Joaquin valley fever. Various immunological tests are available to aid in diagnosis, and a skin test is also available. In disseminated cases of disease, amphotericin B is the most effective agent available. A vaccine is being developed for coccidioidomycosis.
   2. Histoplasmosis
      The soil fungus *Histoplasma capsulatum* causes *histoplasmosis,* or *Darling's disease.* The fungus enters the body by the inhalation of conidia (fungal spores). Diagnosis is made by microscopic identification of small ovoid cells of the organism inside infected human cells. Supportive therapy is used for pulmonary histoplasmosis, and amphotericin B is sometimes effective in treating the disseminated disease.
   3. Cryptococcosis
      Filobasidiella neoformans causes *cryptococcosis,* which is usually characterized by mild symptoms of respiratory infection but can become systemic if large quantities of the spores are inhaled by debilitated patients. Observation of the organism in body fluids and latex agglutination tests can be used to detect the presence of capsular material in body fluids. Flucytosine and amphotericin B can be used in combination to treat systemic disease.
   4. Pneumocystic pneumonia
      Pneumocystis carinii causes *Pneumocystis pneumonia.* Diagnosis is made by finding organisms in biopsied lung tissue or bronchial larvage. Treatment of *Pneumocystis* infection generally involves the administration of a combination of trimethoprim and sulfamethoxazole or pentamidine.
   5. Aspergillosis
      Aspergillus fumigatus or *A. flavus* usually is responsible for *aspergillosis.* Amphotericin B is the drug of choice for invasive infections.

D. Parasitic respiratory diseases
   The lung fluke *Paragominus westermani* causes parasitic respiratory disease. Diagnosis can be made by finding eggs in sputum or by any one of several immunologic tests. The drug praziquantel is effective in treating lung fluke infections.

SELF-TESTS

Continue with this section only after you have read Chapter 21 of your textbook. Write your answers in the appropriate space provided. Correct answers to all questions can be found at the end of the Self-Test section. A score of 80 percent or better is good. If your score is less than 65, reread the chapter.

True/False

Mark T for True, F for False

_____ **1.** Prompt treatment of cases of strep throat is important to avoid the possibility of rheumatic fever.

_____ **2.** Swimmers often contract infections of the ear caused by species of *Hemophilus* due to its resistance to chlorine.

_____ **3.** Since the case rate of whooping cough has declined, there is little need to require the vaccine.

_____ **4.** Chronic alcoholics, drug addicts, and the elderly are at great risk to a bronchial pneumonia caused by *Klebsiella*

_____ **5.** The vaccine for tuberculosis is not used in this country because there are only a few cases per year.

_____ **6.** *Pneumocystic* pneumonia is one of the most commom fungal agents to affect immunodeficient patients, especially those with AIDS.

Multiple Choice

Select the best possible answer.

______ **7.** In cases of tuberculosis:
- **a.** Only the chronically ill are at risk.
- **b.** The causative agent produces a potent exotoxin.
- **c.** The bacteria are phagocytized but not killed.
- **d.** The disease develops rapidly and is usually fatal.

______ **8.** The orthomyxoviruses are characterized by all the following except:
- **a.** The viruses possess hemagglutinins and neuraminidases which are necessary for infectivity.
- **b.** With each antigenic drift, the virus changes completely and would result in a major epidemic.
- **c.** Most viruses in this family are classed as A, B, or C.
- **d.** Amantadine seems to block influenza A replication.

______ **9.** Respiratory infections caused by fungi:
- **a.** are extremely rare in this country.
- **b.** can be detected with skin tests.
- **c.** occur only in the immunocompromised
- **d.** are best treated with antibiotics such as penicillin.

______ **10.** Q-fever:
- **a.** is caused by a chlamydian.
- **b.** has a very high mortality rate.
- **c.** causes a high fever and severe rash.
- **d.** can be transmitted through direct contact or by consumption of raw milk.

Matching

Select the answer from the right-hand side that corresponds to the term or phrase on the left-hand side of the page. An answer may be used more than once. In some cases more than one answer is required.

Topic: Respiratory Diseases

______ **11.** Croup
______ **12.** Associated with toxin-caused disease
______ **13.** Causes middle ear infections
______ **14.** Histoplasmosis
______ **15.** Causes strep throat
______ **16.** Common cold virus
______ **17.** Major cause of lobar pneumonia
______ **18.** Legionnaires' disease
______ **19.** Atypical primary pneumonia
______ **20.** Psittacosis
______ **21.** Pontiac fever
______ **22.** Q fever
______ **23.** A major cause of death among AIDS patients
______ **24.** Whooping cough
______ **25.** Tuberculosis

a. *Haemophilus influenza*
b. rhinoviruses
c. *Staphylococcus aureus*
d. *Corynebacterium diphtheriae*
e. *Streptococcus pyogenes*
f. parainfluenza viruses
g. *Coxiella burnetii*
h. *Histoplasma capsulatum*
i. *Streptococcus pneumoniae*
j. *Chlamydia* species
k. *Legionella*
l. *Pneumocystis carinii*
m. none of the above

Fill-Ins

Provide the correct term or phrase for the following. Spelling counts.

26. A mechanism involving ciliated cells that allows materials in the bronchi, trapped in mucus, to be lifted to the pharynx and spit out or swallowed is ________________ ________________.

27. The ________________ ________________ are modified sebaceous glands that secrete cerumen.

28. Another term for sore throat is ________________.

29. Infections of the sinuses are known as ________________.

30. Club-shaped Gram-positive rods that form metachromatic granules are ________________.

31. Middle ear infections are called ________________ ________________.

32. Another term for the common cold is ________________.

33. The most common cause of colds seem to be ________________, followed closely by viruses in the group ________________.

34. A high-pitched, noisy respiration is known as ________________, while an acute obstruction of the larynx can result in a barking sound called ________________.

35. Pertussis, a highly contagious disease, is also known as ________________ ________________.

36. The word "pertussis" actually means ________________ ________________.

37. The stage of whooping cough that exhibits the worst symptoms is the ________________ stage.

38. Failure to keep the airway open during a case of whooping cough may result in a lack of oxygen resulting in ________________.

39. Solidified fibrin deposits that form in lobar pneumonia may cause ________________.

40. An inflammation of the pleural membranes that causes painful breathing is ________________.

41. *Mycoplasma pneumonia* is also known as ________________ ________________ ________________.

42. Since patients with *Mycoplasma pneumonia* often remain ambulatory, the disease caused by this organism is called ________________ pneumonia.

43. The disease that affected many of the war veterans attending a convention in Philadelphia in 1976 was ________________ disease.

44. A mild form of *legionellosis* that affected a large number of people in a health department in Michigan was ________________ fever.

45. Tuberculosis was once known as ________________.

46. "Cheesy"-appearing lesions in cases of tuberculosis are called ________________ lesions.

47. Cases of tuberculosis that invade other body tissue forming tiny, millet seed lesions are called ________________ tuberculosis.

48. A waxy substance from the cell wall of the mycobacterium that is used in skin testing is called ________________ ________________ ________________.

49. A vaccine for tuberculosis that is made from an attenuated strain is the ________________ strain.

50. Psittacosis is also known as ________________ fever.

51. The Q in Q-fever stands for ________________.

52. Some influenza viruses have an enzyme called ________________ that helps the virus penetrate the mucus layer protecting the respiratory epithelium.

53. A type of antigenic variation in flu viruses that is caused by a reassortment of genes is known as ________________ ________________.

54. RSV infections cause cells in cultures to form multinucleated masses called ________________.

55. *Coccidioides immitis* forms cottony, white ________________ (or colonies) in culture.

56. Another name for histoplasmosis is ________________ disease.

57. The most common cause of pneumonia in AIDS patients is ________________ ________________.

58. Symptoms that follow the recovery of a disease such as diphtheria are called ________________.

Labeling

Questions 59–62: Select the best possible label for each part of the ear: auditory tube, middle ear, cranial VIII nerve, pinna, mastoid area, tympanic membrane, ear hairs, auditory ossicles, inner ear, cochlea, external ear, cartilage, auditory canal of outer ear.

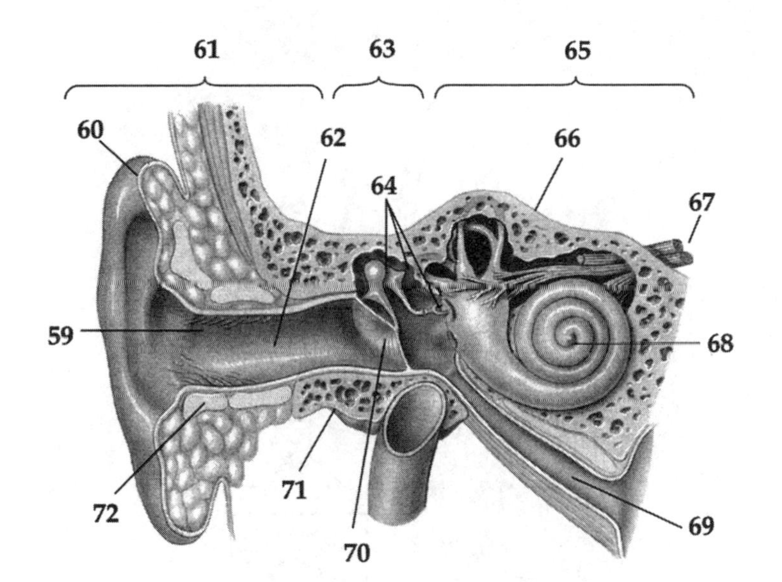

59. ________________
60. ________________
61. ________________
62. ________________
63. ________________
64. ________________
65. ________________

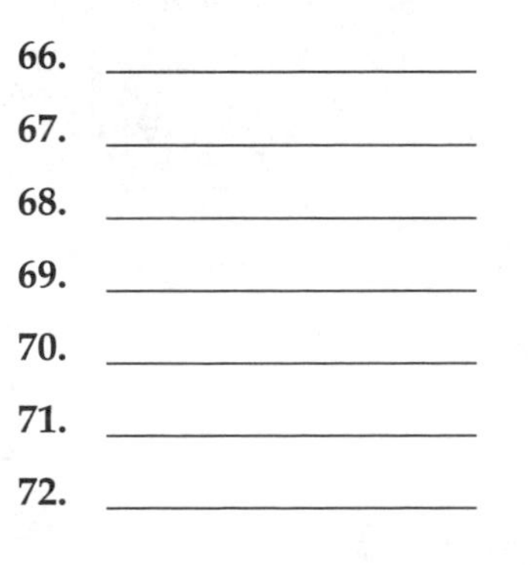

66. ________________
67. ________________
68. ________________
69. ________________
70. ________________
71. ________________
72. ________________

Questions 73–78: Select the best possible label for each component of the upper respiratory system: lung tissue; epiglottis; hyoid; trachea; primary bronchi; larynx.

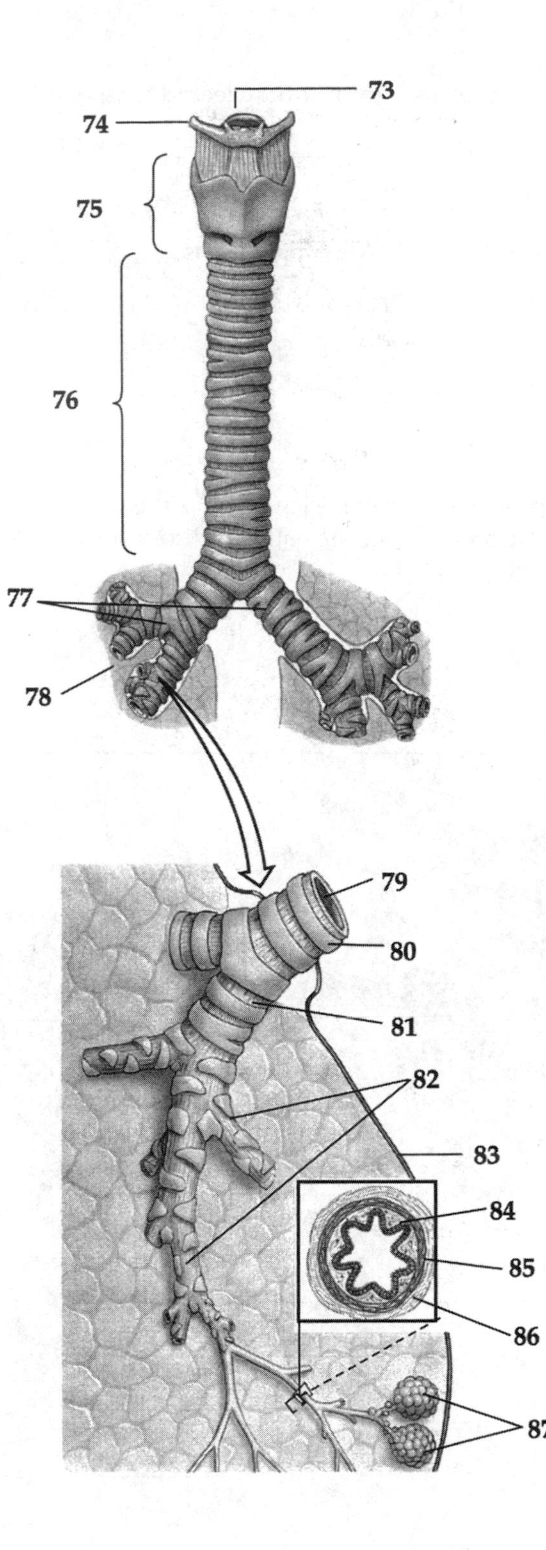

73. ____________________

74. ____________________

75. ____________________

76. ____________________

77. ____________________

78. ____________________

Questions 79–87: Select the best possible label for each component of the lower respiratory system: primary bronchus; secondary bronchus; smooth muscle; cartilage ring; pleura; smaller intrapulmonary bronchi; respiratory epithelium; smooth muscle; bronchiole.

79. ________________

80. ________________

81. ________________

82. ________________

83. ________________

84. ________________

85. ________________

86. ________________

87. ________________

Critical Thinking

88. What bacteria cause upper respiratory infections, and what are the important epidemiologic and clinical aspects of these diseases?

89. What viruses cause upper respiratory infections, and what are the important epidemiologic and clinical aspects of these diseases?

90. What bacteria cause lower respiratory infections, and what are the important epidemiologic and clinical aspects of these diseases?

91. What viruses cause lower respiratory infections, and what are the important epidemiologic and clinical aspects of these diseases?

92. What fungi cause lower respiratory infections, and what are the important epidemiologic and clinical aspects of these diseases?

ANSWERS

True/False

1. **True** If treatment is delayed, the organisms could interact with the immune system and give rise to rheumatic fever.
2. **False** Infections of the ear are commonly caused by species of Pseudomonas that are resistant.
3. **False** Actually the case rate of whooping cough is increasing in many parts of the country. Much of the increase is due to concern about the vaccine's safety.
4. **True** *Since Klebsiella* is a common inhabitant of the gastrointestinal tract, it can readily be aspirated into the lungs. This can result in severe pneumonia.
5. **False** Although the vaccine is not used in this country, it is because of the problem of the recipient developing a positive skin test which would make it more difficult to determine possible cases.
6. **True** *Pneumocystis*, once thought to be a protozoan, causes a severe pneumonia in patients with reduced immune defense systems, especially AIDS patients.

Multiple Choice

7. c; 8. b; 9. b; 10. d.

Matching

11. f; 12. d; 13. a, e, i; 14. h; 15. e; 16. b, f; 17. i; 18. k; 19. k; 20. j; 21. k; 22. g; 23. d; 24. m; 25.m.

Fill-Ins

26. mucociliary escalator; 27. ceruminous glands; 28. pharyngitis; 29. sinusitis; 30. diphtheroids; 31. otitis media; 32. coryza; 33. rhinoviruses, coronaviruses; 34. stridor, croup; 35. whooping cough; 36. violent cough; 37. paroxysmal; 38. cyanosis; 39. consolidation; 40. pleurisy; 41. primary atypical pneumonia; 42. walking; 43. Legionnaires';

44. Pontiac; 45. consumption; 46. caseous; 47. miliary; 48. purified protein derivative; 49. BCG; 50. parrot; 51. query; 52. neuraminidase; 53. antigenic shift; 54. syncytia; 55. mycelia; 56. Darling's; 57. pneumocystis pneumonia; 58. sequelae.

Labeling

59. ear hairs; 60. pinna; 61. external ear; 62. auditory canal of outer ear; 63. middle ear; 64. auditory ossicles; 65. inner ear; 66. temporal bone; 67. cranial VIII nerve; 68. cochlea; 69. auditory tube; 70. tympanic membrane; 71. mastoid area; 72. cartilage.; 73. epiglottis; 74. hyoid; 75. larynx; 76. trachea; 77. primary bronchi; 78. lung tissue; 79. primary bronchus; 80. cartilage ring; 81. secondary bronchus; 82. smaller intrapulmonary bronchi; 83. pleura; 84. respiratory epithelium; 85. smooth muscle; 86. bronchiole; 87. alveoli.

Critical Thinking

88. What bacteria cause upper respiratory infections and what are the important epidemiologic and clinical aspects of these diseases?

Various species of bacteria cause upper respiratory infections and most of these, such as species of *Streptococcus*, *Mycoplasma*, *Staphylococcus*, and *Haemophilus*, are somewhat common inhabitants. Infections related to pharyngitis include laryngitis, epiglottitis, sinusitis, and bronchitis.

These infections are exceedingly common and are easily acquired through inhalation of droplets, especially in winter when people are crowded indoors. Most of these infections can be treated with penicillin or some other broad-spectrum antibiotic; however, complete eradication of the infectious agent or agents is rare. Much depends on the condition of the respiratory system and the host's immune defense system.

Diphtheria is no longer common in the United States because of the use of the diphtheria toxoid (the "D" of DPT shot). Both the organism and its toxin, which is produced by a lysogenic prophage, contribute to the symptoms. These include the formation of a grayish-white pseudomembrane on the pharynx, and massive swelling leading to a symptom called "bullneck." The disease is spread by droplets and can be treated with antitoxin and antibiotics.

Ear infections occur in the middle and outer ear and are very common in young children. Organisms reach the middle ear via the eustachian tube and cause severe pain and swelling. Most infections respond to penicillin or other antibiotic. In severe cases, artificial openings or "tubes" can be made in the eardrum to relieve the pain and fluid buildup.

89. What viruses cause upper respiratory infections and what are the important epidemiologic and clinical aspects of these diseases?

Numerous viral groups affect the upper respiratory tract and are often grouped into the category of "common cold" viruses. The most common agents are found with the rhinoviruses and coronaviruses although parainfluenza and adenoviruses also contribute greatly to the case rate. Pharyngitis, bronchitis, and croup can be caused by parainfluenza viruses.

Common colds are transmitted by fomites, direct contact, and aerosols. Since antibiotics are ineffective, treatment is limited to alleviating symptoms. Vaccines are not available.

Parainfluenza infections range from inapparent to severe croup. Most children have been exposed to these viruses and produced antibodies by age 10. Secretory IgA antibodies seem to be the most protective against these viruses.

90. What bacteria cause lower respiratory infections and what are the important epidemiologic and clinical aspects of these diseases?

Infections of the lower respiratory tract are far less common but often considerably more severe. Two of the greatest killer infections in history, tuberculosis and pneumonia, are among these agents.

Whooping cough, or pertussis, is a highly contagious, worldwide disease. The pertussis vaccine (the "P" of DPT) has decreased its incidence; however, due to the concern about the vaccine's safety, its use declined in the early 1980's with the result of a significant increase in cases. The vaccine is prepared from cell wall material of the organism and is difficult to produce in a completely purified, safe form. However, the number of serious side effects caused by the vaccine (1/1,000,000 doses) is greatly outweighed by the numbers of infants left brain-dead due to the disease.

Classic pneumonia can be lobar or bronchial and is transmitted by respiratory droplets and carriers, including those that actually work in hospitals. Most cases of pneumonia occur in individuals with lowered resistance, such as the elderly, drug addicts, alcoholics, and those under deep anesthesia. Antibiotics can treat the disease; however, since most patients are already in difficulty, the disease often leaves them in a very weakened state. Chronic ulcerative bronchitis caused by *Klebsiella* can be very severe; however, lobar pneumonia caused by *Streptococcus pneumoniae* is more common. Pneumonias caused by *Mycoplasmas* are generally mild and often referred to as "walking" pneumonia because the patient is ambulatory.

Legionnaires' disease appears to be transmitted by aerosols from contaminated water sources, especially improperly installed air conditioning systems. The drug of choice is erythromycin.

Tuberculosis continues to be a worldwide health problem and its incidence in the United States is increasing slowly, especially with the newly-arrived immigrants, with minorities, and with patients with immune deficiencies, such as those with AIDS. It is transmitted by respiratory droplets, and the organisms can persist for years, walled off in lung tissue called tubercles. Numerous species can affect humans, some with devastating effects. Isoniazid is the drug of choice. A vaccine is available but is not used in this country because it always causes the recipient to skin test positive for the disease. Health care workers, especially those working with AIDS patients, are becoming at risk for this disease.

Psittacosis or ornithosis is transmitted to humans from infected birds. In some states, turkeys and chickens (ornithine birds) are the most common carriers whereas in other states, exotic or imported birds such as parrots and cockatiels (psittacine birds) are more likely sources. The disease is caused by a chlamydian and is usually self-limited but can become serious. Tetracyclines are effective in treatment.

Q-fever is caused by a rickettsian and is transmitted by ticks, aerosol droplets, fomites, and infected milk. It varies from other rickettsial diseases in that it does not cause a rash but does cause a high fever. Treatment is with tetracyclines and a vaccine is available to workers with occupational exposure.

91. What viruses cause lower respiratory infections and what are the important epidemiologic and clinical aspects of these diseases?

The most important virus to affect the lower respiratory tract is the influenza virus. The virus has the ability to affect millions by changing its antigenic structure through variations called antigenic shifts and drifts. The virus has surface antigens called hemagglutinins which are responsible for their infectivity and occasionally, neuraminidases, all of which can be changed antigenically to account for their variability.

The disease occurs mainly in December through April and is transmitted in crowded, poorly ventilated conditions, especially in older schools and businesses. The disease can be diagnosed by clinical symptoms and by immunological tests. Vaccines can help prevent the disease but must be correlated with the current antigenic variations that are "expected" during that time period.

The respiratory syncytial and adenoviruses can cause acute respiratory infections which can lead to viral pneumonia. Treatment is for symptoms only.

92. What fungi cause lower respiratory infections and what are the important epidemiologic and clinical aspects of these diseases?

Most fungal infections occur in immunodeficient and debilitated patients. They are usually transmitted by infectious airborne spores, and some can be treated with amphotericin B. Although all of these diseases affect the respiratory system, most exhibit cutaneous and systemic forms as well.

Coccidioidomycosis occurs in the warm, arid regions of the desert Southwest, and histoplasmosis is associated with the Mississippi River Valley regions. The exposure rates of these diseases can be very high but the morbidity is relatively low. Cryptococcosis occurs wherever there are large infected flocks of birds such as pigeons.

Pneumocystis pneumonia occurs primarily in immunodeficient patients, especially those with AIDS. It is diagnosed by finding the organism in biopsied lung tissue.

Lung fluke infections occur primarily in Asia and the South Pacific where infected shellfish are eaten. Thorough cooking can prevent the disease and drugs are available for treatment.

Chapter 22 Oral and Gastrointestinal Diseases

INTRODUCTION

The habitat of the human digestive system is quite complex. Any organism entering the stomach is treated to a bath of hydrochloric acid and enzymes that break down protein. The next stop is the small intestine with a flood of digestive enzymes from the pancreas, bile from the liver, and an environment of no oxygen. These are difficult challenges for the possible rewards of intestinal contents or a meal of our own tissues. Oddly enough, many organisms have adapted their life cycle to even requiring these conditions. Consider, the eggs of the common (over one billion people infected) round worm parasite *Ascaris* require a high concentration of carbon dioxide to germinate. As a large adult of about eight inches in length, the *Ascaris* is quite content with oxygen levels of approximately zero percent. In fact, for this worm the light at the end of the tunnel is likely to be a toilet bowl. It is no wonder that there are more worm infections than there are people on Earth. Many other organisms can survive the journey through the gut and make us sick.

In 1994, an outbreak of bacterial food poisoning occurred on a cruise ship. It was note upon investigation that some people did not get sick. Interviews with those whose intestines were spared drank significant amounts of white wine. Mark Daeschel, at Oregon State University, did experiments to understand the strange outcome of the cruise ship. Red wine did not result in the same result as white wine. More than alcohol had to be at issue. Daeschel constructed artificial stomachs. Each bag would contain pepsin, hydrochloric acid, was at body temperature and contained a paddle to mimic the movement of a normal stomach. *Salmonella* and *E. coli* were placed into the bags and a number the effectiveness of a white wine (Chardonnay) and a typical red wine (pinot noir) were compared. The killing effect of the white wine was about twice as fast as red wine. The explanation as to why white wine is more effective seems related to its higher content of acid. Thus, if someone asks you, what type of wine goes with cruise ships, you know the answer.

STUDY OUTLINE

I. Components of the Digestive System

A. General characteristics

The *digestive system* consists of various organs—mouth, pharynx, esophagus, stomach, and intestines—whereas the accessory organs include the teeth, salivary glands, liver, gallbladder, and pancreas. The digestive system has five major functions: (1) movement; (2) secretion; (3) digestion; (4) absorption;

(5) elimination. Throughout the digestive tract, *mucin*, a glycoprotein in mucus, coats bacteria and prevents their attaching to surfaces.

B. The mouth

The mouth, or *oral cavity*, is lined with a mucous membrane and contains the tongue, teeth, and salivary glands. The mouth is a portal of entry for microorganisms. Each tooth has a *crown* covered with *enamel* above the gum and a *root* covered with *cementum* below the gum. Under these coverings is a porous substance called *dentin*, a central *pulp cavity*, and the *root canals*.

C. The stomach

In the stomach, food mixes with hydrochloric acid and the enzyme pepsin. The lining of the stomach is protected from the acid by viscous mucus.

D. The small intestine

Partially digested food is mixed with secretions from the liver and pancreas in the *small intestine*. The small intestine has *villi, microvilli,* and *Peyer's patches*. Food is then carried directly to the liver. In the liver, capillaries are enlarged to form networks of vessels called *sinusoids* lined with phagocytic *Kupffer cells*.

E. The large intestine

The *large intestine*, or *colon*, joins the small intestine near the appendix and ends at the rectum. *Feces* are created here.

F. Normal microflora of the mouth and digestive system

Over 400 species of microbes have been identified as members of the oral normal microflora. The human esophagus, the stomach, and the first two-thirds of the small intestine contain very few microbes. The last third of the small intestine is characterized by having gram-negative, facultatively anaerobic bacteria plus obligate anaerobes. In the large intestine, the feces are about 50 percent bacteria by weight and volume.

II. Diseases of the Oral Cavity

A. Bacterial diseases of the oral cavity

1. Dental plaque

Dental plaque is a continuously-formed coating of microorganisms and organic matter on tooth surfaces. Plaque formation begins as a pellicle over the tooth surface. Cocci and filamentous bacteria among the normal oral microflora attach to the newly-formed pellicle. Plaque consists of up to 30 different genera of bacteria and their products and offers protection to bacteria in the gingival crevices between the teeth and the gums, including species of *Actinomyces, Veillonella, Fusobacterium,* and sometimes spirochetes, as well as the *streptococci* just noted.

2. Dental caries

Dental caries, or tooth decay, is the chemical dissolution of enamel and deeper parts of the teeth. The combination of sucrose and the action of *S. mutans* on it account for much tooth decay. Dental caries are treated by removing decay and filling the cavity with *resins* or with *amalgam*. Vaccines to prevent dental caries are being developed against strains of *S. mutans*. The use of *fluoride* has been the most significant factor in reducing tooth decay. Another means of preventing dental caries is sealing of the teeth with a type of resin that mechanically bonds to the tooth structure.

3. Periodontal disease

Periodontal disease is a combination of gum inflammation and erosion of periodontal ligaments and the alveolar bone that supports teeth. Periodontal disease is called *gingivitis* in its mildest form and *acute necrotizing ulcerative gingivitis* in its most severe form. Unchecked, periodontal disease can lead to chronic *periodontitis*. Periodontal disease occurs when plaque is allowed to accumulate, resulting in the overgrowth of potentially virulent bacteria. Treatments include antimicrobial mouth rinses, brushing with a mixture of bicarbonate of soda and hydrogen peroxide, surgery to eliminate pockets, and antibiotic therapy in unresponsive or rapidly-progressing cases.

B. Viral diseases of the oral cavity

1. Mumps

Mumps is caused by a paramyxovirus somewhat similar to the measles (rubeola) virus. The virus replicates in the upper respiratory tract, then travels in the blood to the salivary glands and sometimes to other glands and organs, such as the testes (*orchitis*) and the meninges. The MMR vaccine can prevent infection by the mumps virus.

2. Other diseases

Other viral diseases of the oral cavity include *thrush* and *herpes simplex virus*.

III. Gastrointestinal Diseases Caused by Bacteria

A. Bacterial food poisoning

Ingesting food contaminated with preformed toxins causes *food poisoning*. Bacteria that produce toxins responsible for food poisoning include *Staphylococcus aureus* (which causes staphylococcal enterotoxicosis), *Clostridium perfringens, C. botulinum, Bacillus cereus,* and *Pseudomonas cocovenenans* (which causes *bongkrek disease*).

B. Bacterial enteritis and enteric fevers

Enteritis is an inflammation of the intestine. *Bacterial enteritis* is an intestinal infection. When the large intestine is affected, the result is often called *dysentery*. If pathogens spread through the body from the intestinal mucosa and cause systemic infections, the infections are called *enteric fevers*.

1. Salmonellosis

   Salmonellosis is a common enteritis caused by some members of the genus *Salmonella*. About 2000 strains have been identified by their surface antigens and grouped into *serovars*. Because of their ability to invade intestinal tissue and enter the blood, *S. typhimurium* and *S. paratyphi* cause enteric fever, or *enterocolitis*.

2. Typhoid fever

   Typhoid fever, one of the most serious of the epidemic enteric infections, is caused by *Salmonella typhi*. The Widal test is diagnostic of typhoid fever. A live, attenuated oral typhoid vaccine is now available.

3. Shigellosis

   Shigellosis, or *bacillary dysentery*, can be caused by several serovars of *Shigella*. They include *S. dysenteriae* (serovar A), *S. flexneri* (serovar B), *S. boydii* (serovar C), and *S. sonnei* (serovar D). Specific diagnosis of shigellosis involves the swabbing of a bowel lesion during internal examination of the bowel. Treatment involves hydration fluids and possibly a combination of antibiotics—ampicillin, tetracycline, and nalidixic acid.

4. Asiatic cholera

   Vibrio cholerae is the causative organism of *Asiatic cholera*. The enterotoxin, known as *choleragen*, binds to epithelial cells of the small intestine and makes plasma membranes highly permeable to water. Fluid and electrolyte replacement is the most effective treatment for cholera, although tetracycline can be used to reduce the duration of the symptoms. The only available vaccine is not very effective and is not widely used, although a toxoid-type vaccine is being tested.

5. Vibriosis

   An enteritis called *vibriosis* is caused largely by *Vibrio parahaemolyticus*. Once inside the organism, the oganisms colonize the mucosa and release an enterotoxin. The disease usually is not treated, and no vaccine is available.

6. Traveler's diarrhea

   This disorder is caused by pathogenic strains of *E. coli, Shigella, Salmonella, Campylobacter*, rotaviruses, and protozoa such as *Giardia* and *Entamoeba*. Symptoms of traveler's diarrhea vary from mild to severe. The most effective treatment is to keep antidiarrhea medicine available and to use it only after symptoms appear.

7. Other kinds of bacterial enteritis

   Other bacteria that cause enteritis include *Campylobacter jejuni, C. fetus,* and *Yersinia enterocolitica* (which causes *yersiniosis*).

C. Bacterial infections of the stomach, esophagus, and intestines

1. Peptic ulcer and chronic gastritis

   Helicobacter pylori have recently been revealed as a bacterial cause of peptic ulcers and chronic gastritis, and a probable cofactor of stomach cancer. *Peptic ulcers* are lesions of the mucous membranes lining the esophagus, stomach, or duodenum, while *chronic gastritis* is an inflammation of the stomach. Currently available drugs such as Tagament only control, but cannot cure, ulcers. Experimental treatments with antibiotics have reported cure rates as high as 80 to 90 percent, but these drugs must be used very carefully because of the possibility of relapse.

2. Pseudomembranous colitis

   Pseudomembranous colitis, caused by *Clostridium difficile*, is a condition characterized by the formation of a membrane covering on the mucosal surface of the colon.

D. Bacterial infections of the gallbladder and biliary tract

Gallstones, formed from crystals of cholesterol and calcium salts, can block the bile ducts, decreasing the flow of bile and predisposing the individual to *cholecystitis* or *cholangitis*. Diagnosis of gallbladder infection is based on clinical findings of recurring pains (*biliary colic*), nausea, vomiting, chills, fever, and often jaundice due to the absorption of blocked bile in the bloodstream. Despite prompt treatment with antibiotics, gallbladder infections cannot be cured unless the causative obstruction is removed.

IV. Gastrointestinal Diseases Caused by Other Pathogens

A. Viral gastrointestinal diseases

1. Viral enteritis

Rotavirus infection is a major cause of *viral enteritis* among infants and young children. Immunoelectron microscopy and ELISA tests can be diagnostic, while treatment involves restoring fluid and electrolyte balance. Enteritis can also be caused by species of *Enterovirus* and *Clostridium difficile*.

2. Hepatitis

Hepatitis, an inflammation of the liver, is usually caused by viruses. Types of hepatitis include *hepatitis A, hepatitis B, hepatitis C, hepatitis E,* and *hepatitis D*.

B. Protozoan gastrointestinal diseases

1. Giardiasis

The flagellated protozoan *Giardia intestinalis* infects the small intestine of humans, especially children, and causes a disorder called *giardiasis*. Diagnosis is made by microscopic examination and finding cysts of the protozoan in stools. Quinacrine, furazolidine, and metronidazole are used to treat giardiasis.

2. Amoebic dysentery and chronic amebiasis

Entamoeba histolytica causes *amebiasis*, which can appear as *amoebic dysentery* or *chronic amebiasis*. Amoebic infections can be diagnosed by finding trophozoites or cysts in stools, but several stool samples on consecutive days may be needed to find them. Immunoflorescent antibody and ELISA procedures are also available for diagnosis. Metronidazole is used to treat infections.

3. Balantidiasis

Balantidium coli causes a dysentery called *balantidiasis*. Diagnosis is made by finding trophozoites or cysts in fecal specimens. Tetracycline or metronidazole is used to treat the disease.

4. Cryptosporidiosis

Protozoans of the genus *Cryptosporidium* cause *cryptosporidiosis*, which is a self-limiting disease in immunocompetent individuals.

C. Effects of fungal toxins

Fungi produce a large number of toxins, and most come from members of the genera *Aspergillus* and *Penicillium*. Their various effects on humans include loss of muscle coordination, tremors, and weight loss. Some are carcinogenic.

D. Helminth gastrointestinal diseases

1. Fluke infections

The causative agents of fluke infections include the sheep liver fluke *Fasciola hepatica*, the Chinese liver fluke *Clonorchis sinensis*, and a fluke that is common in pigs and humans in the Orient called *Fasciolopsis buski*. Antihelminthic drugs can be used to rid the body of flukes.

2. Tapeworm infections

Human tapeworm infections can be caused by several species, including the pork tapeworm *Taenia solium*, the beef tapeworm *T. saginata*, the dog tapeworm *Echinococcus granulosus*, the insect tapeworm *Hymenolepis nana*, and the broad fish tapeworm *Diphyllobothrium latum*. Diagnosis is made by finding eggs or proglottids in feces. Infections can be treated with niclosamide and other antihelminthic agents.

3. Trichinosis

The small *Trichinella spiralis* roundworm causes *trichinosis*. The disease is difficult to diagnose, but muscle biopsies and immunological tests are sometimes positive. Treatment is directed toward relieving symptoms because the disease cannot be cured.

4. Hookworm infections

Hookworm is most often caused by one of two species of small roundworms—*Ancylostoma duodenale* and *Necator americanus*. Bacterial infection of hookworm penetration sites causes *ground itch*. Diagnosis is made by finding eggs or worms in feces, but samples must be concentrated to find them. Tetrachloroethylene is effective against *Necator* infections, while bephenium hydroxynaphthalate, mebendazole, and several other drugs kill both species.

5. Ascariasis

Ascaris lumbricoides is a large roundworm that causes *ascariasis*. Diagnosis is made by finding eggs or worms in feces. The adult worms can be eradicated from the body by piperazine, mebendazole, and several other drugs, but no treatment is available to rid the body of larvae.

6. Trichuriasis
   Trichuriasis is caused by the whipworm, *Trichuris trichiura*. Infection is diagnosed by finding worm eggs in stools. The drug mebendazole is effective in ridding the intestine of parasites.
7. Strongyloidiasis
   Strongyloidiasis is caused by *Strongyloides stercoralis*. Diagnosis is difficult, so work is in progress to develop a reliable immunological test. The drugs thiabendazole and cambendazole have the greatest effect on the parasites with the fewest undesirable side effects.
8. Pinworm infections
   Pinworm infections are caused by the small roundworm *Enterobius vermicularis*. These infections are diagnosed by finding eggs around the anus. Piperazine or other antihleminthic agents are used for treatment.

SELF-TESTS

Continue with this section only after you have read Chapter 22 of your textbook. Write your answers in the appropriate space provided. Correct answers to all questions can be found at the end of the Self-Test section. A score of 80 percent or better is good. If your score is less than 65, reread the chapter.

True/False

Mark T for True, F for False

_____ **1.** By age 50, nearly everyone has a form of periodontal disease.

_____ **2.** Dental caries is the most common infectious disease in developed countries.

_____ **3.** In cases of bacterial food poisoning, symptoms usually are delayed for a few days until the organisms produce enough toxin.

_____ **4.** Antibiotics should always be given in cases of salmonella food poisonings to reduce the threat of carriers.

_____ **5.** Peanut butter and grains such as rye and wheat could be infected by phallotoxin-producing fungi called *Amanita*.

_____ **6.** Cases of trichinosis have been traced to the consumption of poorly cooked pork, venison, and even horse meat.

_____ **7.** Ascarid worms have been known to wander inside the body to invade other organs and have also crawled up the esophagus and out the mouth.

Multiple Choice

Select the best possible answer

_____ **8.** Select the most correct statement concerning bacterial food poisoning.
- **a.** Most cases can be treated with antibiotics.
- **b.** Staphylococcal food poisons cannot be detected by odor or taste and are heat resistant.
- **c.** Symptoms are usually delayed several days.
- **d.** *Clostridium perfringins* produces a food poisoning that is highly lethal.

_____ **9.** In cases of hepatitis:
- **a.** Exposure to hepatitis A may require shots of antibiotics.
- **b.** Both hepatitis B and C can be prevented by means of vaccines.
- **c.** Although hepatitis A and B are caused by viruses, the hepatitis C variety is actually caused by a species of *Hemophilis*.
- **d.** Hepatitis B occurs only in drug addicts.
- **e.** None of the above are true.

_____ **10.** Amoebic dysentery is characterized by all the following except:
- **a.** The disease is transmitted by cysts.
- **b.** Metronidazole is used in treatment.

c. The parasites cause severe diarrhea but do not invade other tissue.
d. The asexually reproducing form is a trophozoite.

_____ 11. Which of the following viruses is not usually implicated in causing cases of enteritis?
a. rotaviruses
b. Norwalk viruses
c. echoviruses
d. arenaviruses
e. parvoviruses

Matching

Select the answer from the right-hand side that corresponds to the term or phrase on the left-hand side of the page. An answer may be used more than once. In some cases more than one answer is required.

Topic: Disease of the Oral Cavity

_____ 12. Mumps	a. *Candida albicans*
_____ 13. Thrush	b. *herpes simplex* virus
_____ 14. Fever blisters	c. paramyxovirus
_____ 15. Periodontal disease	d. *Streptococcus mutans*
_____ 16. Dental caries	e. *Porphyromonas gingivalis*

Topic: Gastrointestinal Diseases (Part 1)

_____ 17. Food poisoning	a. *Salmonella typhi*
_____ 18. Cryptosporidiosis	b. *Vibrio cholerae*
_____ 19. Traveler's diarrhea	c. *Escherichia coli*
_____ 20. Shigellosis	d. *Clostridium botulinum*
_____ 21. Typhoid fever	e. *Staphylococcus aureus*
_____ 22. Asiatic cholera	f. *Pseudomonas cocovenenans*
_____ 23. Bongkrek disease	g. *Bacillus cereus*
_____ 24. Amebiasis	h. *Salmonella paratyphi*
_____ 25. Botulism	i. *Entamoeba histolytica*
_____ 26. Viral hepatitis	j. *Cryptosporidium* species
_____ 27. Pseudomembranous colitis	k. *Vibrio parahaemolyticus*
_____ 28. Bacillary dysentery	l. *Clostridium difficile*
_____ 29. Enterocolitis	m. none of the above
_____ 30. Vibriosis	
_____ 31. Peptic ulcer	

Topic: Gastrointestinal Diseases (Part 2)

_____ 32. Aflatoxins	a. rotavirus infection
_____ 33. Viral enteritis	b. hepatitis B virus
_____ 34. Hepatitis non-A, non-B	c. *Giardia intestinalis*
_____ 35. Hepatitis D	d. echoviruses
_____ 36. Giardiasis	e. Norwalk virus
_____ 37. Amoebic dysentery	f. hepatitis C virus

_____ 38. Ergot poisoning

_____ 39. Pernicious anemia

g. hepatitis D virus

h. *Entamoeba histolytica*

i. *Aspergillus flavus*

j. *Claviceps purpurea*

k. *Diphyllobothrium latum*

Topic: Helminth Diseases

_____ 40. *Clonorchis sinensis*

_____ 41. *Taenia saginata*

_____ 42. *Trichuris trichiura*

_____ 43. *Fasciola hepatica*

_____ 44. *Fasciolopsis buski*

_____ 45. *Echinococcus granulosus*

_____ 46. *Ancylostoma caninum*

_____ 47. *Taenia solium*

_____ 48. *Ancylostoma duodenale*

_____ 49. *Diphyllobothrium latum*

_____ 50. *Necator americanus*

a. sheep liver fluke

b. pork tapeworm

c. verminous intoxication

d. sheep tapeworm

e. Chinese liver fluke

f. hydatid cysts

g. beef tapeworm

h. fish tapeworm

i. pernicious anemia

j. hookworm

k. ground itch

l. cutaneous larval migrans

m. whipworm

Topic: Helminth Infections, Agents, and Transmission

_____ 51. *Fasciola hepatica*

_____ 52. *Taenia saginata*

_____ 53. *Clonorchis sinensis*

_____ 54. *Enterobius vermicularis*

_____ 55. *Taenia solium*

_____ 56. *Hymenolepis nana*

_____ 57. *Trichuris trichiura*

_____ 58. *Trichinella spiralis*

_____ 59. *Strongyloides stercoralis*

_____ 60. *Diphyllobothrium latum*

_____ 61. *Necator americanus*

_____ 62. *Ancylostoma duodenale*

a. contaminated pork or pork products

b. contaminated water or vegetation

c. contaminated cereals

d. contaminated beef or beef products

e. contaminated fish or fish products

f. contaminated soil

g. contaminated bedclothes and/or inhalation of ova

Fill-Ins

Provide the correct term or phrase for the following. Spelling counts.

63. Movement, secretion, digestion, absorption, and elimination are the five major functions of the _______________ _______________.

64. A glycoprotein in mucus that coats bacteria and prevents their attaching to surfaces is _______________.

65. The sinusoids of the liver are lined with phagocytic ________________ cells, which remove any dead blood cells, bacteria, and toxins from the blood as it passes through the sinusoids.

66. Fingerlike projections of the small intestines are called ________________.

67. ________________ consist of three-fourths water and one-fourth solids and are stored in the rectum until they are eliminated from the body.

68. A coating of microorganisms and organic matter that adheres tightly to the tooth enamel is called ________________.

69. Following tooth brushing, a film of glycoproteins from saliva called ________________ forms on the surface of the teeth.

70. Another word for tooth decay is ________________ ________________.

71. An ion that is added to drinking water to help toughen the teeth against decay is the ________________ ion.

72. A mild form of periodontal disease is ________________.

73. ANUG is also known as ________________.

74. Inflammation of the testes that can occur with mumps is called ________________.

75. An exotoxin that specifically affects the gastrointestinal tract is an ________________.

76. A Polynesian food poisoning associated with a coconut delicacy is ________________ disease.

77. An inflammation of the intestine is called ________________.

78. A disease characterized by the inflammation of the large intestine and severe diarrhea, mucus, pus, and even blood is ________________.

79. Invasion of the intestinal tissue and the blood caused by some species of *Salmonella* is ________________.

80. A famous carrier of typhoid fever in the 1900's was ________________ ________________ ________________.

81. A watery, tissue-laden stool produced in sever cases of cholera is ________________ ________________ ________________.

82. An especially severe strain of cholera with a slower onset is ________________ ________________.

83. An intestinal disorder contracted by people who travel extensively is ________________ ________________.

84. An inflammation of the liver caused by viruses or other agents is called ________________.

85. Hepatitis A is commonly called ________________ hepatitis while hepatitis B is called ________________ hepatitis.

86. Hepatitis C is usually transmitted ________________.

87. A dysentery caused by a large ciliated protozoan is ________________.

88. A fungal toxin produced by species of *Aspergillus* is ________________.

89. A poisoning contracted by eating grains of rye or wheat contaminated with *Claviceps* is ________________.

90. *Fasciola hepatica* is commonly known as the ________________ ________________ ________________.

91. An allergic reaction to toxins produced by certain flukes is known as ________________ ________________.

92. Bladder worms that are found in human tissues are called ________________.

93. Cysts of certain tapeworms that form in tissues are called ________________ cysts.

94. The roundworms of *Ancylostoma* and *Necator* are also called ________________.

95. Larvae of hookworms that cause creeping eruptions following their entrance into the body are called ________________ ________________ ________________.

96. The *Trichuris* worms are known as ________________, due to a long esophagus.

Labeling

Questions 97–115: Select the best possible label for each structure in the digestive system: pharynx; transverse colon; liver; appendix; salivary glands; mouth; sigmoid colon; small intestine; pancreas; esophagus; ascending colon; gallbladder; rectum; large intestine (colon); duodenum; ileum; descending colon; stomach; jejunum.

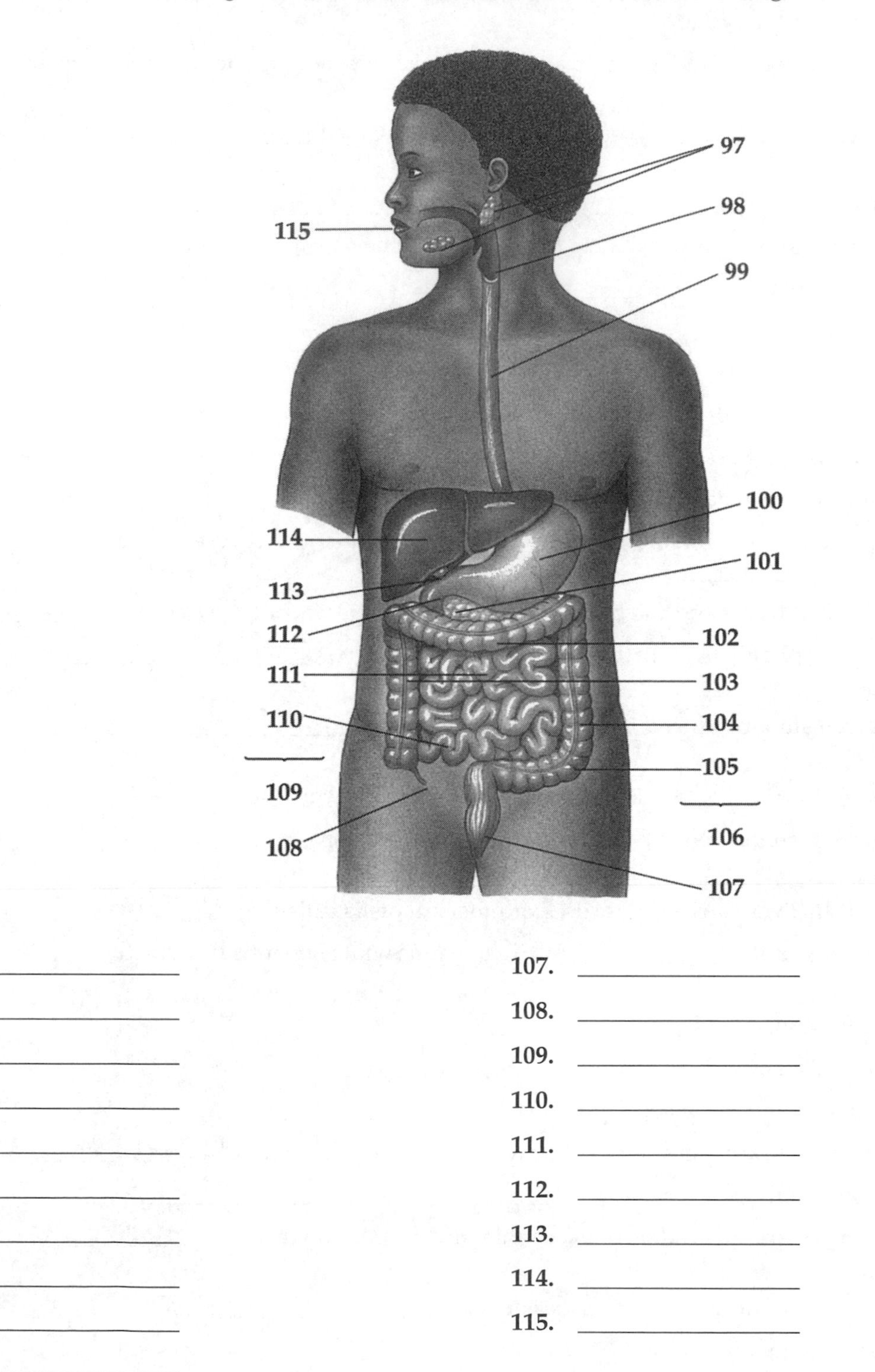

97.	____________	107.	____________
98.	____________	108.	____________
99.	____________	109.	____________
100.	____________	110.	____________
101.	____________	111.	____________
102.	____________	112.	____________
103.	____________	113.	____________
104.	____________	114.	____________
105.	____________	115.	____________
106.	____________		

Questions 116–126: Select the best possible label for each structure in a typical tooth and its surrounding gum: crown; alveolar bone (socket); root canal; cementum; periodontal ligament; root; pulp cavity; gingival sulcus; neck; dentin; enamel.

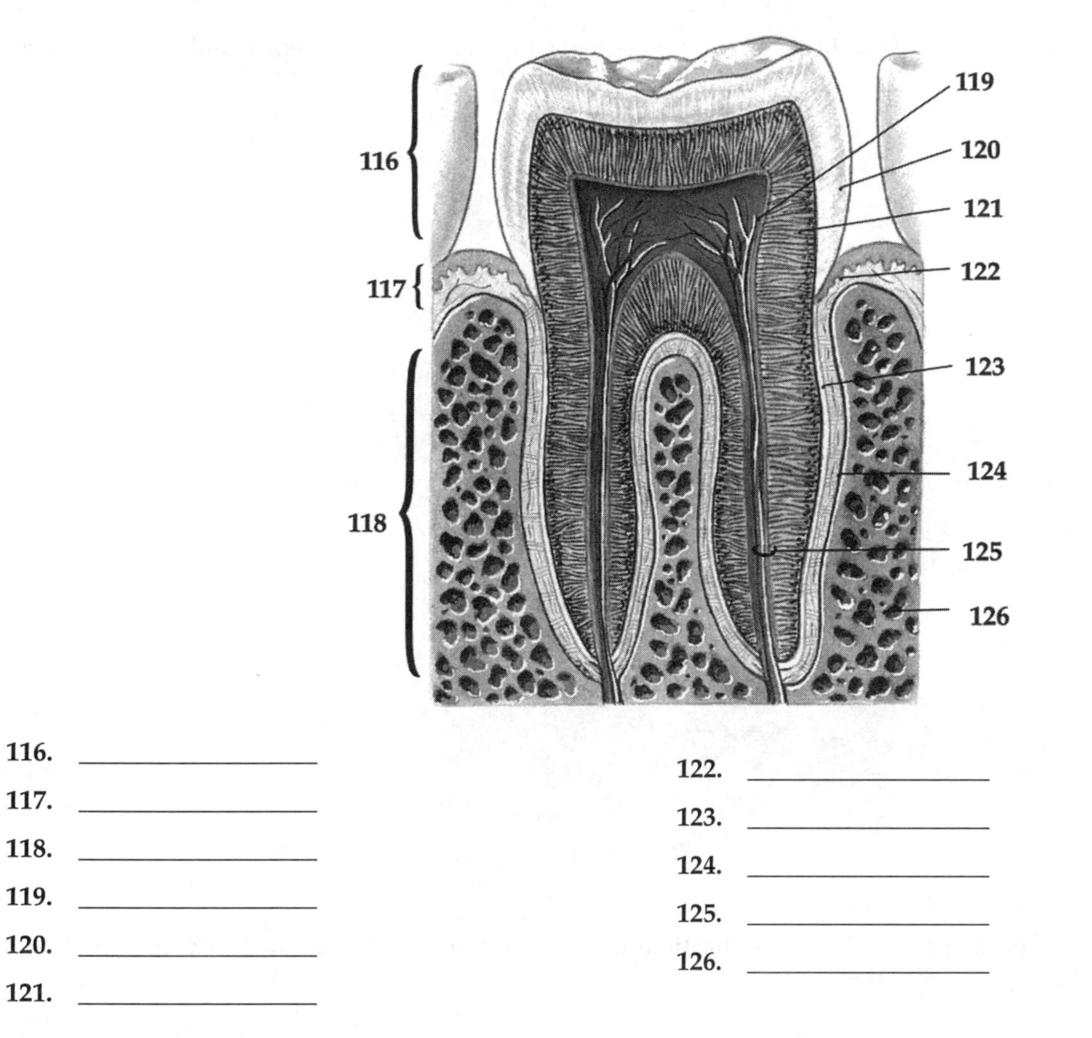

116. ____________________
117. ____________________
118. ____________________
119. ____________________
120. ____________________
121. ____________________
122. ____________________
123. ____________________
124. ____________________
125. ____________________
126. ____________________

Critical Thinking

127. What kinds of pathogens cause diseases of the oral cavity, and what are the important epidemiologic and clinical aspects of these diseases?

128. What bacteria cause gastrointestinal diseases, and what are the important epidemiologic and clinical aspects of these diseases?

129. What other kinds of pathogens cause gastrointestinal diseases, and what are the important epidemiologic and clinical aspects of these diseases?

ANSWERS

True/False

1. **True** Nearly everyone is affected by periodontal disease either with the mild form called gingivitis or the severe form called ANUG.
2. **True** It is the most common disease because the diet contains relatively large amounts of refined sugar.
3. **False** Food poisoning is caused by the ingestion of food contaminated with preformed toxins. Therefore symptoms usually appear within a few hours.
4. **False** Antibiotics should not be given because they could induce carrier states and could contribute to the development of resistant strains.
5. **False** These food products could be poisoned by a toxin called aflatoxin produced by a species of *Aspergillus*. *Amanitas* are poisonous mushrooms.
6. **True** Trichinosis is caused by a small nematode, *Trichinella spiralis*, which is often present in the muscle tissue of hogs, deer, and horses.
7. **True** Ascarid worms are exceptionally large (25–35 cm) and are very active. They have been known to cause abscesses in the liver and other organs during their wandering.

Multiple Choice

8. b; 9. a; 10. c; 11. d.

Matching

12. c; 13. a; 14. b; 15. e; 16. d; 17. e, g; 18. j; 19. c; 20. m; 21. a; 22. b; 23. f; 24. m; 25. d; 26. m; 27. l; 28. m; 29. h; 30. k; 31. m; 32. i; 33. a, d; 34. f; 35. b, g; 36. c; 37. h; 38. j; 39. k; 40. e; 41. b; 42. m; 43. a; 44. c; 45. d, f; 46. l; 47. b; 48. j, k; 49. h, i; 50. j, k; 51. b; 52. d; 53. e; 54. g; 55. a; 56. c; 57. f; 58. a; 59. f; 60. e; 61. f; 62. f.

Fill-Ins

63. digestive sytem; 64. mucin; 65. Kupffer; 66. villi; 67. feces; 68. plaque; 69. pellicle; 70. dental caries; 71. fluoride; 72. gingivitis; 73. trench mouth; 74. orchitis; 75. enterotoxin; 76. bongkrek; 77. enteritis; 78. dysentery; 79. enterocolitis; 80. Typhoid Mary Mallon; 81. rice water stool; 82. El Tor; 83. traveler's diarrhea; 84. hepatitis; 85. infections, serum; 86. parenterally; 87. balantidiasis; 88. aflatoxins; 89. ergot; 90. sheep liver fluke; 91. verminous intoxication; 92. cysticerci; 93. hydatid; 94. hookworms; 95. cutaneous larva migrans; 96. whipworms.

Labeling

97. salivary glands; 98. pharynx; 99. esophagus; 100. stomach; 101. pancreas; 102. transverse colon; 103. ascending colon; 104. descending colon; 105. sigmoid colon; 106. large intestine (colon); 107. rectum; 108. appendix; 109. small intestine; 110. ileum; 111. jejunum; 112. duodenum; 113. gallbladder; 114. liver; 115. mouth; 116. crown; 117. neck; 118. root; 119. pulp cavity; 120. enamel; 121. dentin; 122. gingival sulcus; 123. cementum; 124. periodontal ligament; 125. root canal; 126. alveolar bone (socket).

Critical Thinking

127. **What kinds of pathogens cause diseases of the oral cavity, and what are the important epidemiologic and clinical aspects of these diseases?**
Diseases of the oral cavity can be caused by bacteria and viruses and include dental caries, the most common infectious disease in developed countries. The bacteria that cause disease are commonly found in the oral cavity. Host diseases can be prevented by proper oral hygiene and the use of fluorides and dental sealants.

Dental plaque is a continuously formed coating of microorganisms and organic matter on tooth enamel. Plaque formation is the first step in the process of tooth decay and gum disease. Three factors seem to be important for caries to develop: (1) the presence of caries bacteria, (2) a diet containing relatively large amounts of sugar, and (3) poor oral hygiene or poorly calcified enamel. *Streptococcus mutans* is considered to be the primary causative agent of dental caries. It converts sucrose to sticky dextran material and organic acids which eat away the tooth enamel.

Periodontal disease is caused by a variety of bacteria but especially anaerobic species of *Bacteroides*. It can be prevented by reducing plaque buildup; by proper flossing; and by treatments of peroxide-sodium bicarbonate, mouth rinses, and, as a last resort, surgery. Acute necrotizing ulcerating gingivitis is the most severe form of gingivitis.

The only important viral disease of the oral cavity is mumps caused by a paramyxovirus. It occurs worldwide, mainly in children, but it can be prevented by a vaccine (part of the MMR series). The disease in adults can be more severe, affecting the reproductive organs.

128. What bacteria cause gastrointestinal diseases, and what are the important epidemiologic and clinical aspects of these diseases?

Numerous bacteria affect the gastrointestinal tract. They can be grouped into four categories: those that cause food poisonings; those that cause enteritis and enteric fevers; those that affect the stomach, esophagus, and duodenum; and those that infect the gallbladder and biliary tract.

Bacterial food poisoning is caused by ingesting food containing preformed toxins. Since the tissue damage is due to the action of the toxin, it is considered an intoxication or poisoning rather than an infection. Food poisonings can be prevented by sanitary handling of the food and proper cooking and refrigeration.

The most common type of food poisoning is staphylococcal enterotoxicosis. It usually occurs from eating poorly refrigerated foods, especially dairy and poultry products. The agent usually enters food from human carriers who do not follow proper food-handling procedures. Several other agents cause food poisonings, usually arising from eating undercooked meats and gravies, dairy products, and other foods. Clostridial food poisoning is quite common and causes abdominal cramps, diarrhea, and nausea following the consumption of red meats, turkey, and gravies that have not been properly refrigerated.

Enteritis is an inflammation of the intestines while enteric fevers are systemic diseases caused by pathogens that invade other tissues. All enteritises and enteric fevers are transmitted via the fecal-oral route and can be prevented by good personal hygiene, proper chlorination of drinking water, and good sanitation procedures.

Salmonellosis is often considered a food poisoning because it can be contracted from improperly prepared foods such as chickens and turkeys. It is generally self-limited and should not be treated with antibiotics. Typhoid fever is far more severe and is caused by *Salmonella typhy*. It can be diagnosed by the Widal test and is treated with chloramphenicol. Victims often remain carriers of the disease for years. A vaccine is available but offers limited protection. Shigellosis can be contracted from infected drinking water and can be treated with antibiotics. A more severe tropical form is known as bacillary dysentery. Vaccines are of limited help. Asiatic cholera occurs in Asia and Africa in regions with poor sanitation and inadequate, unchlorinated water.

Treatment requires fluid and electrolyte replacement with antibiotics. Vaccines are of very limited use. Vibriosis tends to be more mild and is common where raw seafood such as oysters is eaten. New strains of vibrios have become lethal. Traveler's diarrhea can be caused by a wide variety of organisms but most commonly by enterotoxigenic strains of *Escherichia coli*. Most cases are debilitating but self-limited.

Recent studies have indicated a bacterial cause of peptic ulcers, chronic gastritis, and other problems of the stomach and duodenum. *Helicobacter pylori* is detected with most patients having these ulcers and following antibiotic therapy, the symptoms rapidly improve.

Infections of the gallbladder and biliary tract are rare because bile destroys most organisms with lipid envelopes. The typhoid bacillus is resistant to bile and can live in the gallbladder and be released in the feces without causing symptoms. Victims of the disease may require surgical removal of the gallbladder. Gallstones blocking the ducts can cause infections, usually by *Escherichia coli,* which can then spread to the bloodstream and to other organs.

129. What other kinds of pathogens cause gastrointestinal diseases, and what are the important epidemiologic and clinical aspects of these diseases?

Several groups of viruses, protozoans and helminths cause gastrointestinal diseases. A few groups of fungi can also produce toxins that cause poisonings.

Most viral gastrointestinal disorders arise from contaminated water or food and are transmitted via the fecal-oral route. The viruses causing hepatitis B, C, D, and E are transmitted parenterally from contaminated blood and other body fluids.

Rotavirus infections cause severe diarrhea and kill many children in underdeveloped countries. Replacement of lost fluids is essential. Hepatitis A (infectious) is usually contracted from infected food handlers but can be prevented by means of a gamma globulin shot. A vaccine is being developed. Hepatitis B (serum) is much more severe and requires extended bed rest and a nutritious diet. Medical and dental personnel are at greatest risk to hepatitis B and should take advantage of the vaccine.

Protozoan gastrointestinal diseases arise from cyst-contaminated food and water via the fecal-oral route and can be prevented by drinking uncontaminated water, and using good personal hygiene and proper sanitation. Diagnosis is based on the observation of the cysts in the concentrated stool and most can be treated with antiprotozoan agents.

Giardiasis is common in children but especially with backpackers since the parasite is maintained in raccoons, deer, and other wild animals. It is one of the most identifiable agents because of its paired nuclei or "eyes." Amoebic dysentery and chronic amebiasis occur worldwide and in some regions are so common that they are considered the rule rather than the exception. Balantidiasis also occurs worldwide but mainly in tropical areas. It is a large parasite and is the only important ciliate to cause disease. Cryptosporoidiosis seems to occur only in immunodeficient patients such as those with AIDS.

Gastrointestinal diseases caused by helminths are mainly acquired in tropical regions and include several kinds of fluke, roundworm, and tapeworm infections. The flukes have interesting life cycles that usually involve snails and crustaceans. They gain access to the host following consumption of vegetation that contains the snails or the larva. Tapeworms are usually contracted by consumption of infected undercooked meats. Roundworms have complex life cycles in which they travel from the intestines following consumption of their eggs, to the heart and lungs and then back to the intestines where they develop into adults. Some, such as the hookworms, can burrow through the skin to cause creeping eruptions. Most worm infections are especially severe in young children. They can be diagnosed by finding the eggs in the stool and can be prevented by good sanitation, by avoiding contaminated water and soil, and by thoroughly cooking foods that might be contaminated.

Fungal toxins can be very poisonous and can result in severe injury and even death. Aflatoxins are potent carcinogens produced by species of *Aspergillus*. Humans and animals can ingest them from moldy grains and raw peanuts. *Claviceps purpura* produces a highly toxic granule called ergot which forms in contaminated grains. Several species of wild mushrooms especially *Amanita* (that are inadvertently picked for dinner) produce a highly toxic poison that causes vomiting, diarrhea, liver damage, hallucinations and even death.

Chapter 23 Cardiovascular, Lymphatic, and Systemic Diseases

INTRODUCTION

The cardiovascular and lymphatic systems may seem a most unlikely habitat for parasites. This is right in the heart (pun intended) of the vertebrate immune defense. In the blood and lymph are circulating antibodies and armies of white blood cells of our immune system. Yet, since the beginnings of recorded history, parasitic diseases of these sites have been every so common. One specific disease that I have studied is thought to be responsible for the evolution of the multiple alleles of human red blood cells and currently infects tens of millions of people, schistomiasis.

This disease occurs with macroscopic worms living in blood vessels of the lower abdomen. Large worms should present multiple antigens to the vertebrate immune system, and yet they can live in our blood vessels for many years. The females lay eggs that are filtered by the liver and spleen and a significant inflammatory response develops resulting in the enlargement of these organs. Infected people are sick, and in the case of the species common in Egypt, *S, haematobium*, even men will commonly develop breast cancer in response. In the second half of the 20th century, when armies from Egypt attacked Israel, the failure of the Egyptian armies is partly due to a high incidence of infected soldiers. Israel does not have this parasite. Even in America, at least ten percent of people who were born in Puerto Rican have this disease.

Eggs of this parasite have been found in the preserved livers from ancient Egypt. It is most likely that when the first humans walked the grasslands of Africa, a schistosome from a related primate began to infect earliest people. The worm is able to evade our immune response by coating itself with carbohydrates that are identical to those found on red blood cells. Thus, our immune system 'thinks' the worms are blood cells. In such a world, there would a selective advantage to evolve a different sugar on the cell coats of red blood cells and to produce an antibody to the sugars on the parasites. In this manner many biologists believe that the ABO system of blood cells evolved. It is for this reason that type O blood is most common. If you remember, people with type O blood have antibodies to type A and B blood. These are very factors that lie on the surface of many schistosome worms.

Many other stories of systemic disease await you in this chapter. Pay special attention to leishmaniaisis. Many of our troops from the Iraq and Afghanistan will return with this disease. It is transmitted by the bite of a sand fly, and our soldiers are largely living in a giant sand box filled with infected flies.

STUDY OUTLINE

I. The Cardiovascular System

A. The heart and blood vessels

The heart lies between the lungs in a tough, membranous *pericardial sac* lubricated by serous fluid. Its wall consists of a thin internal *endocardium*, a thick muscular *myocardium*, and an outer *epicardium*. The heart has two *atria* and two *ventricles*. The blood vessels include *arteries* that receive blood from the heart, *arterioles* that branch from arteries, *capillaries* that branch from arterioles, *venules* that receive blood from capillaries, and *veins* that receive blood from venules and return it to the heart.

B. The blood

The blood consists of plasma (which is more than 90 percent water and contains proteins and electrolytes) and formed elements (cells and cell fragments).

C. The lymphatic system

The lymphatic system consists of a network of vessels, nodes, and other lymphatic tissues, and the fluid *lymph*. The lymphatic system has three major functions: it (1) collects excess fluid from the spaces between body cells (lymphatic circulation); (2) transports digested fats to the cardiovascular system; and (3) provides many of the nonspecific and specific defense mechanisms against infection and disease (lymphoid organs and tissues).

D. Normal microflora of the cardiovascular system

The cardiovascular system does not have any normal resident microflora. When microorganisms do enter the bloodstream, however, they can cause *bacteremia* or *septicemia*.

II. Cardiovascular and Lymphatic Diseases

A. Bacterial septicemia and related diseases

1. Septicemias

Even with antibiotics, septicemia is not easy to treat. Organisms including *Pseudomonas aeruginosa*, *Bacteroides fragilis*, and species of *Klebsiella*, *Proteus*, *Enterbacter*, and *Serratia* cause *septic shock*. Antibiotics often worsen the situation. Diagnosis is made by culturing blood, catheter tips, urine, or other sources of infection.

2. Puerperal fever

Puerperal fever, also called *puerperal sepsis* or *childbed fever*, is caused by *Streptococcus pyogenes*. Streptococci can be isolated from blood cultures to diagnose puerperal fever. Penicillin is effective except against resistant organisms.

3. Rheumatic fever

Rheumatic fever is a multisystem disorder following infection by *Streptococcus pyogenes*. Culturing of streptococci and serological tests are helpful in diagnosis, as is a previous history of streptococcal infection. Prompt treatment of *S. pyogenes* infections with antibiotics before cross-reactive antibodies can form is the only practical way to prevent rheumatic fever. No effective vaccine exists.

4. Bacterial endocarditis

Bacterial endocarditis, or *infective endocarditis*, is caused by many microbes, but most cases are due to bacteria. *Vegetation*, *myocarditis*, and *pericarditis* can be caused by microbial infections. Diagnosis depends on blood cultures and treatment involves penicillin or other antibiotics.

B. Helminthic diseases of the blood and lymph

1. Schistosomiasis

Three species of blood flukes of the genus *Schistosoma* causes *schistosomiasis*. Diagnosis can be made by finding eggs in feces or urine, by intradermal injection of schistosome antigen and measurement of the area of the wheal, or by a complement fixation test. Treatment can involve one of several new drugs, especially praziquantal.

2. Filariasis

Filariasis can be caused by several different roundworms. Repeated infections over a period of years can lead to *elephantiasis*. Filariasis is diagnosed by finding microfilariae in thick blood smears made from blood samples taken at night or by an intradermal test. The drugs diethylcarbamizine and metronidazole are effective in treating the disease.

III. Systemic Diseases

A. Bacterial systemic diseases

1. Anthrax

Bacillus anthracis causes *anthrax*, a zoonosis that affects mostly plant-eating animals, especially sheep, goats, and cattle. Anthrax is diagnosed by culturing blood or examining smears from cutaneous lesions of patients with a history of possible exposure. The disease is treated with penicillin or tetracycline.

2. Plague

Yersinia pestis causes the zoonotic *plague*, which can develop into *bubonic plague*, *septicemic plague*, or *pneumonic plague*. Plague can be diagnosed by fluorescent antibody tests or by the identification of

Y. pestis from stained smears of sputum or from fluid aspirated from lymph nodes. It is treated with streptomycin, tetracycline, or both.

3. Tularemia

   Tularemia, caused by *Francisella tularensis,* is a zoonosis found in more than 100 mammals. It is spread by *transovarian transmission* in ticks. Entry through the skin results in the *ulceroglandular* form of the disease, while bacteremia from lesions can lead to *typhoidal tularemia.* Diagnosis usually involves agglutination tests. Treatment involves streptomycin.

4. Brucellosis

   Several species of *Brucella* are responsible for causing the zoonotic *brucellosis,* which is also called *undulant fever* or *Malta fever.* Brucellosis is diagnosed by serologic tests and treated with tetracycline, streptomycin, gentamicin, or rifampin.

5. Relapsing fever

   A dozen species of the genus *Borrelia* cause *epidemic relapsing fever* and *endemic relapsing fever.* Diagnosis is made by identifying the organism in stained blood smears prepared during a rising fever phase. Tetracycline or chloramphenicol are used to treat relapsing fever. No vaccine is available.

6. Lyme disease

   Borrelia burgdorferi is the causative agent of *Lyme disease.* The disease is usually diagnosed by clinical symptoms after arthritis appears, although an antibody test is available. Lyme disease is treated with antibiotics such as doxycycline and amoxicillin.

B. Rickettsial and related systemic diseases

Rickettsias are named after Howard T. Ricketts, who identified them as the causative agents of typhus and Rocky Mountain spotted fevers. Rickettsial diseases cause skin lesions, *petechiae,* necrosis in organs, fever, headache, extreme weakness, and liver and spleen enlargement.

1. Typhus fever

   Typhus fever occurs in a variety of forms, including *epidemic, endemic (murine),* and *scrub typhus. Brill-Zinsser disease* is a recurrent form of endemic typhus.

2. Rocky Mountain spotted fever

   Rickettsia rickettsii causes Rocky Mountain spotted fever. Prompt antibiotic treatment is important.

3. Other rickettsial infections

   Other rickettsial infections include *rickettsial pox, trench fever, bartonellosis, ehrlichiosis,* and *bacillary angiomatosis.*

C. Viral systemic diseases

1. Dengue fever

   Dengue fever, caused by an arbovirus of the family *Flaviviridae,* has also been called *breakbone fever.* Its main vectors include *Aedes aegypti* and *A. albopictus.* Mosquito control is therefore the primary method of preventing the disease.

2. Yellow fever

   Yellow fever is caused by flavivirus, an arbovirus, and is transmitted by *Aedes aegypti.* Two strains of yellow fever are used to produce vaccines.

3. Infectious mononucleosis

   The herpes virus *Epstein-Barr virus* causes *infectious mononucleosis.* (This virus is also responsible for Burkitt's lymphoma and perhaps chronic fatigue syndrome.) Treatment involves bed rest and antibiotics for secondary infections.

4. Other viral diseases

   Other viral systemic diseases include *filovirus fevers, bunyavirus fevers, arenavirus fevers, colorado tick fever, parvovirus infections, aplastic crisis, fifth disease,* and *coxsackie virus infections.*

D. Protozoan systemic diseases

1. Leishmaniasis

   Three species of protozoa of the genus *Leishmania* can cause *leishmaniasis* in humans. Diagnosis is made by identifying the protozoa in blood smears in kala azar and from scrapings of skin and mucous membrane lesions. Antimony compounds are used to treat both kala azar and skin and mucous membrane lesions.

2. Malaria

   Several species of the protozoan *Plasmodium* are capable of causing *malaria.* The main means of diagnosing malaria is by identifying the protozoa in red blood cells. Chloroquine is the drug of choice for all forms of malaria in the acute stage.

3. Toxoplasmosis
   Toxoplasma gondii can cause *toxoplasmosis*, especially in developing fetuses, in newborn infants, and sometimes in young children. It can be diagnosed by finding parasites in the blood, cerebrospinal fluid, or tissues, by animal inoculation with subsequent isolation of the organisms, or by direct immunofluorescence tests. Pyrimethamine and trisulfapyridine are used in combination to treat toxoplasmosis, but no treatment can reverse permanent damage from prenatal infection.
4. Babesiosis
   Several species of the sporozoan *Babesia* can cause *babesiosis*. Diagnosis is made from blood smears. Chloroquine is the drug of choice for treatment.

SELF-TESTS

Continue with this section only after you have read Chapter 23 of your textbook. Write your answers in the appropriate space provided. Correct answers to all questions can be found at the end of the Self-Test section. A score of 80 percent or better is good. If your score is less than 65, reread the chapter.

True/False

Mark T for True, F for False

_____ **1.** Bacterial endocarditis is generally a life-threatening infection.

_____ **2.** Most cases of Rocky Mountain spotted fever occur in Montana, Wyoming, and Colorado.

_____ **3.** Currently, all malaria parasites respond to the drug chloroquine, and the disease is declining in most countries, including Africa.

_____ **4.** Rheumatic fever is not an infectious disease but actually results from an adverse immune reaction.

_____ **5.** Lyme disease is maintained in the wild animal population such as deer and small mammals and is transmitted by a tick.

_____ **6.** The Epstein-Barr virus appears to exert its effect on lymphocytes.

Multiple Choice

Select the best possible answer.

_____ **7.** In the disease of schistosomiasis:
- **a.** Avoidance of mosquitos can reduce the number of carriers.
- **b.** The parasites are small protozoans that form cysts.
- **c.** The disease primarily affects small rodents and only recently has affected man.
- **d.** The parasites form into cercaria to infect their hosts.

_____ **8.** Select the incorrect statement.
- **a.** Cases of anthrax have occurred from contaminated imported animal hair and skins.
- **b.** Human plague occurs primarily in Europe and Asia with no cases recorded in the United States for the last 10 years.
- **c.** Skinning rabbits could result in tularemia.
- **d.** The agents of brucellosis cause an undulating fever.

_____ **9.** Systemic rickettsial infections are characterized by all the following except:
- **a.** a skin rash.
- **b.** respond slowly to antibiotic therapy.
- **c.** transmitted by arthropod vectors.
- **d.** have a low mortality rate.

_____ **10.** Viral systemic diseases:
- **a.** include dengue fever and yellow fever.
- **b.** are all lethal.
- **c.** generally respond to antibiotics.
- **d.** cannot be prevented by vaccines.

Matching

Select the answer from the right-hand side that corresponds to the term or phrase on the left-hand side of the page. An answer may be used more than once. In some cases, more than one answer may be required.

Topic: Causative Agents of Infectious Diseases (Part 1)

______	**11.** Anthrax	**a.** *Yersinia pestis*
______	**12.** Filariasis	**b.** *Schistosoma* species
______	**13.** Puerperal sepsis	**c.** *Borrelia recurrentis*
______	**14.** Schistosomiasis	**d.** *Brucella abortus*
______	**15.** Bubonic plague	**e.** *Wuchereria bancrofti*
______	**16.** Relapsing fever	**f.** *Streptococcus pyogenes*
______	**17.** Malta fever	**g.** *Borrelia burgdorferi*
______	**18.** Tularemia	**h.** *Rickettsia prowazeki*
______	**19.** Lyme disease	**i.** *Bacillus anthracis*
______	**20.** Epidemic typhus fever	**j.** *Francisella tularensis*

Topic: Causative Agents of Infectious Diseases (Part 2)

______	**21.** Scrub typhus fever	**a.** *Rickettsia typhi*
______	**22.** Trench fever	**b.** *Bartonella bacilliformis*
______	**23.** Endemic typhus fever	**c.** *Rickettsia tsutsugamushi*
______	**24.** Oroya fever	**d.** *Ehrlichia canis*
______	**25.** Ehrlichiosis	**e.** *Rickettsia akari*
______	**26.** Rickettsial pox	**f.** *Bartonella henselae*
______	**27.** Kala azar	**g.** *Bartonella quintana*
______	**28.** Bacillary angiomatosis	**h.** *Leishmania donovani*
______	**29.** Burkitt's lymphoma	**i.** *Plasmodium* species
______	**30.** Infectious mononucleosis	**j.** Epstein–Barr virus
______	**31.** Malaria	**k.** *Toxoplasma gondii*
______	**32.** Babesiosis	**l.** *Babesia bigemina*
______	**33.** Toxoplasmosis	

Topic: Bacterial Disease Transmission (Part 1)

______	**34.** Bacterial endocarditis	**a.** penetration of skin by cercariae
______	**35.** Anthrax	**b.** associated with birth delivery process
______	**36.** Puerperal sepsis	**c.** infected mosquitoes
______	**37.** Septicemia	**d.** invasive medical procedures
______	**38.** Relapsing fever	**e.** zoonosis
______	**39.** Schistosomiasis	**f.** organisms transported from other body sites
______	**40.** Scrub typhus fever	**g.** exposure to contaminated hides, meat, or bones

______ **41.** Filariasis

______ **42.** Brucellosis

______ **43.** Plague

______ **44.** Tularemia

______ **45.** Lyme disease

______ **46.** Rocky Mountain spotted fever

______ **47.** Epidemic typhus fever

______ **48.** Endemic typhus

h. flea bites

i. ingestion of contaminated dairy products

j. body lice

k. ticks

l. mites

Topic: Bacterial Disease Transmission (Part 2)

______ **49.** Bartonellosis

______ **50.** Scrub typhus fever

______ **51.** Trench fever

______ **52.** Rickettsial pox

______ **53.** Ehrlichiosis

______ **54.** Toxoplasmosis

______ **55.** Yellow fever

______ **56.** Dengue fever

______ **57.** Infectious mononucleosis

______ **58.** Bunyavirus fevers

______ **59.** Rift Valley fever

______ **60.** Colorado tick fever

______ **61.** Kala azar

______ **62.** Coxsackie virus infections

______ **63.** Leishmaniasis

______ **64.** Malaria

______ **65.** Babesiosis

a. mites

b. sandflies

c. ticks

d. mosquitoes

e. lice

f. aerosol

g. placental transmission

h. fecal–oral transmission

i. raw or undercooked contaminated meat

Topic: Signs and Symptoms of Disease

______ **66.** Filariasis

______ **67.** Septicemia

______ **68.** Schistosomiasis

______ **69.** Plague

______ **70.** Bacterial endocarditis

______ **71.** Rheumatic fever

______ **72.** Dengue fever

______ **73.** Localized leishmaniasis

a. dermatitis from cercaria, allergic reactions to eggs, tissue damage to urinary bladder

b. inflammation and blockage of lymph ducts, leading to elephantiasis

c. mitral valve damage due to immunological response

d. formation of buboes

e. inflammation and vegetation of heart valves

f. septic shock due to endotoxins

______ 74. Malaria

g. oriental sore, and skin and mucous membrane lesions

h. periods of high fever associated with release of parasites from blood cells; relapses can occur

i. severe bone and joint pain

Topic: Properties of Disease Agents

______ 75. *Schistosoma japonicum*

______ 76. *Yersinia pestis*

______ 77. *Streptococcus pyogenes*

______ 78. *Francisella tularensis*

______ 79. *Brucella* species

______ 80. *Bacillus anthracis*

______ 81. *Wuchereria bancrofti*

______ 82. *Borrelia recurrentis*

______ 83. *Rickettsia typhi*

______ 84. *Plasmodium* species

______ 85. *Leishmania* species

______ 86. *Ehrlichia canis*

______ 87. *Bartonella henselae*

a. gram-positive coccus

b. nematode

c. gram-positive spore former

d. gram-negative rod

e. fluke

f. gram-negative coccobacillus

g. gram-negative spiral

h. protozoan

Fill-Ins

Provide the correct term or phrase for the following. Spelling counts.

88. The ________________ ________________ consists of the heart, blood vessels, and blood.

89. The heart lies between the lungs in a tough, membranous ________________ ________________ lubricated by serous fluid.

90. The ________________ gland is a multilobed lymphatic organ located beneath the sternum that releases T cells into the blood around the time of birth.

91. The ________________, located in the upper left quadrant of the abdominal cavity, is the largest of the lymphatic organs and is similar to the lymph nodes.

92. The Peyer's patches of the small intestine are examples of ________________ ________________, or unencapsulated areas filled with lymphocytes.

93. Microorganisms occasionally enter the bloodstream of all humans, causing a transient ________________.

94. Blood poisoning, or ________________, occurs when microbes grow and multiply in the bloodstream.

95. A drop in blood pressure along with blood vessel collapse due to bacterial infection can result in ________________ ________________.

96. Puerpéral sepsis due to beta-hemolytic streptococci is also known as ______________________________ __________________.

97. Fibrous adhesions that form on heart valves are often called __________________.

98. Schistosomiasis is also known as __________________.

99. Gross enlargements of limbs caused by blockage of lymphatic vessels may be called __________________.

100. A respiratory form of anthrax is __________________ disease.

101. The agent that transmits plague from animals to humans is the __________________.

102. Painfully enlarged lymph nodes ,especially of the groin and armpit are __________________.

103. If plague bacteria move from the lymphatics to the circulatory system, they cause a form of the plague called __________________ plague because of its hemorrhagic symptoms.

104. In ticks, __________________ transmission allows the pathogen that is present in the eggs to be transmitted from one generation to another.

105. A septicemia of tularemia can allow the pathogen to cause a __________________ __________________ that resembles typhoid fever.

106. Infections caused by species of *Brucella* cause the disease brucellosis, __________________ fever, or __________________ fever.

107. Since the disease of brucellosis occurs in animals, it can be called a __________________.

108. The epidemic form of relapsing fever is transmitted by a __________________.

109. Lyme disease is now known to be transmitted by a __________________.

110. Pinpoint-sized hemorrhages most common in skin folds that are caused by rickettsia are called __________________.

111. Recrudescent typhus, which is a recurrent typhus infection, is also known as __________________ - __________________ disease.

112. Scrub typhus is also known as __________________ fever and is transmitted by a __________________.

113. The vector that transmits Rocky Mountain spotted fever is the __________________.

114. The severe bone and joint pain associated with dengue fever gives reason to call this disease __________________ fever.

115. Yellow fever is transmitted by a __________________.

116. A cancer caused by the same virus that causes infectious mononucleosis is ______________________________ __________________.

117. A parvovirus may be the cause of a symptom of sickle-cell anemia called ______________________________ __________________.

118. Visceral leishmaniasis is also known as __________________.

119. The cutaneous form of leishmaniasis is known as __________________.

120. One form of malaria causes large numbers of red blood cells to lyse, resulting in a condition called __________________ fever.

Labeling

Questions 121–127: Select the best possible label for each structure in the lymphatic system: spleen; tonsil; lymph node; lymphatic vessel; thoracic (left lymphatic) duct; thymus gland; right lymphatic duct. *Use the spaces provided for your answers.*

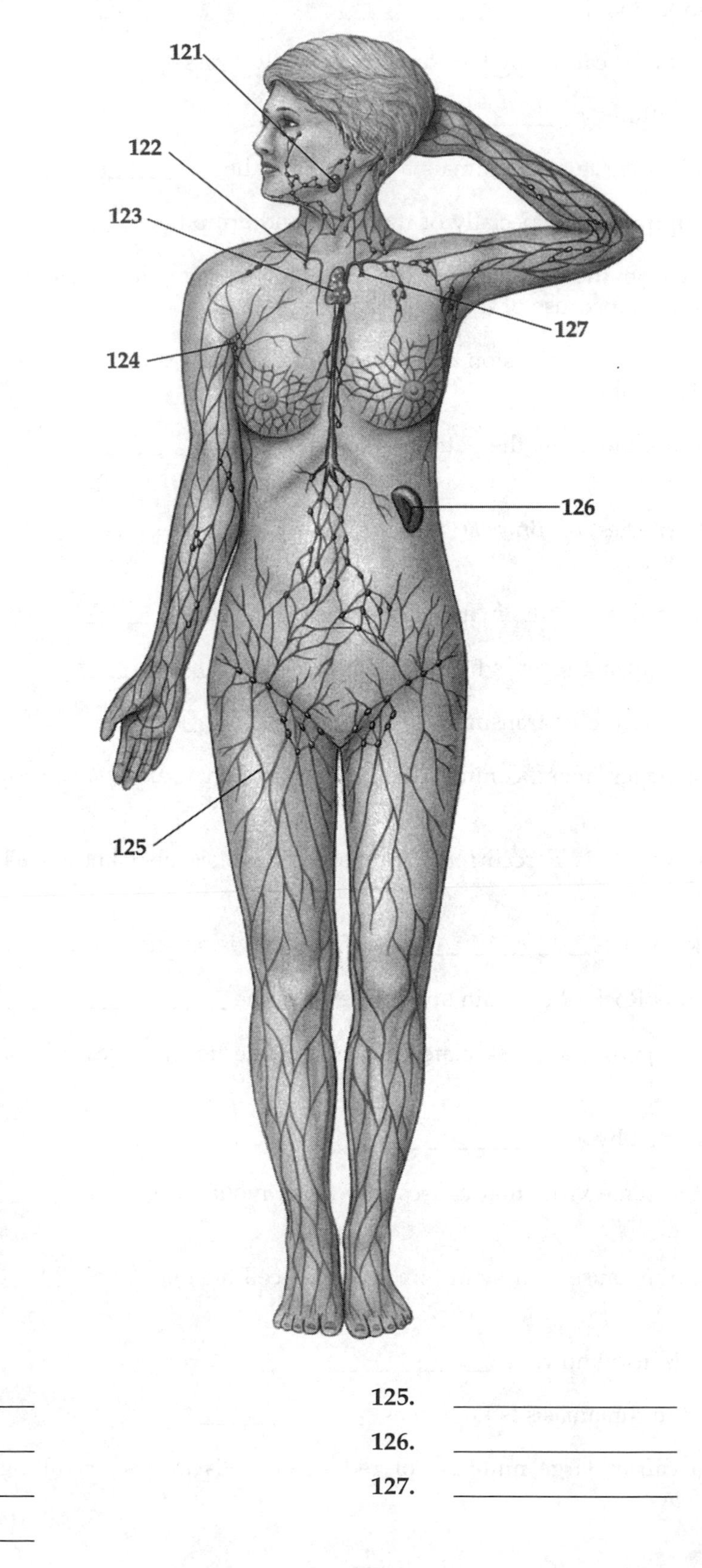

121. ____________________

122. ____________________

123. ____________________

124. ____________________

125. ____________________

126. ____________________

127. ____________________

Critical Thinking

128. What pathogens cause bacterial septicemias and related diseases, and what are the important epidemiologic and clinical aspects of these diseases?

129. What pathogens cause parasitic diseases of the blood and lymph, and what are the important epidemiologic and clinical aspects of these diseases?

130. What pathogens cause bacterial systemic diseases, and what are the important epidemiologic and clinical aspects of these diseases?

131. What pathogens cause rickettsial systemic disease, and what are the important epidemiologic and clinical aspects of these diseases?

132. What pathogens cause viral systemic diseases, and what are the important epidemiologic and clinical aspects of these diseases?

133. What pathogens cause parasitic systemic diseases, and what are the important epidemiologic and clinical aspects of these diseases?

ANSWERS

True or False

1. **True** Acute endocarditis is a rapidly progressive disease that can destroy heart valves and cause death in only a few days. Subacute endocarditis, although not as rapidly fatal, still may cause serious valve damage.
2. **False** Most cases of Rocky Mountain spotted fever actually occur in the Eastern part of the United States, especially in the Appalachian mountains, probably because of a higher tick population and because of a lack of awareness of the disease itself.
3. **False** Many species of Plasmodia are becoming more resistant to quinine derivatives. Furthermore, the disease is still endemic in most tropical areas, especially in Africa, and may be increasing in other countries, even the United States.
4. **True** Rheumatic fever appears to be a sequelae of infections caused by beta-hemolytic *Streptococcus pyogenes*. There may be genetic predisposition for the disease.
5. **True** Lyme disease is caused by the spirochaete, *Borrelia bourgdorferi.* The agent is transmitted by the deer tick, *Ixodes dammini.*
6. **True** The EBV seems to infect certain kinds of mature B lymphocytes and replicates in the nucleus of the cell. Its DNA is replicated faster than cellular DNA.

Multiple Choice

7. d; 8. b; 9. d; 10. a.

Matching

11. f; 12. e; 13. f; 14. b; 15. a; 16. c; 17. d; 18. j; 19. g; 20. h; 21. c; 22. g; 23. a; 24. b; 25. d; 26. e; 27. h; 28. f; 29. j; 30. j; 31. i; 32. l; 33. k; 34. f; 35. g; 36. b; 37. d; 38. j, k; 39. a; 40. l; 41. c; 42. i; 43. e, h; 44. e, k; 45. k; 46. k; 47. j; 48. h; 49. b; 50. a; 51. e; 52. a; 53. c; 54. g, i; 55. d; 56. d; 57. f; 58. b, d; 59. b; 60. c; 61. b; 62. h; 63. b; 64. d; 65. c; 66. b; 67. f; 68. a; 69. d; 70. e; 71. c; 72. i; 73. g; 74. h; 75. e; 76. d; 77. a; 78. f; 79. d; 80. c; 81. b; 82. g; 83. g; 84. h; 85. h; 86. f; 87. f.

Fill-Ins

88. cardiovascular system; 89. pericardial sac; 90. thymus; 91. spleen; 92. lymphoid nodules; 93. bacteremia; 94. septicemia; 95. septic shock; 96. childbed fever; 97. vegetations; 98. bilharzia; 99. elephantiasis; 100. woolsorters'; 101. flea; 102. buboes; 103. septicemic; 104. transovarian; 105. typhoidal tulermia; 106. undulant, Malta; 107. zoonosis; 108. louse; 109. tick; 110. petechiae; 111. Brill-Zinsser; 112. tsutsugamushi; 113. tick; 114. breakbone; 115. mosquito; 116. Burkitt's lymphoma; 117. aplastic crisis; 118. kala azar; 119. oriental sore; 120. blackwater.

Labeling

121. tonsil; 122. right lymphatic duct; 123. thymus gland; 124. lymph node; 125. lymphatic vessel; 126. thoracic (left lymphatic) duct; 127. spleen.

Critical Thinking

128. **What pathogens cause bacterial septicemia and related diseases, and what are the important epidemiologic and clinical aspects of these diseases?**
Several different types of normal flora bacteria cause septicemia and related diseases, especially members of the *Streptococci* and *Staphylococci*. Before the advent of antibiotics, these infections were usually fatal. Today, they are still serious and difficult to treat. Diagnosis of septicemia is made by culturing the blood, catheter tips, urine, or other specimens.

    Rheumatic fever is actually an immunologic reaction to constant, untreated infections caused by *Streptococcus pyogenes*. Antibiotic treatment will not reverse existing damage, but can prevent further destruction of the heart valves. Bacterial endocarditis also follows infections caused by species of *Streptococci* which gain access to the blood following tooth extractions and teeth cleaning. Normally these organisms are removed from circulation immediately; however, if the victim has existing heart valve damage, the bacteria can colonize and eventually form massive vegetations.

129. What pathogens cause parasitic diseases of the blood and lymph, and what are the important epidemiologic and clinical aspects of these diseases?

Several species of schistosome flukes and a filarial worm cause serious systemic infections.

Schistosomiasis or bilharzia is acquired by the aquatic larvae penetrating the skin of a victim and developing in the liver or bladder. The disease is diagnosed by finding the eggs in the stool. Praziquantel can be used for treatment and the disease can be prevented by eradicating the infected snails (rather hard to do) or by avoiding snail-infested water. Most cases occur in Africa, South America, and Asia; however, the disease is prevalent on the island of Puerto Rico. A vaccine has not yet been developed.

Filariasis or elephantiasis is transmitted by mosquitoes and diagnosed by finding the microfilaria in the blood. It is treated with diethylcarbamazine or metronidazole and may require surgical drainage. Avoidance of infected mosquitoes is the best prevention.

130. What pathogens cause bacterial systemic diseases, and what are the important epidemiologic and clinical aspects of these diseases?

Several groups of bacteria cause serious systemic diseases, some of which are commonly found in the United States.

Anthrax is a zoonosis that affects mostly animals and rarely afflicts humans except in countries such as Africa and Asia. It is acquired through spores from infected domestic animals, especially the hides and hair products. U.S. Customs agents have kept infected, imported products from entering this country. The disease is diagnosed by blood cultures or from smears of lesions. Treatment is obtained with antibiotics and the disease can be prevented by vaccinating the animals and reducing the risk of exposure to the agent. Unfortunately, anthrax has been considered as a biological germ warfare weapon.

Plague is also a zoonosis that is maintained in the small animal population throughout the world, including the Western part of the United States. It is transmitted by fleas that fed off of infected rodents and can be diagnosed by stained smears and antibody tests. The disease can be treated with broad spectrum antibiotics but still has a high mortality rate. A vaccine is available for those entering endemic areas.

Tularemia, or rabbit fever, is a zoonosis that is maintained in the small animal population and is transmitted by direct contact, inhalation, ingestion, and through insect bites. It can be diagnosed by blood tests and treated with streptomycin. Taxidermists and small game hunters are at great risk for the disease. It can be prevented by avoiding the infected animals and by wearing protective gloves and clothing. The vaccine is not too effective.

Brucellosis, or relapsing fever, is a zoonosis that is maintained in the domestic large animal population, including cows, pigs, goats, and even dogs. It is an occupational hazard for those in the meat-packing industry and is acquired through the skin, by inhalation, by ingestion, and by drinking infected raw milk. It is diagnosed by serological tests, treated with prolonged antibiotic therapy, and prevented by avoiding contact with infected animals and contaminated milk products.

Relapsing fever is a zoonosis maintained in the small animal population and is transmitted by lice and ticks. It can be diagnosed from blood smears, treated with antibiotics, and prevented by avoiding the vectors.

Lyme disease is a zoonosis maintained in the wild animal population, especially deer, and is transmitted by ticks. It is diagnosed by clinical signs and by antibody tests. It can be treated with antibiotics such as doxycycline and prevented by avoiding tick bites. The disease appears to be spreading across the country, but its increase may be due to better identification of the symptoms.

131. What pathogens cause rickettsial systemic disease, and what are the important epidemiologic and clinical aspects of these diseases?

Several species of rickettsia cause serious systemic disease. Rocky Mountain spotted fever and murine typhus are the only two that occur with regularity in the United States. The others occur more often in Europe, Asia, and other parts of the world. The organisms are small Gram-negative, highly infectious, intracellular parasites that can not be cultivated artificially. They are all transmitted by arthropod vectors, including ticks, mites, lice, and fleas. Mosquitoes have not been found to transmit them. They all respond to tetracyclines but even in treated cases, the mortality rate can be high. They are diagnosed by serological tests. Clinical symptoms are all very similar, with a high fever, chills, and a rash being the most common.

Murine typhus is transmitted by fleas and is maintained in the rodent population. Southern Texas is the endemic region for the disease.

Epidemic typhus is more of historical note in that it has changed the course of world history several times by infecting armies such as Napoleon's in 1812. It's transmitted by body lice and is common in unsanitary, overcrowded conditions.

Rocky Mountain spotted fever is carried and maintained in the tick and small animal population and is actually distributed more in the Eastern part of the United States than in the Rocky Mountains.

132. What pathogens cause viral systemic diseases, and what are the important epidemiologic and clinical aspects of these diseases?

Several viruses cause serious systemic diseases, most of which occur with little regularity in the United States. Since antibiotics are of no value, prevention is the best medicine. Vaccines are available for some of these diseases but the best method is to avoid those diseases that are transmitted by vectors.

Dengue or breakbone fever occurs worldwide and is beginning to occur in the United States. It is transmitted by a mosquito and can be diagnosed by serologic tests. Mosquito control seems to be the best means of prevention.

Yellow fever is most common in tropical areas of Central and South America and Africa, is diagnosed by symptoms and serological lab tests, and can be prevented by a vaccine.

Infectious mononucleosis seems to affect those who have not acquired the virus early in life. Children in underdeveloped countries and those that attend public schools seem to have the best opportunity of early exposure and the buildup of antibodies to the virus. Those without this exposure seem to be affected more severely by the virus when they get infected during their teen and young adult years. The disease is diagnosed through clinical symptoms and is only treated symptomatically. Chronic fatigue may be a complication.

Filovirus, bunyavirus, and arenavirus fevers seem to occur predominately in Africa and Central and South America. Parvoviruses may cause aplastic anemia but most commonly affect dogs and cats. Animal vaccines are available. Coxsackie viruses cause a variety of rare infections of the heart, pharynx, and meninges. There is no treatment, vaccine, or prevention available.

133. What pathogens cause parasitic systemic diseases, and what are the important epidemiologic and clinical aspects of these diseases?

Of all the parasitic systemic diseases, only two, malaria and toxoplasmosis, occur with regularity in the United States.

Leishmaniasis occurs in tropical and subtropical countries where the tiny sandflies are present. A cutaneous and visceral form can occur, with each responding poorly to medication which includes antimony and arsenic. The disease is diagnosed from blood smears or scrapings from the lesions. By reducing the sandfly population and eliminating the rodent reservoirs, the disease could be controlled.

Malaria is the world's greatest public health problem and affects millions worldwide. Most cases in the United States come from people who have been in endemic areas; however, with the increase in the number of active carriers, the disease has begun to actually occur in many states, especially in the South. The disease is transmitted by *Anopheles* mosquitoes and is diagnosed by identifying the parasites in blood smears. Active disease is treated with quinine derivatives; however, some strains are now becoming resistant. Great efforts have been made to find better mosquito-control methods, new forms of chemotherapy, and an effective vaccine. At present, the mosquito and the parasite are winning the battle.

Toxoplasmosis is a zoonosis that is maintained in the rodent population but has infected most domestic cats. The disease agent is now transmitted from the feces of these cats and is most dangerous to pregnant women, especially the fetus, and infants who get in contact with the infected materials. It is diagnosed by finding the parasites in body fluids or tissues and is treated with pyrimethamine and trisulfapyridine. However, once the fetus is infected, the parasites may cause serious damage before it can be treated. Therefore, pregnant women should avoid the "kitty litter" box and any soil contaminated with cat feces.

Babesiosis is not very common but is transmitted by ticks. It can be diagnosed from blood smears and treated with chloroquine.

Chapter 24 Diseases of the Nervous System

INTRODUCTION

The nervous system is perhaps the most protected part of our body. The endothelium that lines the blood vessels that vascularize the central nervous system are designed to prevent microbes from invading. Unfortunately, parasites are very cleaver and many viruses, bacteria, and protozoa are up to the challenge. Once a parasite penetrates the nervous system there is little immune response to protect the host. In some part of the world, diseases of the central nervous system can be a major cause of death and economic stagnation. A case in point is the country of Brazil and South American sleeping sickness or Chagas' disease. This disease may have been the caused of the illness and death of perhaps the most important biologist to have every lived, Charles Darwin.

During his voyage with the ship Beagle, Darwin spent much time in South America. Insects also fascinated Darwin, and when he found a type of cone-nosed kissing bug his interest was most focused. Darwin noted that when these bugs take a blood meal their abdomen greatly swells as they engorge themselves with host blood. He found many of these bugs and placed them on his arm to watch them swell. This may not have been a good idea. These insects transmit the protozoan, *Trypanosoma cruzi* that cause Chagas' disease. The parasites invade muscle and nerve cells and can result in a chronic disease. When Darwin returned to England, he never worked more than four hours a day. It is not impossible that he suffered from Chagas' disease. Today, large numbers of rural poor people in Brazil live in homes that do not exclude these bugs and are infected when quite young. By the time they get to young adulthood, the illness prevents them from experiencing the prime of their lives. The economic development of Brazil is severely hampered. No vaccines or effective treatment exists. People suffer from a slow chronic disease. Often so many heart cells are killed that the victims require cardiac pacemakers. This is a great stress on the relatively poor health care system of Brazil. Clearly, it is one of the most compelling of the 'forgotten' disease of the world.

STUDY OUTLINE

I. Components of the Nervous System

A. General characteristics

The ability of the *nervous system* to control body functions at the same time depends on many *neurons*, or *nerve cells*. Structurally, the nervous system has two components: the *central nervous system (CNS)* and the *peripheral nervous system (PNS)*. Other important components include *nerves, ganglia,* and *meninges.*

B. Normal microflora of the nervous system

The nervous system does not have normal microflora.

II. Diseases of the Brain and Meninges

A. Bacterial diseases of the brain and meninges

1. Bacterial meningitis

Bacterial meningitis is an inflammation of the *meninges,* the membranes that cover the brain and spinal cord. Most cases are acute but some are chronic. Meningitis is diagnosed by culturing cerebrospinal fluid. Antibiotic treatment varies with the causative organism. Types of bacterial meningitis include *meningococcal meningitis, Haemophilus meningitis,* and *Streptococcus meningitis.*

2. Listeriosis

Listeria monocytogenes causes another kind of meningitis called *listeriosis,* which is responsible for infection in kidney transplant patients and for many cases of fetal damage.

3. Brain abscesses

Microorganisms can cause *brain abscesses* if they can reach the brain from head wounds or via blood from another site. The masses can be detected by CAT scans or X-rays, and causative agents can be identified by serologic tests and by culturing cerebrospinal fluid. Antibiotic treatment can be sufficient in very early stages but surgical drainage or removal of abscesses is usually necessary later on.

B. Viral diseases of the brain and meninges

1. Viral meningitis

Enteroviruses and the mumps virus are the usual causative agents of *viral meningitis,* which is usually a self-limiting and nonfatal disease.

2. Rabies

The *rabies virus* is an RNA-containing rhabdovirus. A sample from a brain or skin biopsy can be stained by the immunofluorescent antibody test to identify rabies virus antigen before a patient dies. Hyperimmune rabies serum can be introduced into and around a thoroughly cleaned wound in hopes of neutralizing viruses before they reach the nervous system, where they are beyond the reach of antibodies. Interferon and a series of vaccine injections can also be used in treatment.

3. Encephalitis

Encephalitis is an inflammation of the brain caused by a variety of togaviruses or by a flavivirus. Four types are: (1) *eastern equine encephalitis;* (2) *western equine encephalitis;* (3) *Venezuelan equine encephalitis;* and (4) *St. Louis encephalitis.* Encephalitis sometimes can be diagnosed by isolating the causative agent from cell cultures or mice inoculated with blood or spinal fluid. Serological methods can be used to identify antibodies at any time during and following the illness. Treatment only alleviates symptoms.

4. Other viral diseases of the brain and meninges

Other viral diseases of the brain and meninges include *herpes meningoencephalitis* and *polyomavirus infections* (including *progressive multifocal leukoencephalopathy*).

III. Other Diseases of the Nervous System

A. Bacterial nerve diseases

1. Hansen's disease

Mycobacterium leprae causes *Hansen's disease,* the currently preferred name for *leprosy.* It is diagnosed by PCR or by finding the organism in acid-fast-stained smears and in scrapings from lesions or biopsies. The disease is treated with dapsone and rifampin.

2. Tetanus

Tetanus is caused by the obligately anaerobic, gram-positive, spore-forming *Clostridium tetani.* Tetanus toxoid vaccine given prior to injuries protects against the toxin. Antitoxin and antibiotics are given to nonimmunized patients when injuries are treated.

3. Botulism

Botulism is caused by *Clostridium botulinum.* It is diagnosed based on clinical symptoms and history, with confirmation based on a demonstration of the toxin in serum, feces, or food remains. Treatment with a polyvalent antitoxin is started immediately.

B. Viral nerve diseases: poliomyelitis

Poliomyelitis is caused by three strains of polioviruses (picornaviruses). It is diagnosed when the virus is isolated from pharyngeal swabs or feces, cultured, and studied for its cytopathic effects. Treatment alleviates symptoms and vaccines are now available.

C. Prion diseases of the nervous system
Prions cause *transmissible spongiform encephalopathies*, including *kuru, Creutzfeldt-Jakob disease, Gerstmann-Strassler disease, scrapie, transmissible mink encephalopathy, chronic wasting disease, and mad cow disease.*

D. Parasitic diseases of the nervous system

1. African sleeping sickness
African sleeping sickness, or *trypanosomiasis*, is caused by protozoan blood parasites of the genus *Trypanosoma*. Diagnosis is made by finding the trypanosome parasite in the blood. Pentamidine, suramin, and melarsoprol are used in treatment.
2. Chagas' disease
Chagas' disease is caused by *Trypanosoma cruzi*. Diagnosis involves: (1) finding the parasites in the blood during fever in acute cases; (2) using *xenodiagnosis*; or (3) allowing patients to be bitten by uninfected lab-reared bugs and examining the bugs later for trypanosomes. No effective treatment is available.

SELF-TESTS

Continue with this section only after you have read Chapter 24 of your textbook. Write your answers in the appropriate space provided. Correct answers to all questions can be found at the end of the Self-Test section. A score of 80 percent or better is good. If your score is less than 65, reread the chapter.

True/False

Mark T for True, F for False

_____ **1.** Major contributing factors in the epidemics caused by meningococci have been stress, overcrowding, and a decrease in resistance.

_____ **2.** The most common cause of heart attacks in Central and South America is the trypanosome of Chagas' disease.

_____ **3.** Fortunately, rabid wild animals are easy to spot because they always exhibit symptoms of aggressive behavior, foaming at the mouth, and irregular gait.

_____ **4.** Leprosy or Hansen's disease occurs primarily in Asia, Africa, and South America and is only seen in the United States as a result of imported cases.

_____ **5.** Symptoms of botulism include respiratory paralysis, slurred speech, and blurred vision.

_____ **6.** Listeriosis is a disease that can cross the placenta of pregnant women and infect the fetus.

_____ **7.** Virtually all prion-associated diseases exhibit a filamentous protein called amyloid and significantly high inflammatory response.

_____ **8.** Because of the abundance of oxygen, most brain abscesses tend to be caused by aerobes.

Multiple Choice

Select the best possible answer.

_____ **9.** In cases of polio:
a. The Salk vaccine is most commonly used in the United States.
b. Transmission is by the fecal–oral route.
c. All patients contracting polio must live in an "iron lung" for the rest of their lives.
d. The disease is maintained in the wild animal population.

_____ **10.** Which of the following is not true concerning rabies?
a. Bats can actually carry the virus.
b. A human diploid fibroblast vaccine is currently available and requires only a few injections.
c. The virus seems to be confined to the North American continent.
d. The observation of Negri bodies in the brain of a suspected animal is a positive sign of rabies.

______ **11.** Diseases caused by trypanosomes:
- **a.** are easily treated with antibiotics.
- **b.** do not occur in the United States.
- **c.** first gain access to the blood stream via insect bites and then travel to affect other tissue.
- **d.** are now controlled by effective vaccines.

______ **12.** Tetanus:
- **a.** can be prevented by means of a live attenuated virus vaccine.
- **b.** affects the neuromuscular junction and blocks neural transmission.
- **c.** is a fastidious gram-negative rod which forms an endotoxin.
- **d.** is commonly found in the soil, especially if is enriched with manure.
- **e.** causes serious sequelae such as brain damage following the disease.

______ **13.** Select the most incorrect statement concerning encephalitis.
- **a.** The most reliable method of identification of the viral agent requires cultivation of the virus in tissue cultures.
- **b.** The main reservoir of most equine encephalitis viruses seems to be the swamp bird population.
- **c.** The St. Louis encephalitis variety is transmitted mostly among English sparrows, mosquitoes, and humans.
- **d.** Vaccines are available to protect horses but are not generally used on humans.
- **e.** All of the above are true.

Matching

Select the answer from the right-hand side that corresponds to the term or phrase on the left-hand side of the page. An answer may be used more than once. In some cases more than one answer is required.

Topic: Diseases of the Brain and Meninges

______ **14.** *Listeria monocytogenes*
______ **15.** RNA rhabdovirus
______ **16.** Herpes simplex virus
______ **17.** *Neisseria meningitidis*
______ **18.** Togavirus
______ **19.** Flavivirus
______ **20.** *Haemophilus influenzae* type B
______ **21.** Polyomavirus
______ **22.** JC virus

- **a.** bacterial meningitis
- **b.** Waterhouse-Friderichsen syndrome
- **c.** hydrophobia
- **d.** western equine encephalitis
- **e.** rabies
- **f.** listeriosis
- **g.** progressive multifocal leukoencephalopathy
- **h.** St. Louis encephalitis
- **i.** herpes meningoencephalitis

Topic: Microbial Diseases of the Nervous System

______ **23.** Lockjaw
______ **24.** Wound botulism
______ **25.** Hansen's disease
______ **26.** Botulism
______ **27.** Tetanus neonatorum
______ **28.** Leprosy
______ **29.** African sleeping sickness
______ **30.** Chagas' disease
______ **31.** Kuru

- **a.** *Mycobacterium leprae*
- **b.** *Clostridium tetani*
- **c.** *Trypanosoma cruzi*
- **d.** *Clostridium botulinum*
- **e.** *Trypanosoma brucei gambiense*
- **f.** prions

Topic: Disease Transmission

______	**32.** Listeriosis	**a.** warmblooded animal bites
______	**33.** Poliomyelitis	**b.** contaminated foods
______	**34.** WEE	**c.** head wounds
______	**35.** Rabies	**d.** mosquitoes
______	**36.** Chagas' disease	**e.** contamination of deep wounds or cuts
______	**37.** Brain abscess	**f.** fecal–oral transmission
______	**38.** St. Louis encephalitis	**g.** tsetse flies
______	**39.** African sleeping sickness	**h.** reduviid bugs
______	**40.** Botulism	
______	**41.** Tetanus	
______	**42.** Wound botulism	
______	**43.** Brain abscesses	

Topic: Characteristics of Nervous System Diseases

______	**44.** Hansen's disease	**a.** Negri bodies
______	**45.** Rabies	**b.** lepromas
______	**46.** African sleeping sickness	**c.** oligodendrocyte destruction
______	**47.** Infant botulism	**d.** hydrophobia
______	**48.** Mad cow disease	**e.** floppy baby
______	**49.** Chagas' disease	**f.** causative agent evades host's defenses
______	**50.** Progressive multifocal leukoencephalopathy	**g.** pseudocyst formation
______	**51.** Scrapie	**h.** amyloid production
		i. spongy brain

Topic: Diagnosis of Nervous System Diseases

______	**52.** Brain abscesses	**a.** cerebrospinal fluid culture
______	**53.** Rabies	**b.** IFAT
______	**54.** Hansen's disease	**c.** lepromin skin test
______	**55.** Poliomyelitis	**d.** acid-fast staining
______	**56.** African sleeping sickness	**e.** PCR
______	**57.** Chagas' disease	**f.** tissue culture isolations
		g. blood smears

Topic: Treatment, Prevention, and/or Control

______	**58.** *Haemophilus influenzae* infection	**a.** hyperimmune globulin
______	**59.** African sleeping sickness	**b.** Hib vaccine
______	**60.** Chagas' disease	**c.** Pasteur treatment and/or modifications
______	**61.** EEE	**d.** elimination of insect vector
______	**62.** Poliomyelitis	**e.** tetanus toxoid

______	63. Rabies	f.	elimination of honey from diet of children under 1 year
______	64. VEE	g.	Sabin vaccine
______	65. Tetanus	h.	Salk vaccine
______	66. Infant botulism		

Fill-Ins

Provide the correct term or phrase for the following. Spelling counts.

67. The ________________ ________________ ________________ consists of the brain and the spinal cord.
68. The ________________ ________________ ________________ is composed of nerves that supply all parts of the body.
69. Aggregations of nerve cell bodies in the PNS are called ________________.
70. Phagocytes of the nervous system called ________________ cells can destroy invaders that reach the brain and spinal cord.
71. Thick-walled capillaries of the brain that limit entry of substances into brain cells form the ________________-________________ ________________.
72. An inflammation of the meninges is called ________________.
73. A complication of meningococcal infections where the cocci invade all parts of the body is ________________ - ________________ syndrome.
74. Among adults, ________________ ________________ is the most common cause of meningitis.
75. Diagnosis of rabies can be made on the observation of ________________ bodies in neural brain cells.
76. A common symptom of rabies in humans is ________________.
77. The most severe form of encephalitis is ________________ ________________ encephalitis.
78. The form of encephalitis that primarily affects horses is ________________ ________________ encephalitis.
79. The preferred name of leprosy is ________________ disease.
80. The two forms of leprosy are ________________ and ________________.
81. The type of vaccine material given to prevent tetanus is tetanus ________________.
82. A form of tetanus that affects the umbilical cord of neonates is tetanus ________________.
83. Infant botulism seems to be transmitted by feeding infants ________________.
84. The vaccine currently used to prevent polio is the oral ________________ vaccine.
85. A condition that affects survivors of polio and causes muscle weakness or paralysis is called ________________ ________________.
86. A special form of ________________ - ________________ disease is Gerstmann-Strassler disease.
87. The cells that regulate the passage of materials from the blood to neurons are called ________________.
88. The neurofibrillary tangles that have been found at autopsy in the brains of Alzheimer's patients are also called ________________.
89. The fly that transmits African sleeping sickness is the ________________.
90. American trypanosomiasis is also called ________________ disease.
91. When small animals are inoculated with blood from patients and observed for disease symptoms, the technique is called ________________.

Labeling

Questions 92–99: Select the best possible label for each structure in the nervous system: peripheral nervous system; cerebellum; brain stem; ganglion; peripheral nerves; central nervous system; spinal cord; cerebrum. *Use the blanks provided for your answers.*

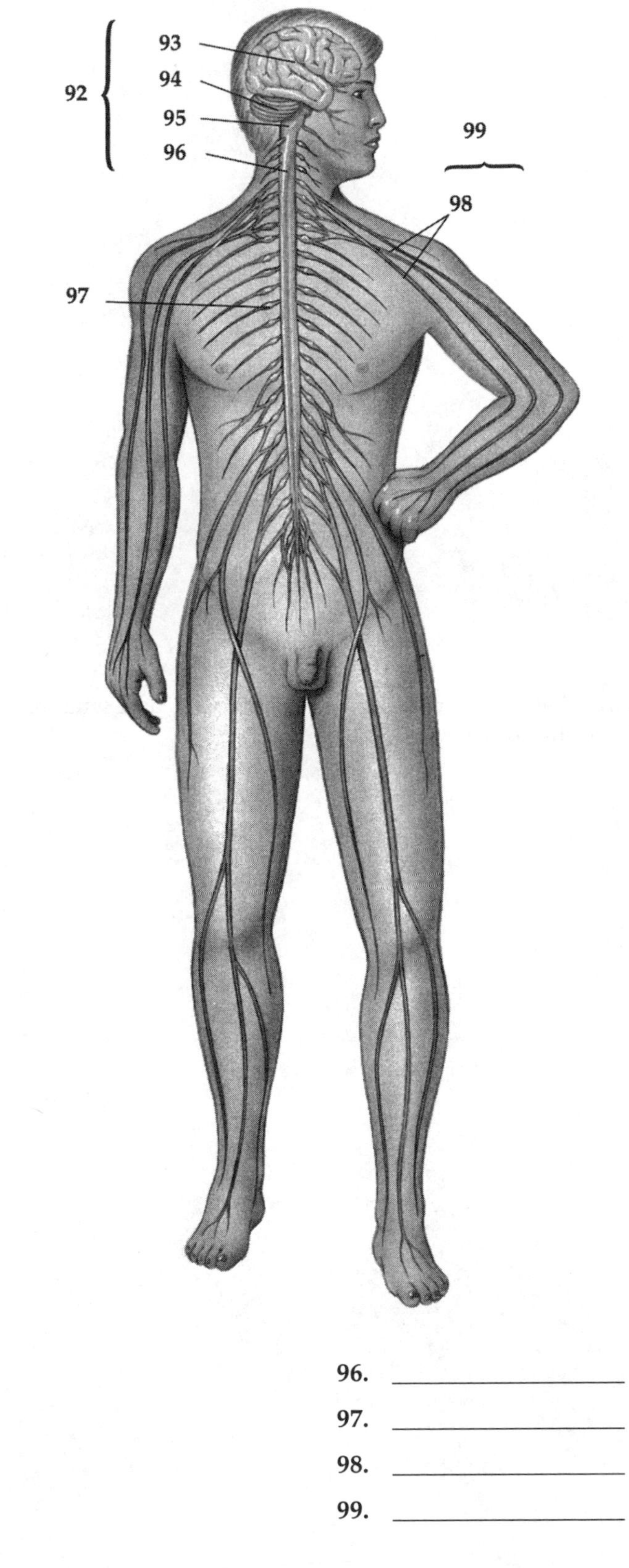

92. ____________________

93. ____________________

94 ____________________

95. ____________________

96. ____________________

97. ____________________

98. ____________________

99. ____________________

Critical Thinking

100. What pathogens cause bacterial diseases of the brain and meninges, and what are the important epidemiologic and clinical aspects of these diseases?

101. What pathogens cause viral diseases of the brain and meninges, and what are the important epidemiologic and clinical aspects of these diseases?

102. What pathogens cause bacterial nerve diseases, and what are the important epidemiologic and clinical aspects of these diseases?

103. What pathogens cause viral nerve diseases, and what are the important epidemiologic and clinical aspects of these diseases?

104. What pathogens cause parasitic diseases of the nervous system, and what are the important epidemiologic and clinical aspects of these diseases?

ANSWERS

True/False

1. **True** *Neisseria meningitidis* has caused about 3000 cases of meningitis per year for the last 10 years in the United States. The disease was the leading cause of death from infectious disease among service men due in large part to the stress and overcrowding during boot camp.
2. **True** *Trypanosoma cruzi* is the flagellated protozoan that is credited for causing the greatest number of heart attacks in the southern hemisphere. It is transmitted by the triatomid bug, common to those areas. It forms pseudocysts which rupture and cause inflammation and tissue necrosis.
3. **False** Rabid animals are often difficult to detect since many animals release the virus in the saliva 3–6 days before exhibiting symptoms. In addition, many rabid wild animals are often "friendly" and quite approachable.
4. **False** Leprosy does occur in the United States and especially in many of the Southern states. Increases in reported cases may be due to better diagnosis. Fortunately, most cases respond to treatment.
5. **True** Botulism is a neuroparalytic disease with sudden onset and rapidly progressing paralysis. The toxin prevents the release of acetylcholine, resulting in the symptoms.
6. **True** *Listeria monocytogenes* can be transmitted in unpasteurized milk and cheese products. Once ingested, the agent can cross the placenta of a pregnant woman and infect the fetus, resulting in spontaneous abortion, stillbirth, or neonatal death.
7. **False** Although amyloid is found in prion-associated patients, most lack any inflammatory response.
8. **False** Most brain abscesses are caused by multiple species and anaerobes are just as likely to cause infection as aerobes.

Multiple Choice

9. b; 10. c; 11. c; 12. d; 13. a.

Matching

14. a, f; 15. c, e; 16. i; 17. a, b; 18. d; 19. h; 20. a; 21. g; 22. g; 23. b; 24. d; 25. a; 26. d; 27. b; 28. a; 29. e; 30. c; 31. e; 32. b; 33. f; 34. d; 35. a; 36. h; 37. c; 38. d; 39. g; 40. b; 41. e; 42. e; 43. c; 44. b; 45. a, d; 46. f; 47. e; 48. h, i; 49. g; 50. c; 51. h, i; 52. a; 53. b; 54. c, d, e; 55. f; 56. g; 57. g; 58. b; 59. d; 60. d; 61. d; 62. g, h; 63. c; 64. d; 65. a, e; 66. f.

Fill-Ins

67. central nervous system; 68. peripheral nervous system; 69. ganglia; 70. microglial; 71. blood-brain barrier; 72. meningitis; 73. Waterhouse-Friderichsen; 74. *Streptococcus pneumoniae*; 75. Negri; 76. hydrophobia; 77. eastern equine; 78. Venezuelan equine; 79. Hansen's; 80. tuberculoid, lepromatous; 81. toxoid; 82. neonatorum; 83. honey; 84. Sabin; 85. postpolio syndrome; 86. Creutzfeldt-Jakob; 87. astrocytes; 88. plaques; 89. tsetse; 90. Chagas'; 91. xenodiagnosis.

Labeling

92. central nervous system; 93. cerebrum; 94. cerebellum; 95. brain stem; 96. spinal cord; 97. ganglion; 98. peripheral nerves; 99. peripheral nervous system.

Critical Thinking

100. What pathogens cause bacterial diseases of the brain and meninges, and what are the important epidemiologic and clinical aspects of these diseases?

A wide variety of bacteria can infect the meninges, with certain ones affecting each age level. Most cases of meningitis are acute but some are chronic.

These agents can be acquired from carriers or from endogenous organisms. Meningococcal meningitis tends to be the most severe, especially during epidemics, and affects primarily adults. *Haemophilus* meningitis is the most common and affects primarily children. Vaccines are available to protect children from *Haemophilus* meningitis and to control *Streptococcus pneumoniae* in the elderly. Antibiotic treatment varies with the causative agent.

Listeriosis can be transmitted by improperly processed dairy products and can cross the placenta to affect the fetus. It can also affect immunodeficient patients.

Brain abscesses can arise from wounds or as secondary infections caused by anaerobes. Antibiotics should be used as early as possible to control the infection or surgery may be required.

101. What pathogens cause viral diseases of the brain and meninges, and what are the important epidemiologic and clinical aspects of these diseases?

Viral diseases of the brain and meninges are usually very severe and often life-threatening. Since antibiotics are of no value, the patient can only be given supportive therapy and hope. Vaccines are available for rabies and for encephalitis but the latter can only be used for horses.

Aseptic (nonbacterial) meningitis, although not mentioned in the text, is caused by certain enteroviruses and occasionally by the mumps virus. It is usually a severe but self-limited infection since there is little therapy available.

Rabies has a worldwide distribution and is difficult to control because of the large number of small mammals such as skunks and bats that can serve as reservoirs. The disease can be diagnosed by the finding of Negri bodies in neural cells and by the IFAT. The disease can be treated by thoroughly cleaning the bite wound, injecting hyperimmune rabies serum, and by giving the human diploid cell vaccine. It can be prevented by immunizing all pets and people who may be at risk and by avoiding contact with wild animals, especially the very "friendly" or unafraid ones.

Encephalitis is an inflammation of the brain caused by a variety of togaviruses. It is transmitted by mosquitoes, often from horses, and seems to be maintained in the wild bird populations. It can be diagnosed by culturing blood or spinal fluid in cell cultures or mice. The St. Louis variety seems to affect mostly humans, whereas the other varieties affect mostly horses.

Herpes meningoencephalitis often follows a generalized herpes infection and is especially severe.

102. What pathogens cause bacterial nerve diseases, and what are the important epidemiologic and clinical aspects of these diseases?

Several bacteria specifically affect the central and/or peripheral nervous system by damaging or destroying the nerve cells directly or by blocking neural impulses.

Hansen's disease, or leprosy, affects millions of people worldwide including the United States. In fact, the disease appears to be increasing in the Southern states, especially Texas, probably because of better detection. Many patients are asymptomatic for years after infection and no diagnostic test, including the lepromin skin test, can detect every case. Dapsone and rifampin are effective in treating most cases so few individuals ever have need to be placed in a "leper" colony. Vaccines are not available. Recently, the Texas armadillo was found to harbor the exact same leprosy bacillus as affects humans. Perhaps the "love affair" with armadillos may need revision.

Tetanus represents a true infection and toxemia since the organisms (usually spores) must enter a deep wound and germinate to develop the disease. It is treated with antitoxin and antibiotics and prevented by means of the tetanus toxoid (the "T" of DPT shot).

Botulism is normally a food poisoning acquired from ingesting foods containing the preformed neurotoxin. It is treated with a polyvalent antitoxin. Most cases of botulism arise from improperly home canned foods, especially those that are neutral in pH. Cases of infant botulism contracted from honey contaminated with the endospores and wound botulism following contamination of a wound have been recorded.

103. What pathogens cause viral nerve diseases, and what are the important epidemiologic and clinical aspects of these diseases?

Only a few viruses specifically affect the nerve tissue. The most important one is the poliovirus. Prior to the development of vaccines in the 1950's, polio was a common and dreaded disease. Today only a few religious groups opposed to immunization and groups of illegal immigrants who have not received the vaccine are at risk for the disease. Diagnosis is made from cultures and by immunologic methods. Treatment alleviates symptoms, but patients with the paralytic strains of the virus may require use of the "iron lung." Both the injectable Salk and the oral Sabin vaccine are available and each has advantages and disadvantages. Worldwide use of the vaccine could eradicate the disease. A few cases of postpolio syndrome have occurred and may be related to overuse of previously polio-weakened muscles.

104. What pathogens cause parasitic diseases of the nervous system, and what are the important epidemiologic and clinical aspects of these diseases?

The trypanosomes are the most important parasites to affect the nervous system. They also do great damage to the cardiovascular system as well. Both are transmitted by vectors and are difficult to treat.

African sleeping sickness occurs in equatorial Africa and is transmitted by the tsetse fly. It is diagnosed by finding the trypanosomes in the blood and can be treated with pentamidine and other drugs but they tend to be very toxic. Without treatment, the disease becomes chronic and the victim lapses into a coma and dies.

Chagas' disease occurs from southern United States to all but the southernmost part of South America and is transmitted by various species of triatomid or "kissing" bugs. It is diagnosed by finding the parasites in the blood and does not respond well to any treatment. It is the leading cause of cardiovascular illness in Central and South America.

Chapter 25 Environmental Microbiology

INTRODUCTION

In many of our essays, we have focused on the concept that a person is like a world. There is a famous scientist who believes the opposite is true; that is, the Earth is like an organism. In this 'Gaia' hypothesis the planet is a self-regulating system and interactions between the world of the living and the nonliving result in a stable planet. In his view, Lovelock believes that bacteria may be the most important players in creating a world that we can live with. Bacteria are the most ancient form of life on Earth, with a history of almost four billion years. Clearly, for about half of the history of life on Earth, the only life was bacterial life. Lovelock feels that in most early stages of the planet it was bacteria that permitted the planet to hold on to its water and to evolve into the world we see today instead to a dead planet like Venus. Whether or not he is correct in this hypothesis, bacteria engage in so many of the biogeochemical cycles that life as we know it could not exist without them. Microbes make soil an environment in which crops can grow. Bacteria breakdown human waste in treatment plants. When oil spills, microbes degrade them into nontoxic compounds. Microbes maintain the ecological health of our skin and intestines. Indeed, it is really their world and they are kind enough to share it with us. Many of us would do well to appreciate them more and express less zeal to destroy them with antibacterial soap and the like.

STUDY OUTLINE

I. Fundamentals of Ecology

A. Ecology defined

Ecology is the study of the relationships among organisms—the *biotic factors*—and their environment—the *abiotic factors*. An *ecosystem* comprises all the organisms in a given area together with surrounding abiotic and biotic factors.

B. The nature of ecosystems

Ecosystems are organized into various biological levels. The *biosphere* is the region of the earth inhabited by living organisms. An ecological *community* consists of all the kinds of organisms that are present in a given environment. *Indigenous organisms* are organisms that are native to the given environment. *Nonindigenous organisms* are temporary inhabitants of an environment. Microorganisms make up a *microenvironment*.

C. Energy flow in ecosystems

Energy flow in ecosystems starts at the *producers*, continues with the *consumers*, and ends with the *decomposers*.

II. Biogeochemical Cycles

A. General characteristics

Biogeochemical cycles are mechanisms by which water and elements that serve as nutrients are recycled.

B. The water cycle

The *water cycle*, or *hydrologic cycle*, is the process by which water is recycled through precipitation, ingestion by organisms, respiration, and evaporation.

C. The carbon cycle

The *carbon cycle* is the process by which carbon from atmospheric carbon dioxide enters living and nonliving things and is recycled through them.

D. The nitrogen cycle and nitrogen bacteria

The *nitrogen cycle* is the process by which nitrogen moves from the atmosphere through various organisms and back into the atmosphere. Nitrogen bacteria fall into one of three categories according to the roles they play in the nitrogen cycle: (1) nitrogen-fixing bacteria; (2) nitrifying bacteria; and (3) denitrifying bacteria.

E. The sulfur cycle and sulfur bacteria

The *sulfur cycle* is the process by which sulfur moves cyclically through an ecosystem. The various sulfur bacteria can be categorized according to their roles in the sulfur cycle. These roles include sulfate reduction, sulfur reduction, and sulfur oxidation.

F. Other biogeochemical cycles

Phosphorus and other elements also move through ecosystems cyclically.

III. Air

A. Microorganisms found in air

Though microorganisms do not grow in air, spores are carried in air, and vegetative cells can be carried on dust particles and water droplets in air.

B. Methods for determining microbial content of air

Airborne microbes can be detected by collecting those that happen to fall onto an agar plate or in a liquid medium. A special centrifuge-like air-sampling instrument provides a better measure of airborne microbes.

C. Methods for controlling microorganisms in air

Chemical agents, radiation, filtration, and laminar airflow all can be used to control microorganisms in the air.

IV. Soil

A. Components of soil

Soil is divided into a number of layers, or *soil horizons*, including the *topsoil*, *subsoil*, and *parent material*. The most abundant inorganic components of soil are pulverized rocks and minerals. Soil also contains water and gases like carbon dioxide, oxygen, and nitrogen. *Humus*, or nonliving organic matter, is constantly changing as organisms die and as decomposers degrade complex molecules to simpler ones. In addition to microorganisms, soil also contains the root systems of many plants, a variety of invertebrate animals, and a few burrowing reptiles and mammals.

B. Microorganisms in soil

All the major groups of microorganisms—bacteria, fungi, algae, and protists, as well as viruses—are present in soil, but bacteria are more numerous than all other kinds of microorganisms. Factors like moisture, oxygen concentration, pH, and temperature affect soil microorganisms. Soil microorganisms that are decomposers are important in the carbon cycle and the nitrogen cycle because of their ability to decompose organic matter.

C. Soil pathogens

Soil pathogens are primarily plant pathogens, but a few can affect humans (like those from the genus *Clostridium*) and other animals (like *Bacillus anthracis*).

V. Water

A. Freshwater environments

Freshwater environments include surface water, such as lakes, ponds, rivers, and streams, and groundwater that runs through underground strata of rock. Although groundwater contains few microorganisms, surface water contains large numbers of many different kinds of microorganisms. Which organisms are present depends on the temperature and pH of the water, dissolved minerals, the depth of sunlight penetration, and the quantity of nutrients in the water.

B. Marine environments
Large numbers of many different kinds of microorganisms live in the ocean. The temperature and pH of the water, dissolved minerals (especially salt), the depth of sunlight penetration, the oxygen concentration, and the quantity of nutrients in the water affect the microorganisms living in the ocean.

C. Water pollution
Water is considered polluted if a substance or condition is present that renders the water useless for a particular purpose. The major types of pollutants are organic wastes (such as sewage and animal manures), industrial wastes, oil, radioactive substances, sediments from soil erosion, and heat. Human pathogens in water supplies usually come from contamination of water with human feces; *Escherichia coli* is used as an *indicator organism* because its presence in water indicates that the water is contaminated with fecal material.

D. Water purification
Purification procedures for human drinking water are determined by the degree of purity of the water at its source. Usually, water undergoes settling, *flocculation*, *filtration*, and *chlorination*. Water is then tested for purity by looking for *coliform bacteria* using the *multiple-tube fermentation method* (involving the *presumptive test*, the *confirmed test*, and the *completed test*), the *membrane filter method*, or the *ONPG and MUG test*.

VI. Sewage Treatment

A. General characteristics
Sewage is used water and the wastes it contains. Completed sewage treatment involves three steps: *primary*, *secondary*, and *tertiary treatment*.

B. Primary treatment
In *primary treatment*, physical means are used to remove solid wastes from sewage.

C. Secondary treatment
In *secondary treatment*, biological means (actions of decomposers) are used to remove solid wastes that remain after primary treatment. Secondary treatment involves *trickling filter systems*, *activated sludge systems*, and *sludge digesters*.

D. Tertiary treatment
In *tertiary treatment*, chemical and physical means are used to produce an effluent of water pure enough to drink.

E. Septic tanks
Septic tanks are underground tanks for receiving sewage, where solid materials settle out as sludge that must be pumped periodically.

VII. Bioremediation

A. Process that uses microorganisms to transform harmful substances into less toxic or nontoxic compounds
B. Used to treat oil spills like the Exxon Valdez spill of 1989.

SELF-TESTS

Continue with this section only after you have read Chapter 25 of your textbook. Write your answers in the appropriate space provided. Correct answers to all questions can be found at the end of the Self-Test section. A score of 80 percent or better is good. If your score is less than 65, reread the chapter.

True/False

Mark T for True, F for False

______ **1.** Denitrification is a very useful process because it helps to remove nitrates which are extremely toxic to plants.

______ **2.** Although microorganisms do not grow in air, spores of others can be transported on dust particles and water droplets.

______ **3.** As water becomes warmer it can support an increasing amount of oxygen and hence a greater variety of microbes.

______ **4.** The presence of coliforms generally indicates fecal pollution.

______ **5.** Following secondary treatment of sewage, the resultant water is cleaned of all organic material and is safe to drink.

______ **6.** Sulfate-reducing bacteria are among the oldest life forms.

Multiple Choice

Select the best possible answer.

______ **7.** The steps necessary to purify water are (in order):
a. chlorination, flocculation, filtration, settling
b. flocculation, chlorination, filtration, settling
c. settling, flocculation, filtration, chlorination
d. settling, filtration, flocculation, chlorination

______ **8.** Select the most correct statement.
a. Compared to fresh water, the ocean is much more variable in temperature and pH.
b. Marine organisms must tolerate varying degrees of salinity which changes with the depth.
c. Nutrient concentrations vary in ocean water, often depending on depth and proximity to the shore.
d. Terrestrial and freshwater environments occupy more of the earth's surface than marine environments.

______ **9.** Water is considered polluted when:
a. The BOD is very low.
b. A substance or condition is present that renders the water useless for a particular purpose.
c. The producers outnumber the consumers.
d. Heterotrophic microbes are present.
e. All of the above are true.

______ **10.** In terms of water pollution all the following are true except:
a. Heat can act as an important water pollutant.
b. Excessive plant nutrients can lead to overgrowth of undesirable algae.
c. Sediments can deplete the oxygen content of water and reduce visibility.
d. Inorganic chemicals can alter the acidity of water.
e. All of the above are true.

Matching

Select the answer from the right-hand side that corresponds to the term or phrase on the left-hand side of the page. An answer may be used more than once. In some cases more than one answer is required.

Topic: Fundamentals of Ecology

______ **11.** Region of earth inhabited by living organisms

______ **12.** The role of an organism in the ecosystem

______ **13.** Gaseous envelope surrounding the earth

______ **14.** Physical features of the environment

______ **15.** Organisms always found in a given environment

______ **16.** The earth's crust

______ **17.** Groups of organisms of the same species

______ **18.** Temporary inhabitants of an environment

______ **19.** Hydrosphere

______ **20.** Physical location of an organism

a. habitat
b. earth's water supply
c. niche
d. indigenous
e. atmosphere
f. biosphere
g. lithosphere
h. populations
i. abiotic factors
j. nonindigenous

Topic: Biogeochemical Cycles

______ 21. Photosynthesis

______ 22. Reduction of N_2 to NH_3

______ 23. Transpiration

______ 24. Nitrogenase

______ 25. Carbon dioxide

______ 26. *Rhizobium*

______ 27. Legumes

______ 28. *Thiobacillus*

______ 29. Conversion of inorganic phosphates to orthophosphate

______ 30. Reduction of SO_4^{2-} to H_2S

______ 31. Oxidation of ammonia to nitrate

a. water cycle

b. carbon cycle

c. nitrogen cycle

d. sulfur cycle

e. phosphorous cycle

Topic: Organisms in Soil

______ 32. *Cytophaga*

______ 33. *Bacillus anthracis*

______ 34. *Methanosarcina*

______ 35. *Clostridium perfringens*

______ 36. *Clostridium botulinum*

a. methane producers

b. cellulose digester

c. gas gangrene cause

d. anthrax cause

e. botulism cause

Topic: Water

______ 37. Sunlit water away from the shore

______ 38. Between limnetic zone and lake sediment

______ 39. Shoreline

______ 40. Sediment

______ 41. Nutrient enrichment

______ 42. Oxygen required for organic waste decomposition

______ 43. Passage of water through beds of sand

______ 44. The addition of alum to precipitate suspended colloids

a. littoral zone

b. profundal zone

c. limnetic zone

d. benthic zone

e. BOD

f. eutrophication

g. flocculation

h. filtration

Topic: Water Purity and Sewage Treatment

______ 45. Skimmers remove oily substances

______ 46. ONPG test

______ 47. Use of flocculating substances to precipitate solids

______ 48. Spraying of sewage over bacteria-coated rocks

______ 49. Membrane filter method

______ 50. Screens remove large pieces of debris

______ 51. Use of sludge digester

______ 52. Multiple-tube fermentation method

______ 53. Denitrifying bacteria convert nitrates to nitrogen gas

______ 54. An activated sludge system agitates and aerates sewage effluent

a. test for coliform detection

b. primary (sewage) treatment

c. secondary (sewage) treatment

d. tertiary (sewage) treatment

Fill-Ins

Provide the correct term or phrase for the following. Spelling counts.

55. The study of the relationships among organisms and their environment is called ________________.
56. Native organisms always found in a given environment are ________________.
57. Organisms that can produce energy from the sun are called ________________.
58. Microorganisms act as ________________ by obtaining their energy from wastes of producers and dead bodies.
59. Mechanisms by which recycling of chemicals such as carbon and nitrogen occurs are termed ________________ cycles.
60. The biogeochemical cycle that recycles water is the water or ________________ cycle.
61. In the carbon cycle, the atmospheric form is ________________.
62. In the nitrogen cycle, the atmospheric form is ________________.
63. The process by which bacteria reduce atmospheric nitrogen into ammonia is ________________ ________________.
64. The nitrogen-fixing enzyme that bacteria must have to fix nitrogen is ________________.
65. Some microbes fix nitrogen by acting in a relationship with plants called ________________.
66. Plants of the pea family that often enter a relationship with bacteria to fix nitrogen are called ________________.
67. The process by which ammonia is oxidized to nitrites or nitrates is ________________.
68. The process by which nitrates are reduced to nitrous oxide or nitrogen gas is ________________.
69. A sulfur-containing gas with a rotten egg smell is ________________.
70. Sulfate converted to hydrogen sulfide is called ________________ ________________.
71. Sulfur-oxidizing bacteria can create environmental problems due to the formation of ________________ ________________, which ________________the pH.
72. A type of air-flow hood that suctions air away from the opening and filters it before expelling it is the ________________ flow hood.
73. All of the organic nonliving components of soil are collectively known as ________________.
74. In swamp gas, the main carbon-containing product is ________________.
75. The main human pathogens of soil are in the genus ________________.
76. The main producers of the ocean are ________________.
77. Water that is fit for human consumption is termed ________________ water.
78. BOD refers to ________________ ________________ ________________.
79. ________________ is the excessive growth of algae that often follows the nutrient enrichment of water.
80. The presence of *Escherichia coli* in water indicates that it has been contaminated by ________________ ________________.
81. The process of adding alum to water, which acts to precipitate out suspended colloids, is ________________.
82. Testing for water involves three stages: ________________, ________________, and ________________.
83. The inoculation of water samples into lactose broth tubes represents the ________________ test.
84. When samples of positive lactose tubes are plated onto EMB plates, the test is called ________________.
85. Used water with the wastes it contains is called ________________.

86. In the treatment of sewage, the use of physical means to remove solid wastes from sewage is ________________ treatment.

87. The use of biological means to remove residual solid wastes is ________________ treatment.

88. Spreading sewage over a bed of rocks is a treatment system called ________________ ________________.

89. Sludge that floats to the surface of water in the treatment process is known as ________________.

90. An anaerobic chamber in which sludge from both primary and secondary treatments is digested is called a ________________ ________________.

91. When chemical and physical means are used to produce an effluent pure enough to drink, the treatment is known as ________________ treatment.

Labeling

Questions 92–96: Select the best possible label for each step in the water cycle: (a) transpiration and evaporation from plants, animals, lakes, land; (b) seepage; (c) evaporation from oceans; (d) subsurface runoff; (e) precipitation returns moisture to land.

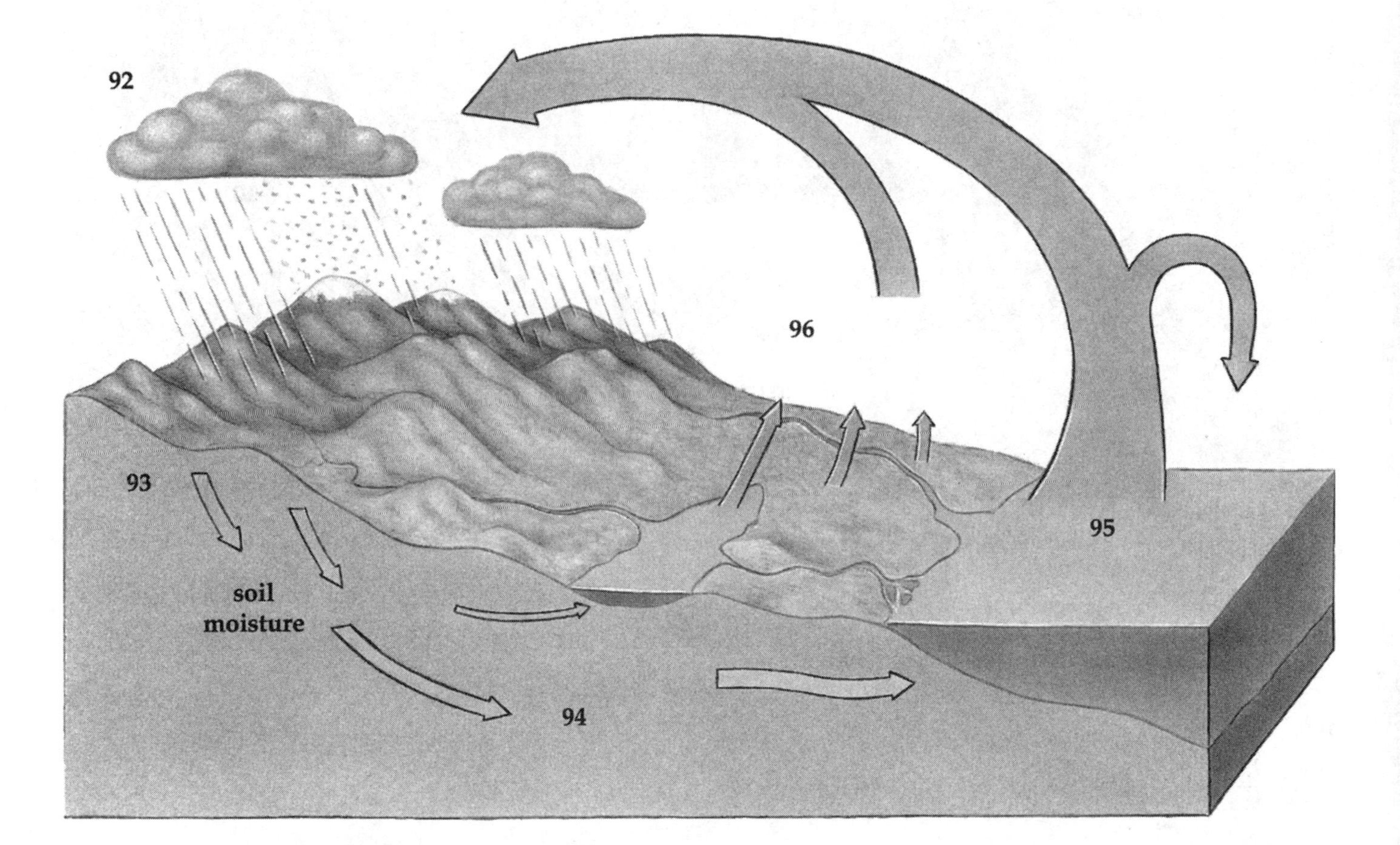

92. ________________

93. ________________

94. ________________

95. ________________

96. ________________

Questions 97–105: Select the best possible label for each step in the nitrogen cycle: (a) industrial fixation; (b) denitrifying bacteria; (c) ammonia; (d) fixation by nitrogen-fixing bacteria; (e) reservoir of nitrogen in atmosphere; (f) fertilizers; (g) nitrates in soil; (h) plant and animal waste and remains; (i) atmospheric fixation.

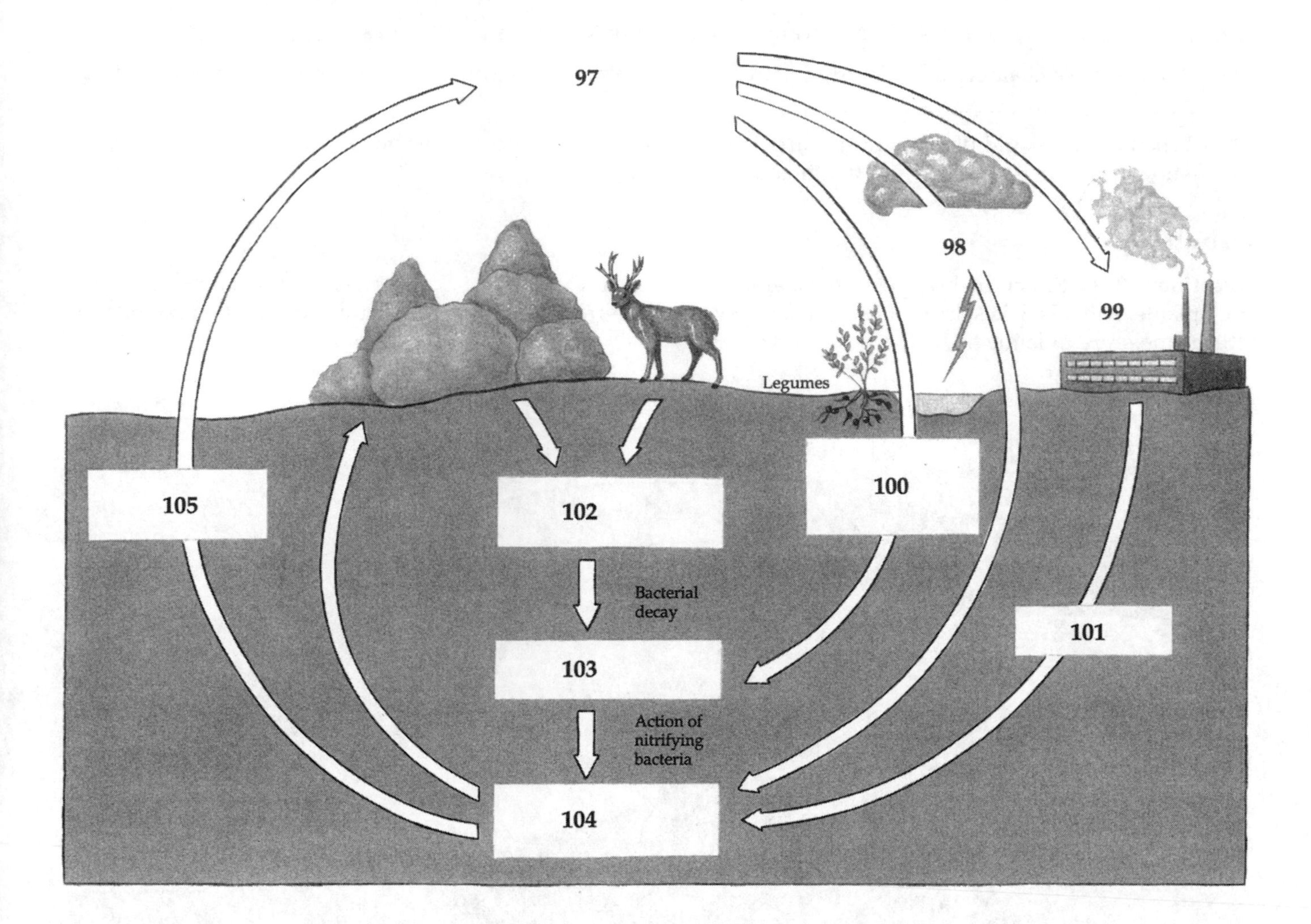

97. ______________

98. ______________

99. ______________

100. ______________

101. ______________

102. ______________

103. ______________

104. ______________

105. ______________

Questions 106–109: Select the best possible label for each soil horizon: bedrock; subsoil; topsoil; parent material.

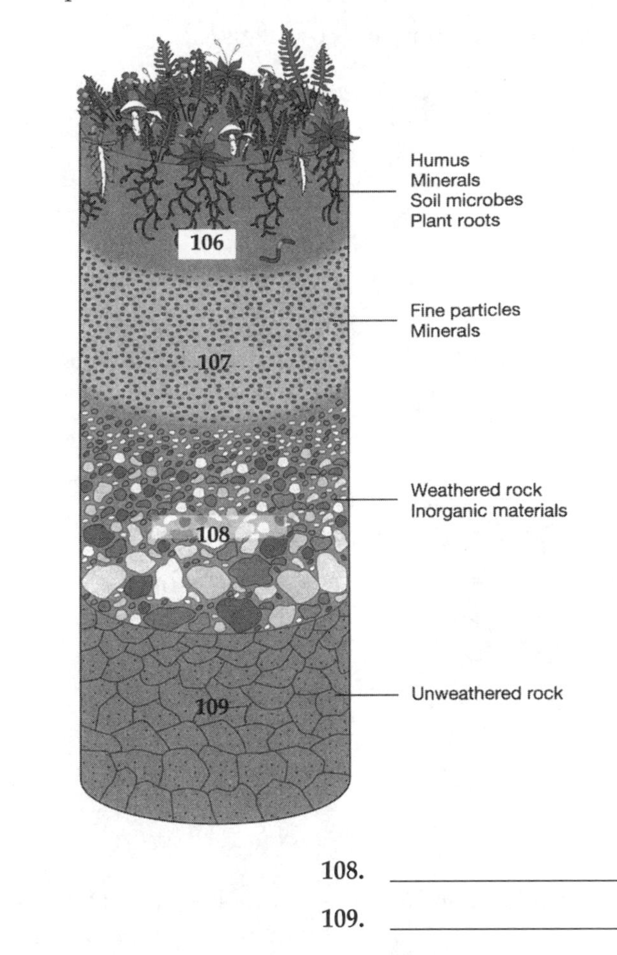

106. ____________________

107. ____________________

108. ____________________

109. ____________________

Questions 110–113: Select the best possible label for each zone of a typical lake or pond: littoral zone; profundal zone; limnetic zone; benthic zone.

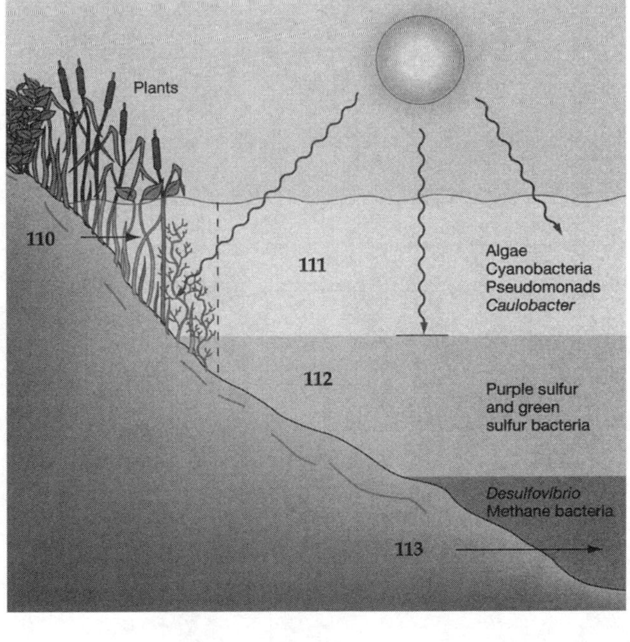

110. ____________________

111. ____________________

112. ____________________

113. ____________________

Critical Thinking

114. What is ecology, and how does energy flow in ecosystems?

115. Why is recycling important, and how are water and carbon recycled?

116. What other biogeochemical cycles exist, and what roles to microorganisms play in them?

117. What kinds of microorganisms are found in air, and how are they detected and controlled?

118. What kinds of microorganisms are found in soil, and what are their roles in biogeochemical cycles and as pathogens?

119. How do freshwater and marine environments, and their microorganisms, differ?

120. How do water pollution and waterborne pathogens affect humans?

121. How is water purified, and how is it tested to determine purity?

122. What is sewage, and what processes are involved in primary, secondary, and tertiary sewage treatment?

ANSWERS

True/False

1. **False** Nitrates are very useful for plant growth and their removal would interfere with their metabolism.
2. **True** Endospores of bacteria, pollen of plants, and asexual spores of molds are commonly carried by wind currents, dust, and water droplets in the air. During dust storms and heavy pollution the microbial count can be very high.
3. **False** Moderate temperatures can support a wide variety of microbes. As water becomes very warm (above 50°) it is more difficult to support useful organisms and especially fish because oxygen will begin to come out of the solution.
4. **True** Coliforms including *Escherichia coli* are those gram-negative, non-spore-forming aerobic or facultative anaerobic bacteria that ferment lactose and produce acid and gas. They are commonly associated with the GI tract.
5. **False** The effluent from secondary treatment contains only 5 to 20% of the original quantity of organic matter. It can be discharged into streams but can cause problems and is not safe to drink.
6. **True** Sulfate reducers may be up to 3 billion years old and are found in anaerobic environments. They reduce sulfate to hydrogen sulfide.

Multiple Choice

7. c; 8. c; 9. b; 10. e.

Matching

11. f; 12. c; 13. e; 14. i; 15. d; 16. g; 17. h; 18. j; 19. b; 20. a; 21. b, c; 22. c; 23. a; 24. c; 25. b; 26. c; 27. c; 28. d; 29. e; 30. d; 31. c; 32. b; 33. d; 34. a; 35. c; 36. e; 37. c; 38. b; 39. a; 40. d; 41. f; 42. e; 43. h; 44. g; 45. b; 46. a; 47. b; 48. c; 49. a; 50. b; 51. b, c; 52. a; 53. d; 54. c.

Fill-Ins

55. ecology; 56. indigenous; 57. producers; 58. decomposers; 59. biogeochemical; 60. hydrologic; 61. CO_2; 62. N_2; 63. nitrogen fixation; 64. nitrogenase; 65. symbiosis; 66. legumes; 67. nitrification; 68. denitrification; 69. H_2S; 70. sulfate reduction; 71. sulfuric acid, lowers; 72. laminar; 73. humus; 74. CH_4; 75. *Clostridium*; 76. phytoplankton; 77. potable; 78. biological oxygen demand; 79. eutrophication; 80. fecal matter; 81. flocculation; 82. presumptive, confirmed, completed; 83. presumptive; 84. completed; 85. sewage; 86. primary; 87. secondary; 88. trickling filter; 89. bulking; 90. sludge digester; 91. tertiary.

Labeling

92. e; 93. b; 94. d; 95. c; 96. a; 97. e; 98. i; 99. a; 100. d; 101. f; 102. h; 103. c; 104. g; 105. b; 106. topsoil; 107. subsoil; 108. parent material; 109. bedrock; 110. littoral zone; 111. limnetic zone; 112. profundal zone; 113. benthic zone.

Critical Thinking

114. What is ecology, and how does energy flow in ecosystems?

Ecology is the study of relationships among organisms and their environment which include interactions of organisms with physical features and with each other. An ecosystem includes all the living or biotic factors and the nonliving or abiotic factors of an environment. Microorganisms can be native or indigenous to the environment or temporary, nonindigenous inhabitants. All the living organisms in an ecosystem make up a community.

Energy in an ecosystem flows from the ultimate source, the sun, to the producers and then to the consumers. Decomposers, which are mostly microorganisms, obtain their energy from digesting dead bodies and wastes of other organisms. This insures that all nutrients are recycled.

115. Why is recycling important, especially with regard to water and carbon atoms?

Although energy is continuously available from the sun, nutrients must be recycled to insure their availability. Dead organisms and wastes must be recycled by decomposers or too much matter would be incorporated into this material and life would soon become extinct.

The water or hydrologic cycle insures that water is continuously recycled as all living organisms use this as part of their metabolism.

The carbon cycle insures that carbon in both an inorganic and an organic form is always available to living organisms. Carbon is the basic building block for all living organisms.

116. What other biogeochemical cycles exist, and what roles do microorganisms play in them?

The nitrogen cycle insures that nitrogen in both an inorganic and an organic form is continuously recycled from the atmosphere through various organisms and the soil and back to the atmosphere. All living organisms including microorganisms require nitrogen for their metabolism and thus become essential parts of the nitrogen cycle.

Nitrogen bacteria fall into three categories according to the roles they play in the nitrogen cycle. These are nitrogen fixers, nitrifying bacteria, and denitrifying bacteria.

Nitrogen fixation is the reduction of atmospheric nitrogen to ammonia and is accomplished by free-living aerobes and anaerobes and also by nitrogen fixers in combination with plants called legumes. *Rhizobium* is an organism that has established a symbiotic relationship with legumes such as soybean and alfalfa to fix atmospheric nitrogen. This helps the bacteria and the plant gain the nitrogen they need. These plants can then be rotated with corn plants that extract nitrogen from the soil.

Nitrification is the conversion of ammonia to nitrites and nitrates. *Nitrosomonas* converts ammonia to nitrites and *Nitrobacter* converts the nitrites to nitrates. This helps to provide plants with needed amounts of nitrates which they use in their metabolism.

Denitrification is the conversion of nitrates to nitrous oxide and nitrogen gas. It primarily occurs in waterlogged soils and is a detrimental process for plants because it removes the useful nitrates.

Sulfur is another important element that is continuously recycled. Various bacteria carry out important parts of this cycle including sulfate reduction, sulfur reduction, and sulfur oxidation.

Phosphorus also is recycled. Again, microorganisms become essential parts of the cycle to insure that organic and inorganic components are returned to useful forms.

117. What kinds of microorganisms are found in air, and how are they detected and controlled?

Microorganisms do not grow in air, but spores and vegetative cells can be transmitted by air currents or on dust particles and water droplets. The kinds and numbers of microorganisms vary tremendously in different environments.

Air can be sampled by exposing agar plates to the air or by drawing air over the surface of an agar plate or liquid medium. The organisms can then be identified and studied.

Microorganisms in the air can be controlled by chemical agents such as triethylene glycol, by UV radiation, by filtration, and by unidirectional airflow such as a laminar airflow system.

118. What kinds of microorganisms are found in soil, and what are their roles in biogeochemical cycles and as pathogens?

Soil consists of inorganic and organic components. The inorganic components include rocks, minerals, water, and gases. The organic components include humus and microbes of all taxonomic groups such as bacteria, fungi, algae, viruses, and protists.

Various physical factors affect these microbes. These include moisture, oxygen concentration, pH, and temperature.

Microorganisms also alter the characteristics of their environment as they are effective decomposers and they release wastes, some of which can be used by other organisms. They represent important members of the carbon and nitrogen cycles.

Soil pathogens affect mainly plants and insects and rarely affect humans. The main human pathogens found in soil are members of the genus *Clostridium*. These include those that cause tetanus, botulism, and gas gangrene.

119. How do freshwater and marine environments, and their microorganisms, differ?

Freshwater environments are characterized by low salinity and a great variability in temperature, pH, and oxygen concentration. Because of these extremes, a wide diversity of microorganisms may be found. Most prefer moderate temperatures and a neutral pH environment. Aerobic bacteria are abundant where oxygen is plentiful and anaerobic bacteria accumulate where oxygen is depleted.

Marine environments are characterized by high salinity, and smaller variabilities in temperature, pH, and oxygen concentrations. As the depth increases, pressure increases and sunlight penetration decreases. Microbes from all taxonomic groups are found; however, they must adjust to the higher salinity. Photosynthetic microbes are found near the surface, heterotrophs in the middle and lower strata, and decomposers in the bottom sediments.

120. What is water pollution and what are some of the waterborne pathogens that affect humans?

Water is polluted if a substance or condition is present that renders the water useless for a particular purpose. Thus, the concept of water pollution is a relative one, depending both on the nature of the pollutants and the intended uses of the water. Water may be too polluted for drinking but may be safe for swimming or boating.

Many human pathogens can be transmitted in water. These include: *Salmonella*, *Shigella*, *Vibrio's*, *Escherichia*, *Giardia*, Hepatitis A viruses, and many others.

121. How is water purified, and how is it tested to determine purity?

Purification procedures for human drinking water are determined by the degree of purity of the water at its source. If the source is very contaminated, the procedures may be more elaborate. The usual procedures involve flocculation of suspended matter, filtration through beds of sand to remove bacteria, and chlorination to insure the water is safe to drink.

Tests for purity are designed to detect coliform bacteria, especially *Escherichia coli* which is always associated with the human intestinal tract. These tests include the multiple-tube fermentation test and the membrane filter method.

122. What is sewage, and what processes are involved in primary, secondary, and tertiary sewage treatment?

Sewage is used water and the wastes it contains. It is usually about 99.9% water and about 0.1% solid or dissolved wastes, which include household wastes such as human feces, detergents, and grease; industrial wastes such as acids and other chemicals; and wastes carried by rainwater that enters the sewers.

Sewage treatment involves three main processes. Primary treatment is the removal of solid wastes by physical means; secondary treatment is the removal of organic matter by the action of aerobic bacteria; and tertiary treatment is the removal of all organic matter, nitrates, phosphates, and any surviving microorganisms by physical and chemical means. Most treatment plants utilize only the primary and secondary processes.

Chapter 26 Applied Microbiology

INTRODUCTION

This is our last chapter and I hope by this time, microbiology leaves you with a good taste in your mouth (another pun intended!). I often teach microbiology in our evening division and students spend at least four hours and miss their dinner at home. I have one or two students from each class prepare a snack for our break and of course it stresses at least one food that contains a microbe. Each week students bring in foods that contain baked yeast (bread), live bacteria (yogurt), or preserved seaweed (nori).

While applied microbiology is the last chapter in our textbook, it was my first job in microbiology. I tested food intended for schoolchildren in New York City. My task was to determine how many *E. coli* organisms were present per gram of each food. I stated how many, because all the prepared foods I tested had at some *E. coli*, and it is really true what people say about meatloaf! I was only concerned with *E. coli* because it is a measure of fecal contamination and one may assume that an array of other more nasty organisms might be coming along for the ride. I began with frozen food items and used a sterile drill tip to create one gram of sterile food shavings per Petri dish. The best part of the job was I would select only one food idem to test per carton of food. Thus, the rest of the meat patties or hamburgers we would split up and take home to eat. Yes, I knew what was in the patties but I also knew that cooked bacteria are a good source of vitamins and protein. We all eat more of them we would like to think and I hope by now you know that we are the better for it.

STUDY OUTLINE

I. **Microorganisms Found in Food**

A. Grains

If stored under moist conditions, grains can easily become contaminated with molds and other microorganisms. Insects, birds, and rodents also transmit microbial contaminants to grains. Examples of contaminating microbes include the mold *Claviceps purpurea*; *Rhizopus nigricans*; several species of *Penicillium, Aspergillus*, and *Monilia*; *M. sitophila*; and *Bacillus* species.

B. Fruits and vegetables

Fruits and vegetables easily become contaminated with organisms from soil, animals, air, irrigation water, and equipment used to pick, transport, store, or process them. Examples of contaminating microbes include *Pseudomonas fluorescens, Salmonella, Shigella, Entamoeba histolytica, Ascaris, Erwinia carotovora, Phytophthora infestans, Pseudomonas syringae, Fusarium, Rhizopus, Leuconostoc, Lactobacillus, Monilia fructicola,* and *Penicillium expansum.*

C. Meats and poultry

Meat animals arrive at slaughterhouses with numerous and varied microorganisms in the gut and feces, on hides and hoofs, and sometimes in tissues. Common diseases identified in slaughterhouses are abscesses, pneumonia, septicemia, enteritis, toxemia, nephritis, pericarditis, and lymphadenitis. Examples of contaminating microbes of meats and poultry include *Cladosporium herbarum, Rhizopus, Mucor, Pseudomonas mephitica, Clostridium, Trichanella spiralis, Lactobacillus, Salmonella, Clostridium perfringens, Staphylococcus aureus, Penicillium,, Sporotrichum,* and *Salmonella pullorum.*

D. Fish and shellfish

Fresh fish abound with microorganisms. Several species of enteric bacteria and clostridia, enteroviruses, and parasitic worms are commonly found on fresh fish. Shellfish, such as oysters and clams, carry many of the same organisms as fish, including *Salmonella typhimurium* and *Vibrio cholerae.* Among crustaceans, shrimp are extremely likely to be contaminated. Crabs also carry *Vibrio cholerae, Clostridium botulinum, Cryptococcus,* and *Candida.*

E. Milk

Modern mechanized milking and milk handling have greatly reduced the microbial content of raw milk. However, breeding diary cattle for increased milk production has resulted in exceptionally large udders and teats that easily admit bacteria. Most microorganisms in freshly drawn milk are *Staphylococcus epidermidis* and *Micrococcus,* but *Pseudomonas, Flavobacterium, Erwinia,* and some fungi also can be present. Other examples of contaminated microbes include *Escherichia coli, Acinetobacter johnsoni, Mycbacterium bovis, Brucella, Staphylococcus aureus, Salmonella, Streptococcus lactobacillus,* and *Lactobacillus.*

F. Other edible substances

Other edible substances subject to microbial contamination include sugar, spices, condiments, tea, coffee, cocoa, raw sugar-cane juice, maple sugar sap, honey, condiments, carbonated beverages, freshly-harvested coffee beans, and tea leaves.

II. Preventing Disease Transmission and Food Spoilage

A. General characteristics

Diseases acquired from food are due mainly to the direct effects of microorganisms or their toxins, but they can result from microbial action on food substances. Industrialization has increased the spread of foodborne pathogens. Bacterial pathogens that can be transmitted through food include *Klebsiella pneumoniae; Mycobacterium tuberculosis;* those that cause anthrax, brucellosis, Q fever, and listeriosis; *Yersinia enterocolitica;* and *Erysipelothrix rhusiopathiae.* Enteroviruses, echoviruses, coxsackie viruses, poliomyelitis viruses, and hepatitis A viruses are among the viruses frequently transmitted through food. Microbes that can be transmitted through milk include *Mycobacterium bovis, Brucella* species, *Listeria monocytogenes, Coxiella burnetii, Bacillus anthracis,* and *Leuconostoc cremoris.*

B. Food preservation

Methods of food preservation described in the text include canning using moist heat; refrigeration and freezing; lyophilization and drying; the use of radiation; the use of chemical additives; and the use of antibiotics.

C. Pasteurization of milk

In both *high-temperature short-time (HTST) pasteurization* (or *flash pasteurization*) and *low-temperature long-time (LTLT) pasteurization* (or *holding method*), vegetative cells of pathogens likely to be found in milk are destroyed and the organisms that can cause souring are decreased in number.

D. Standards regarding food and milk production

Because food and milk production are carefully regulated by federal, state, and local laws, consumers are better protected in the United States than in many other countries.

III. Microorganisms as Food and in Food Production

A. Algae and fungi as food

Among microorganisms, yeasts show great promise for increasing our food supplies. Algal culture is another promising avenue for increasing human food supplies, since the use of algae as human food shortens the food chain. Even some bacteria are used as food. If technical difficulties can be overcome and the products made acceptable as human food, yeasts, algae, and some bacteria can increase the world food supply. But, at best, the use of microorganisms as food can buy only a little time to allow humans to control their own numbers.

B. Food production

In modern food production, specific organisms are purposely used to make a variety of foods, including bread, cheeses and other dairy products, vinegar, sauerkraut, pickles, olives, soy sauce, and other soy products.

IV. Beer, Wine, and Spirits

Beer and wine are made by fermenting sugary juices; spirits, such as whiskey, gin, and rum, are made by fermenting juices and distilling the fermented product.

V. Industrial and Pharmaceutical Microbiology

A. General characteristics

Industrial microbiology deals with the use of microorganisms to assist in the manufacture of useful products or to dispose of waste products. *Pharmaceutical microbiology* is a special branch of industrial microbiology concerned with the manufacture of products used in treating or preventing disease.

B. Useful metabolic processes

The production of complex molecules and metabolic end products in amounts that are commercially profitable requires the manipulation of microbial processes. Industrial microbiologists thus manipulate organisms in the following ways: (1) by altering nutrients available to the microbes; (2) by altering environmental conditions; (3) by isolating mutant microbes that produce excesses of a substance because of a defective regulatory mechanism; and (4) by using genetic engineering to program organisms to display particular synthetic capabilities.

C. Problems of industrial microbiology

Problems faced by industrial microbiologists include: (1) adapting processes so that they will be profitable in a large-scale industrial setting; (2) keeping industrial microorganisms from being so extensively modified that their products may be useless or even toxic to the organisms; and (3) isolating and purifying the product with or without killing the organisms.

VI. Useful organic products

A. Simple organic compounds

Simple organic compounds such as solvents (like ethanol, butanol, acetone, and glycerol) and organic acids (like acetic, lactic, and citric acids) can be manufactured with the aid of microorganisms.

B. Antibiotics

Industrial microbiologists work diligently to find ways to cause organisms to make their particular antibiotic in large quantities. They are also always looking for ways to increase potency, improve therapeutic properties, and make antibiotics more resistant to inactivation by microorganisms.

C. Enzymes

All enzymes used in industrial processes are synthesized by living organisms. Examples of industrial enzymes include proteolytic enzymes and enzymes used in paper production.

D. Amino acids

Microbial production of amino acids has become a commercially successful industry. Examples of microbially-synthesized amino acids include lysine, glutamic acid, phenylalanine, aspartic acid, and tryptophan.

E. Other biological products

Vitamins, hormones, and single-cell proteins are the major categories of other biologically useful products of industrial microbiology.

VII. Microbiological Mining

As the availability of mineral-rich ores decreases, methods are needed to extract materials from less concentrated sources. This need spawned the new discipline known as *biohydrometallurgy*, or the use of microbes to extract metals from ores.

VIII. Microbiological Waste Disposal

Sewage-treatment plants are prime examples of microbiological waste-disposal systems, but microbes are also used in disposing of chemical pollutants and toxic wastes (as in bioremediation, or the use of microorganisms to dispose of some chemical wastes).

SELF-TESTS

Continue with this section only after you have read Chapter 26 of your textbook. Write your answers in the appropriate space provided. Correct answers to all questions can be found at the end of the Self-Test section. A score of 80 percent or better is good. If your score is less than 65, reread the chapter.

True/False

Mark T for True, F for False

______ **1.** The skins of most plant foods are very inhibitory to microbial agents.

______ **2.** Since eggs have very hard shells, they are rarely associated with any microbial agents and are not considered sources of infections.

______ **3.** Foods that contain artificial sweeteners do not retard growth by osmotic pressures and therefore required more elaborate methods to inhibit microbes.

______ **4.** The manufacture of vinegar requires the action of vinegar eels to produce the acid.

______ **5.** Wines can be made from any fruit.

______ **6.** By adding more sugar to a wine, the alcohol content can be increased to as much as 40%.

______ **7.** All enzymes used in industrial processes are synthesized by living organisms.

______ **8.** Flat sour spoilage of canned foods occurs when bacterial endospores germinate, grow, and spoil the food but do not cause the cans to bulge with gas.

Multiple Choice

Select the best possible answer.

______ **9.** The use of nitrates and nitrites in food:
- **a.** should be stopped immediately because when heated, they are converted to formaldehyde.
- **b.** are effective in controlling diseases such as botulism in sausages.
- **c.** cause meats to turn a dull brown color.
- **d.** retard the maturation of fruits and resist spoilage.

______ **10.** Foods that are irradiated:
- **a.** become slightly radioactive.
- **b.** can be sterilized.
- **c.** can be stored at room temperature indefinitely.
- **d.** will cause considerable color changes in the product.

______ **11.** Which of the following is not true concerning the addition of antibiotics to animals used for food?
- **a.** Humans may become allergic to antibiotics given to animals.
- **b.** The antibiotics might interfere with the activities of microbes such as those essential in fermenting milk and cheese.
- **c.** The antibiotics might be relied on instead of good sanitation.
- **d.** The antibiotics might reduce the number of resistant strains of bacteria thus making the foods safer to eat.
- **e.** All of the above are true.

______ **12.** In the production of cheeses:
- **a.** Hard cheeses ripen in about two weeks.
- **b.** Species of *Clostridium* are commonly used to produce the best flavors in cheddar and Roquefort.
- **c.** Hard cheeses are often very large and are ripened by microbial action.
- **d.** Soft cheeses are ripened by the action of salt.

Matching

Select the answer from the right-hand side that corresponds to the term or phrase on the left-hand side of the page. An answer may be used more than once. In some cases more than one answer is required.

Topic: Microorganisms Found In Grains and Bread

______ **13.** Common bread mold		**a.** *Claviceps purpurea*
______ **14.** Source of aflatoxins		**b.** *Aspergillus* species

_____ 15. Ergot poisoning

_____ 16. Bread mold contaminant

_____ 17. Used in bread making

_____ 18. Gives bread a stringy texture

c. *Saccharomyces cerevisiae*

d. *Rhizopus nigricans*

e. *Bacillus* species

f. *Monilia sitophila*

Topic: Microorganisms Involved in Food Spoilage

_____ 19. Sour milk

_____ 20. Eggshell contamination

_____ 21. Bone stink

_____ 22. Potato blight

_____ 23. Brown rot

_____ 24. Bacterial soft rot of leafy vegetables

_____ 25. Patulin production

_____ 26. Soft rot of tomatoes

_____ 27. Green discoloration of refrigerated meat

_____ 28. Whiskers on meat

_____ 29. Bulging cans

a. *Phytophthora infestans*

b. *Streptococcus lactis*

c. *Penicillium*

d. *Clostridium* species

e. *Erwinia carotovora*

f. *Fusarium* species

g. *Monilia fructicola*

h. *Pseudomonas*

i. *Rhizopus* species

Topic: Pathogenic Organisms Spread by Food and Milk

_____ 30. *Staphylococcus aureus*

_____ 31. *Escherichia coli*

_____ 32. *Clostridium botulinum*

_____ 33. *Clostridium perfringens*

_____ 34. *Bacillus cereus*

_____ 35. *Vibrio cholerae*

_____ 36. *Erysipelothrix rhusiopathiae*

_____ 37. *Yersinia enterocolitica*

a. causes food poisoning

b. causes botulism

c. causes cholera

d. causes traveler's diarrhea

e. causes Adirondack disease

f. causes erysipeloid

Topic: Food Preservation

_____ 38. Preserves natural food flavor better than canning

_____ 39. Causes foods to become soft upon thawing

_____ 40. Moist heat under pressure

_____ 41. Freeze-drying

_____ 42. Uses temperature around—10°C

_____ 43. Poor penetrating power

_____ 44. Most common food preservation method

_____ 45. Stops microbial growth but does not kill all organisms on food

_____ 46. Good penetrating ability

_____ 47. Lower pH to prevent pathogens from growing

_____ 48. May interfere with microbes essential for cheese making

a. canning

b. refrigeration

c. freezing

d. drying

e. lyophilization

f. ultraviolet light

g. gamma rays

h. chemical additives

i. antibiotics

Topic: Food Production

______ 49. Buttermilk
______ 50. Yogurt
______ 51. Bread making
______ 52. Acidophilus milk
______ 53. Koumiss
______ 54. Bulgarian milk
______ 55. Kefir

a. *Saccharomyces cerevisiae*
b. *Streptococcus cremoris*
c. *Streptococcus thermophilus*
d. *Lactobacillus bulgaricus*
e. *Lactobacillus acidophilus*
f. *Streptococcus lactis*
g. *Leuconostoc citrovorum*

Topic: Fermentation Products and the Organisms That Produce Them

______ 56. Pickles
______ 57. Sauerkraut
______ 58. Vinegar
______ 59. Soy Sauce
______ 60. Olives
______ 61. Poi
______ 62. Sufu
______ 63. Beer and wine

a. *Acetobacter aceti*
b. *Leuconostoc* species
c. *Pediococcus* species
d. *Lactobacillus* species
e. pseudomonads
f. *Saccharomyces* species
g. *Mucor*

Topic: Industrially Useful Organic Products and the Organisms That Produce Them

______ 64. Acetic acid
______ 65. Lactic acid
______ 66. Proteases
______ 67. Amylases
______ 68. Lipase
______ 69. Lactase
______ 70. Invertase
______ 71. Lignin-digesting enzymes
______ 72. Amino acids

a. *Aspergillus* species
b. *Clostridium aceticum*
c. *Lactobacillus delbrueckii*
d. *Acetobacterium woodii*
e. *Trichoderma* species
f. *Kluyveromyces* species
g. *Saccharomyces* species
h. *Phanerochaete* species
i. *Corynebacterium glutamicum*
j. *Sacharomycopsis* species

Fill-Ins

Provide the correct term or phrase for the following. Spelling counts.

73. Raw grains contaminated with the mold *Claviceps purpurea* cause ______________ ______________.

74. A common disease identified in sheep and lambs in slaughterhouses is ______________, or an inflammation of the lymph nodes.

75. Several species of ______________ cause putrefaction called bone stink deep in the tissues of large carcasses.

76. A parasitic worm that is found in raw pork samples is ______________ ______________, the cause of trichinosis.

77. Since ______________ survives on eggshells and can enter broken eggs or be deposited with bits of shell in foods, eggs and foods that contain them should be cooked thoroughly.

78. Among crustaceans, ______________ are extremely likely to be contaminated; some studies have shown that over half of the breaded variety on the market contain in excess of 1 million bacteria per gram.

79. The contaminating microorganism that can give milk a fecal flavor is _________________ _________________.

80. When *Streptococcus lactis* and species of *Lactobacillus* release enough lactic acid to bring the pH of milk below 4.8, the milk proteins coagulate and the milk is said to have _________________.

81. Honey should not be given to infants, since it is a potential source of _________________ _________________ toxins that can cause "floppy baby syndrome."

82. The most common method of food preservation is _________________, or the use of moist heat under pressure.

83. Spoilage due to the growth of thermophilic anaerobic endospores sometimes does not cause cans to bulge with gas; such spoilage is called _________________ _________________ spoilage.

84. Refrigeration at temperatures of about _________________ is suitable for preserving foods only for a few days, since refrigeration does not prevent the growth of psychrophilic organisms that can cause food poisoning.

85. Freezing, which prevents the growth of most microorganisms but does not destroy them, involves storage of foods at temperatures of about _________________.

86. Currently, _________________, or freeze-drying, is employed in the food industry almost exclusively for the preparation of instant coffee and dry yeast for breadmaking.

87. _________________ radiation, such as gamma rays, has great penetrating ability and is antimicrobial.

88. In the United States, only the anticlostridial agent _________________ may be used in milk.

89. Milk that is heated to 71.6°C for 15 seconds is pasteurized by the _________________ pasteurization process.

90. The use of _________________ as human food shortens the food chain.

91. The microbe used to leaven bread is _________________ _________________.

92. Coagulated milk proteins form into a cheesy mass called _________________, and the liquid that remains is _________________.

93. Vinegar contains the organic acid _________________ acid.

94. Cabbage that is allowed to ferment in the presence of species of *Lactobacillus* and *Leuconostoc* is called _________________.

95. A food common in the South Pacific that is made from the fermented roots of the taro plant is _________________.

96. A soft curd of soybeans is _________________.

97. _________________ is a process that separates alcohol and other volatile substances from solid and nonvolatile substances.

98. _________________ microbiology deals with the use of microorganisms to assist in the manufacture of useful products or to dispose of waste products.

99. _________________ microbiology is a special branch of industrial microbiology concerned with the manufacture of products used in treating or preventing disease.

100. Animal feed consisting of microorganisms is called _________________ - _________________ _________________.

101. In a _________________ _________________, fresh medium is introduced at one end and medium containing the product is withdrawn at the other.

102. The antibiotic industry came into being in the 1940s with the manufacture of _________________.

103. _________________ are important in industrial processes because of their ability to act on a certain substrate and yield a certain product.

104. A reaction in which one compound is converted to another by enzymes in cells is called _________________.

105. The use of microbes to extract metals from ores is known as _________________.

106. _________________ is defined as the use of microorganisms to dispose of some chemical wastes.

Critical Thinking

107. What kinds of microorganisms are found in different categories of food?

108. What diseases can be transmitted in food, and how can these diseases be avoided?

109. How can food spoilage and disease transmission be prevented, and what standards relate to these problems?

110. How are microorganisms used as food or in the making of food products?

111. How are microbes used in the manufacture of beer, wines, and spirits?

112. How can microbes be used in industry, and what problems are associated with their use?

113. What is the role of microorganisms in the manufacture of simple organic compounds, antibiotics, enzymes, and other biologically useful substances?

114. How are microorganisms used in mining?

115. How are microorganisms used in waste disposal?

ANSWERS

True/False

1. **True** Plant skins are often very tough and contain waxes and may even release antibiotic substances to inhibit microbial invasion.
2. **False** Most eggs are free of contamination; however, some bacteria such as pseudomonads and some fungi can grow on eggshells. *Salmonella* can also survive on eggshells and can be involved in disease transmission.
3. **True** Artificial sweeteners cannot retard growth by osmotic effects as foods containing natural sugars. High sugar content can create a hypertonic condition.
4. **False** Vinegar is made from ethyl alcohol by the action of acetic acid bacteria and not eels.
5. **True** Wine is made from juice extracted usually from grapes although any fruit including nuts and certain blossoms can be used. Normally, sulfur dioxide is used to kill any wild yeasts that may lurk on any of the fruits being used for fermentation.
6. **False** Once the alcohol content reaches 12 to 15 percent, it poisons the yeasts carrying out the fermentation. Extra alcohol is added after the wine is made in order to increase the alcohol content.
7. **True** Enzymes are made of protein and are produced by protein-synthetic mechanisms found in living organisms. A few enzymes are extracted from plants but most are produced by bacteria.
8. **True** When spores germinate and produce gas, the resultant condition is known as thermophilic anaerobic spoilage.

Multiple Choice

9. b; 10. b; 11. d; 12 c.

Matching

13. d; 14. b; 15. a; 16. f; 17. c; 18. e; 19. b; 20. c; 21. d; 22. a; 23. g; 24. e; 25. c; 26. f; 27. h; 28. i; 29. d; 30. a; 31. d; 32. b; 33. a; 34. a; 35. c; 36. f; 37. e; 38. c; 39. c; 40. a; 41. e; 42. c; 43. c; 44. a; 45. d; 46. g; 47. h; 48. i; 49. b, f, g, h; 50. c, d; 51. a; 52. e; 53. d, f; 54. d; 55. d, f; 56. b, c; 57. b, d; 58. a; 59. c, f; 60. b, d; 61. e; 62. g; 63. f; 64. b, d; 65. c; 66. a; 67. a; 68. j; 69. e, f; 70. g; 71. h; 72. i.

Fill-Ins

73. ergot poisoning; 74. lymphadenitis; 75. *Clostridium*; 76. *Trichinella spiralis*; 77. *Salmonella*; 78. shrimp; 79. *Escherichia coli*; 80. soured; 81. *Clostridium botulinum*; 82. canning; 83. flat sour; 84. 4°C; 85. −10°C; 86. lyophilization; 87. ionizing; 88. nisin; 89. HTST; 90. algae; 91. *Saccharomyces cerevisiae*; 92. curd, whey; 93. acetic; 94. sauerkraut; 95. poi; 96. tofu; 97. distillation; 98. industrial; 99. pharmaceutical; 100. single-cell protein; 101. continuous reactor; 102. penicillin; 103. enzymes; 104. bioconversion; 105. biohydrometallurgy; 106. bioremediation.

Critical Thinking

107. What kinds of microorganisms are found in different categories of food?
Anything that people eat or drink is food for microorganisms. Since foods are derived from plants or animals that are associated with the soil, they will automatically have soil organisms on them. Fortunately most of these are not pathogenic but they can cause food spoilage. As long as the food is properly prepared, handled, and stored there is little chance for illness or spoilage.

Since harvested grains are normally dry there is little chance for microbial contamination. If they become moist because of improper storage, contamination especially by molds can occur. Flour, however, is purposely inoculated with yeasts to make bread.

Fruits and vegetables are usually protected by their resistant skins; however, if they are not properly stored, they will succumb to soft rot and mold damage.

Meats, poultry, fish, and seafood contain many kinds of microorganisms, some of which cause zoonoses. Because they are neutral in pH, they must be handled and prepared very carefully to keep dangerous pathogens from developing. Poultry often is contaminated with *Salmonella, Clostridium perfringens*, and *Staphylococcus aureus*. Fish and shellfish can be contaminated with several kinds of bacteria such as *Vibrio's* and viruses such as hepatitis A.

Milk can contain organisms from the cows, milk handlers, and the environment. Unfortunately, many of these bacteria can be potentially dangerous and, as a result, have required the pasteurization of milk and milk products.

Other substances such as sugar can support the growth of microbes but they are normally killed during the refining process. Spices are often used to mask the effects of microbes in foods; however, they can be the source of these microbes. Syrups or carbonated beverages can be contaminated with molds, and even tea, coffee, and cocoa can be subject to these molds if not kept dry.

108. What diseases can be transmitted in food, and how can these diseases be avoided?
Numerous types of diseases such as shigellosis, salmonellosis, botulism, and various food poisonings can be transmitted in food and milk Good sanitation and proper food handling, preparation, and storage greatly reduce the chances of getting food-borne diseases.

109. How can food spoilage and disease transmission be prevented, and what standards relate to these problems?
The most important factor in preventing spoilage and disease transmission in food and milk is cleanliness in handling. Other factors include common-sense rules such as proper refrigeration, prompt use of fresh foods, and proper processing and storage.

Methods of food preservation include canning, refrigeration, freezing, lyophilization, drying, ionizing radiation, and the use of chemical additives such as sulfur dioxide, salt, and nitrates and nitrites. Most of these have been described in Chapter 10. Canning is the most common method of food preservation and involves the use of moist heat under pressure. If properly done, canning destroys all harmful spoilage microbes, prevents spoilage, and avoids any hazards of disease transmission.

Milk and milk products are rendered safe for human consumption by pasteurization or sterilization. This is done to avoid diseases such as tuberculosis, salmonellosis, brucellosis, and Q-fever.

Certain standards regarding food and milk production are maintained by federal, state, and local laws. Many states have banned the sale of raw milk to avoid the threat of milk-borne diseases.

110. How are microorganisms used as food or in the making of food products?

The rapid rise in world population is greatly increasing the demand for new and inexpensive sources of human food. Various microorganisms, especially yeasts, have been developed to provide some of the proteins and vitamins that would be required. Problems associated with this development include the expensive equipment and the difficulty in getting people to accept the food as part of their diet. Algae has also been used as a protein food supplement and can be readily grown in lakes and on sewage. Problems include the danger of viral contamination and the lack of its acceptability as a food.

Yeasts are commonly used in food production to leaven bread and to ferment wines and beer. Certain bacteria are used to make dairy products such as buttermilk, sour cream, yogurt, and cheeses. In cheesemaking, the whey of milk is discarded and microorganisms ferment the curd and impart flavors and texture to the product.

Many other foods are produced by microbial action and include vinegar, sauerkraut, pickles, olives, poi, soy sauce, and other soy products. At the next opportunity, go to the grocery store and make a list of all the foods that are affected in some way by microbial action. The list will be quite lengthy.

111. How can microbes be used in industry, and what problems are associated with their use?

Industrial microbiology deals with the use of microorganisms to assist in the manufacture of useful products such as simple organic compounds and alcoholic beverages, or to dispose of waste products such as garbage and waste paper. Pharmaceutical microbiology deals with the use of microorganisms in the manufacture of medically useful products such as antibiotics, vitamins, and hormones.

The production of complex molecules and metabolic end products in commercially profitable quantities usually requires the manipulation of microbial processes. These manipulations include altering the nutrients, or the environmental conditions, and isolating mutants and/or modifying them by genetic engineering.

Problems include adapting small-scale processes into large-scale commercial processes and developing techniques that are necessary for recovery of the products.

112. How are microbes used in the manufacture of beer, wines, and spirits?

Beer and wine are made by fermenting sugary juices; spirits, such as whiskey, gin, and rum, are made by fermenting juices and distilling the fermented product. Strains of *Saccharomyces* yeasts are used as the fermenters for all alcoholic beverages. The yeasts produce ethyl alcohol, carbon dioxide, and other substances such as acetic and butyric acids and tannin which add much to the flavor of the beverage.

113. What is the role of microorganisms in the manufacture of simple organic compounds, antibiotics, enzymes, and other biologically useful substances?

Microorganisms are capable of producing a wide variety of useful organic compounds including alcohols, acetone, glycerol, and organic acids. Currently, microbes play a limited industrial role; however, they may become more important in the future because of advances in genetic engineering.

Miccrobes such as *Streptomyces*, *Penicillium*, *Cephalosporin*, and *Bacillus* are greatly involved in producing a wide variety of antibiotics. New strains are continuously being discovered that are capable of producing potential substances. Some antibiotics such as the penicillins have beta-lactam rings that can be modified in the laboratory to produce more effective agents.

Various enzymes including proteases and amylases have been extracted from microorganisms such as *Aspergillus* and *Bacillus* and are produced commercially. Many of these are used in detergents, drain cleaners, enrichment agents of foods, and in the manufacture of paper.

Vitamins and hormones are now made by manipulating organisms so that they produce excessive amounts of the products. Single-cell proteins consist of whole organisms such as *Candida* that are grown in simple foods to produce large quantities of protein. They are used primarily as animal feed.

114. How are microorganisms used in mining?

Since the availability of mineral-rich ores has decreased, various methods must be used to extract the minerals from less concentrated sources. Microbes have been used to extract copper from low-grade ores as well as small quantities of iron, uranium, arsenic, lead, zinc, cobalt, and nickel.

115. How are microorganisms used in waste disposal?

Microbes are very important components in sewage treatment plants. A few organisms have been found to degrade toxic wastes and certain types of plastics. Microbes have also been used to degrade oil products lost during spills. Research is under way to identify and develop other uses for microbes.